D1498611

THE COMPLETE BOOK OF
HOME
WELDING

JOHN TODD

TAB BOOKS
Blue Ridge Summit, PA

FIRST EDITION
THIRD PRINTING

Printed in the United States of America

Library of Congress Cataloging in Publication Data

Todd, John (John T.)
The complete book of welding techniques.

Includes index.
1. Welding—Amateurs' manuals.   I. Title.
TS227.T594 1986   671.5′2   86-5927
ISBN 0-8306-0317-4
ISBN 0-8306-2717-0 (pbk.)

TAB BOOKS offers software for sale. For information and a catalog, please contact TAB Software Department, Blue Ridge Summit, PA 17294-0850.

Questions regarding the content of this book should be addressed to:

Reader Inquiry Branch
TAB BOOKS
Blue Ridge Summit, PA 17294-0214

Cover photography courtesy of HOBART BROTHERS COMPANY.

Contents

Acknowledgments

Howard Adams
Bart Attebery
DuWayne Arneson
Charles Burkart
Gene Brady
Earl Bowick
Dick Earle
Rob Eytchison
Bob Hall
Lori Hatcher
Bob Hereford
Kent Kammerer
Jeff Laing
Tom McGowan
John and Sheila Morris
Darrell Nicholson
Rodger Squirrell
Kim Tabor
Mary Todd
Jack Uchida
Kelley Waller
Bob Wilson

Introduction

LOOK OUT! HERE COMES THE DETERMINED home craftsman. If he or she wants to build or repair something, you step back or get involved. But don't fight it, join in the fun with these people who are never quite satisfied until they can do it themselves.

Home craftsmen are now welding. Home welding saves money and provides a practical avenue for creative expression. What's more, home welders do quality work without extensive training and without major changes to their workshops. They have found today's welding technology to be sufficiently advanced so that welding is truly a do-it-yourself activity.

This book is about welding and having fun. Building and repairing with durable welded constructions is satisfying. There is an elusive magic to working with molten weld metal: perhaps in the idea of practicing ancient arts by modern means. Follow the footsteps of the artisans and blacksmiths of earlier days and build lasting metal fabrications with home welding. It is fun and you can do it yourself.

Just a few years ago, the time needed to develop even mediocre skills was reason enough to discourage home welding. Also, nearly all equipment was designed for industrial use; extensive (and expensive) electrical supply changes were needed to accommodate such heavy duty equipment. Home welding with the versatile wire feed processes was only a dream. Serving the home craftsman was not a priority for welding equipment manufacturers.

The first home welders were limited mostly to oxyacetylene (or "gas") welding (Fig. I-1). Some did electric arc welding with the Shielded Metal Arc Welding (SMAW) process and coated *electrodes* (Fig. I-2). In this process, an arc is formed between the electrode and plate to be welded; the melting electrode becomes the filler. SMAW (also called "stick" welding) gave many home craftsmen new abilities. Others, however, found it—with its long and disheartening learning curve—a demanding, skill-intensive process. Creative hobbyists sometimes adapted commercial processes and gear to their lighter needs. Fortunately, it is no longer difficult to be a home welder.

The practicality of home welding has greatly improved with new arc welding equipment like that shown in this book. The new and already popular *single-phase wire feeder* redefines welding as a bona-fide do-it-yourself skill. (Single-phase is the type of electrical power supplied to residential areas; industrial plants often use three-phase power.) The impact of these units parallels that of the personal home computer (PC) that gives amateurs computing powers that formerly belonged only to professionals.

Developments with smaller, more efficient welding units promise even more advantages to the home craftsman. Those with some welding experience find his or her capabilities broadened with new affordable equipment and supplies. Even aluminum and stainless steel welding are now realistic home shop activities.

Home welders routinely use several electric arc processes along with oxyacetylene. These include Shielded Metal Arc Welding (SMAW), Gas Metal Arc Welding (GMAW), also called "Mig" or several other slang names, and Flux Cored Arc Welding (FCAW), with "Dual Shield" and "Innershield" variations. Little is said in this book about it, but a few home welders use Gas Tungsten Arc Welding (GTAW), also called "Tig" or "Heliarc,"

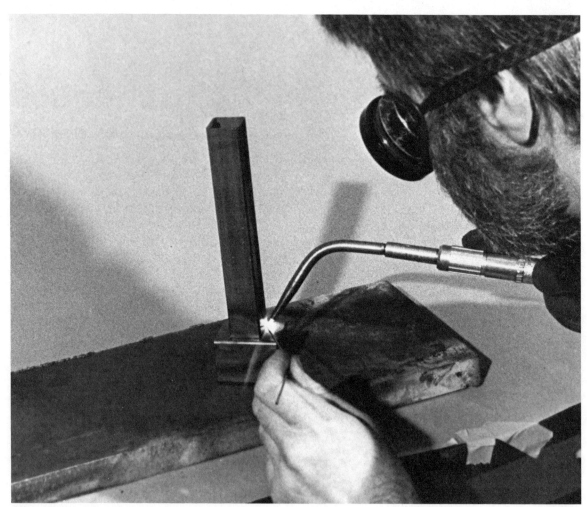

Fig. I-1. Until recently, home welding was done almost entirely with the use of an oxyacetylene flame as shown here.

Fig. I-2. Home welders now use nearly all of the electric arc welding processes along with oxyacetylene. Today, home welding projects are almost unlimited. (Courtesy of Kent Kammerer Photography, Seattle.)

for welding delicate aluminum and stainless steel assemblies.

For years, hobbyists have spent rejuvenating hours working with wood. There is real joy in moving from an idea to building a useful and graceful object. The desire to recapture this joy can develop into a healthy sort of compulsion. Many home woodworkers are extending their activities to include building with metal.

Everyone recognizes that wood and metal are different, but many working skills transfer between these substances. The possible beautiful combinations of metals, wood, and other materials are limited only by the imagination.

Metal sculpture is often done with welding as the main or entire creative tool. Welding is an easy transition for many artists accustomed to sculpting with wood, clay, or stone. It is hoped that this book provides some smoother, wider, and more interesting avenues for the creative home craftsman.

How to Use This Book

N O ONE REALLY KNOWS, BUT THE WORLD might be better off if everyone had the time and inclination to read technical books from cover to cover; do-it-yourselfers seldom have the time or desire. Home shopwork is irresistible fun with an understandable eagerness to get right into it. We probably ought to know better, but reading just enough to get started seems to be a standard approach to doing things in the shop. In spite of smug sayings like, "When all else fails, read the directions," or, "There is always enough time to do it over, but seldom enough time to do it right"—we still are anxious to start building.

This is your book about home welding to be used any way you want. It is hardly a textbook (there are no teacher's guides, chapter questions, or student study guides), but it contains a lot of easy-to-locate information. There is method to its content and the order in which it appears. To get the most use of your time, take a moment or two now to check out the plan of the book.

By all means, look over the table of contents and notice the definitions in Appendix A. Depending on your background, you might find it helpful to refer often to this section. Also, flip forward and backward as you use the book. Words of special significance are italicized as they first appear. If you are in a big hurry, read Appendix A and all available safety information about the process you are using. As you find time, read more. Suggestions for further reading are in Appendix C.

Welded fabrications are all around us. Most welded products are not too complex to be made in the home shop. The only common welded product that I emphatically advise the amateur to avoid is a pressure vessel. Projects are shown in many photographs, but this book is not a collection of blueprints and step-by-step instructions. Hopefully it is more a springboard for you to take your own pictures and sketch your own plans of whatever weldable catches your fancy.

Get out and see what others are doing with home welding. It is amazing how much can be learned in a short time from someone else's experiences. Most of all—have fun.

Chapter 1

Who is Doing Welding Now? Who Could Be?

T HE FIRST CONTACT WITH WELDING FOR many people takes place when they engage the services of a welder to repair a valued broken metal object. Just seeing someone welding often provides enough stimulation to inspire the do-it-yourselfer to try it for him/her self. Once on this course, there is little chance of turning back. There is no shortage of things to fix with welding. But welding is also used to fabricate useful and beautiful new items—which is right up the home craftsman's alley. Given the broad cross section of the population now involved with do-it-yourself home and vehicle repairs, it follows that the term *home welder* should include a diverse and growing number of individuals.

General Home Handymen. Many homeowners and farmers have, of necessity, become versatile repair people to fix anything from a leaking pipe to a sagging backporch. Maybe that porch could be better secured to the house with steel brackets—easy to cut and weld right at home. While at it, the home craftsman might also make a steel railing for that porch. Perhaps a few decora-

tive scrolls and flower pot holders could be incorporated in the design. A welded steel-framed canopy or awning might be other items to dress up the porch. Decorative and functional security grills and gates are other popular projects. And so it goes—once the imagination starts, there seems to be no end to home welding projects.

Vehicle Enthusiasts. People with all kinds of special interest vehicles sooner or later become acquainted with welding projects as Figs. 1-1 to 1-4 show. Welding hobbyists are also involved with tractors, race cars, restoring steam locomotives, airplanes, and every kind of boat. Working with vehicles often requires making a weld repair or fabricating an original (or duplicate) part. Special workstands, and holding and lifting devices can be welded at home. Custom tools and clamping devices like pullers, engine stands, and lifting aids are simple welded fabrications.

Home Energy Conservationists. The home craftsman can use welding to make interesting and economically attractive home energy projects—including wood heat (Fig. 1-5) and even

Fig. 1-1. Little deuce coupe. This completely reliable vehicle was brought to its present condition with the use of home welding.

Fig. 1-2. Lots of shiny chrome, but beneath lies a basic welded steel fabrication. These bumpers are popular with off-the-road enthusiasts.

Fig. 1-3. Truck racks are basic welded fabrications.

Fig. 1-4. Almost any type of vehicle can be made using home welding. This garden tractor with a dump bed has prevented many backaches.

Fig. 1-5. Just a couple of years ago this Majestic wood stove was found upside down in a gully. It was rusted and rat-infested. Today it is a showpiece. Many parts and plates were duplicated using home welding techniques.

Fig. 1-6. An electromechanical exercise, this homemade solar tracker involved home welding.

Fig. 1-7. This antique irrigation pump engine was discarded with a cracked block. It was repaired with oxyacetylene welding and a lot of elbow grease.

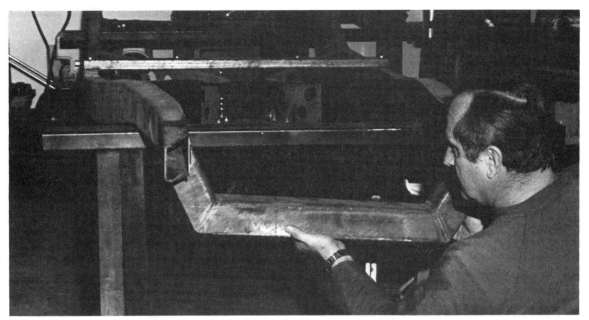

Fig. 1-8. A better-than-new frame for a 1935 Ford is made using welding techniques.

solar energy (Fig. 1-6). Projects made by home welders include log lifters, log splitters, and even wood stoves. These help with cost of home heating. A carriage for hauling logs out of the brush and a trailer to bring fire wood home can also be welded in the home shop. Almost all chain saw attachments are basic welding fabrications. Articles on building such fabrications can be found in periodicals like *Mother Earth News, The Stabilizer,* and *Farm Journal.*

WELDING FOR (ALMOST) EVERYBODY

Recreational welding is used by people involved with: antique restoration and/or replication (Figs. 1-7 and 1-8), home construction and repair, outdoor activities, model building, gold prospecting, piping systems, art projects, and even scientific endeavors. The applications seem endless because those able to weld at home find themselves equipped with another dimension of creative capabilities.

Chapter 2

Projects You Can Do at Home

AS ALREADY MENTIONED, THE AVERAGE CONsumer first thinks of welding when he or she discovers a broken favorite metal object. Industrial metal fabricators were in the same situation in the early 1900s. The need for repairs, and welding's respectable response to that need, prompted the start of commercial welding. Maintenance and repair is still nearly one-half of industrial welding. The other half is, of course, joining metal for new fabrications. Your home projects also fall into these categories of new constructions or maintenance and repair.

Home welding projects include thousands of items both large and small. Here are but a few:

- Storage racks, tables, and furniture
- Wrought iron work in every imaginable application
- Spiral and straight staircases
- Trailers for boats, general utility, motorbikes, snowmobiles, horses, etc.
- Automobile frames including the new generation of kit cars
- Tow bars, engine hoists, ramps, and safety stands
- Barbeques
- Boating hardware
- Router tables and table saw extensions
- Band saws
- Bicycle storage and servicing devices
- Fireplaces and equipment
- Playground equipment

A HOME WELDING PROJECT SAMPLER

This chapter has a sampling of welded projects suitable for the home shop (Figs. 2-1 through 2-9). Most are new fabrications but there is no shortage of items needing weld repair; you will have little trouble finding them. The projects shown have been successfully fabricated at home. While some are more sophisticated than others, little if any, "outside contracting" was used.

Projects ideas appear throughout the book. In fact, Chapter 17 walks you through the fabrication of four projects involving representative operations to construct many more of your own selection and/or design.

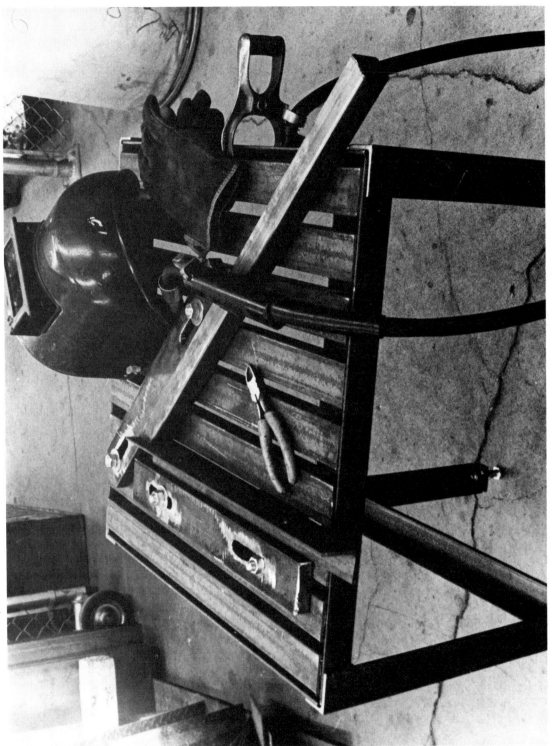

Fig. 2-1. This versatile work table serves as a locating jig for lining up parts for welding. A length of angle bar is fitted to square tubing. Any number of attachments could be fabricated for use with such a table.

Fig. 2-2. A 30-gallon oil drum is the basis for this wood heater.

Fig. 2-3. This cast iron clamping lever for a lathe tailstock was repaired with braze welding.

Fig. 2-4. The use of bolted connections permits easy transporting to the assembly site for this satellite disk.

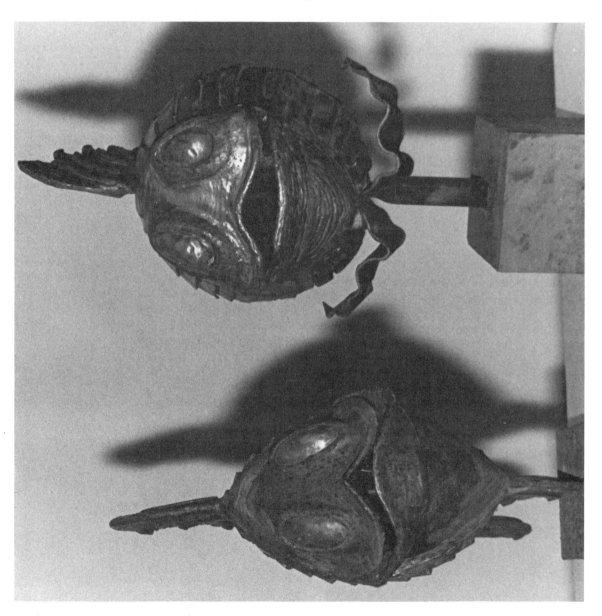

Fig. 2-5. Grinning fish sculptures.

Fig. 2-6. A few lengths of pipe and some flame cut plate were used to form this strong bumper and guard.

Fig. 2-7. The home mechanic will find an engine overhaul stand to be extremely useful.

Fig. 2-8. This press was fabricated mainly from channel bar and uses a standard hydraulic jack.

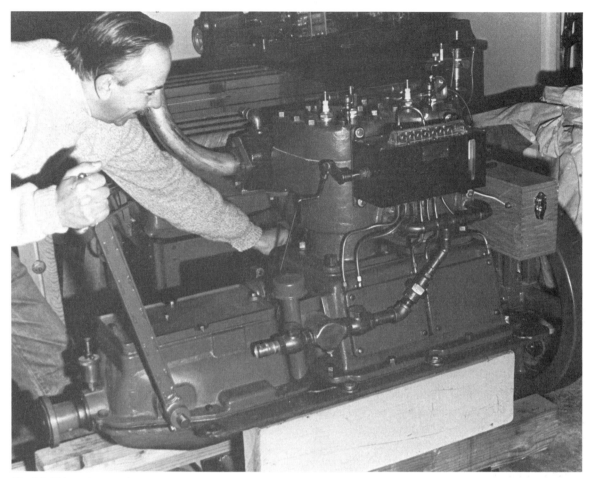

Fig. 2-9. This antique engine was restored to mint condition in a home shop using welding, brazing, and soldering techniques.

Chapter 3
Getting Started

WELDING IS FUN, PRACTICAL, AND SAFE. Why, knowing how to weld can even make you a local hero. However, some practical matters, if overlooked, could lead to safety hazards, disgruntled neighbors, and disappointment. Among these are: your aptitudes, workplace safety and practicality, expenses, and use of time.

YOUR APTITUDES

Let's face it; welding is a manipulative skill, like playing a musical instrument. Not everyone can be a concert musician, nor can everyone be a potential great nuclear piping system welder. But being the world's finest has little to do with having a lot of amateur fun. Welding is no exception. Luckily, modern welding equipment can make those of us who have something less than super talents look pretty good. Also, equipment designed for certain processes is much more "user friendly" than others. You will need to find out what equipment and processes fit you best.

Everyone has different strengths and welding equipment can be expensive, so try before you buy. It would be equally unwise to buy expensive musical instruments or a shopful of specialty welding gear without a clue as to your specific talents.

One proven way to explore your aptitudes is to visit an evening welding class at a local school. These are always popular with do-it-yourself welders. Talk with the instructor and discuss your interest in becoming a more versatile home craftsman. Your visit should give you a good idea of what can be done in a class. Hopefully it will encourage you to sign up for a few weeks of evening training sessions which should answer most questions about your talents and give you some good ideas about welding at home.

Some home welders have tested their aptitude by running their first bead under the eye of a friend with welding equipment. This might not be encouraging. However well-intentioned, if the experienced person lacked either patience or the knack to convey skill training, this could be a negative experience. Avoid putting a great deal of stock in such informal "auditions."

Another way to assess your aptitudes is by inspecting the equipment at a welding supply house. With a little advance warning, many dealers will set up equipment for demonstration. Some stores have a special demonstration area.

When you look at equipment, have a list of questions. At this point, you have nothing to lose except some uncertainties. You might find either good advice or excessive salesmanship. Avoid impulse buying and get other opinions concerning what you intend to do with your welding. Rush, and you could spend a lot on equipment that gathers dust from lack of use.

Take the same cautious approach whenever you decide to expand your skills into new areas of welding requiring additional equipment. Your equipment must match you.

A SAFE PLACE TO WORK

A workplace must be safe from both obvious and subtle hazards. Arc flash, fire danger, grinding chips, slag, sparks, and pungent fumes are easily noticed as dangers. Less conspicuous hazards include depletion and/or displacement of breathing oxygen, unsensed metallic fumes, and even noise and smoke.

They might at first seem severe, but respiratory threats from welding are no more drastic than those faced by a woodworker or a cook. The social use of a table saw or Skilsaw involves precautions from dust, fumes, fire, noise, and unplanned flying objects. Besides you, the welfare of bystanders and neighbors must also be considered. Have a fire extinguisher handy at your work site. If there is an extra fire danger, post a firewatch assistant to stand by with an extinguisher.

As just mentioned, practical safeguards for potential onlookers must be arranged. The welding arc can cause eye damage if watched for a minute or so from even 35-40 feet away. However, a wood chip in the eye is at least equally dangerous. Use common sense. Your hood is down during the actual "arc-on" welding time, so you can't watch for someone wandering into your workplace. It should be secured with a gate or a door.

Learn from construction pros on a crowded downtown site. A few sheets of cheap plywood make an excellent portable shield from arc flash and flying objects. Later, the plywood can be used elsewhere. Black plastic sheeting effectively blocks the arc when placed over a window or other avenue for escape of the flash.

Welding fabrication, with hammering, grinding, sawing, and dropping things, is noisy. Limits to your activities must be observed in the early morning and evening hours. Considering others is standard procedure for carpentry work and tinkering in general. Certain operations like mowing the lawn involve noise but, with planning, courtesy is always possible. The same is true with welding.

Fresh air must always flow through your welding area. Most operations produce harmful fumes and/or use up breathing oxygen. The oxyacetylene flame consumes more oxygen than that supplied through the torch. It is also easy to overheat bronze filler rod—causing toxic fumes from a flareup of the zinc component.

Besides getting fresh air to an otherwise closed workspace, make provisions to remove smoke as it is generated. A few minutes to arrange a small electric fan, a bit of sheet metal, and some wire could save you and others considerable discomfort.

A sure way to discourage fire from welding sparks in a wooden garage or outbuilding is to install interior drywall (or plasterboard). Great cosmetic efforts are not needed; the drywall joints need only be sealed. Drywall also makes your shop more secure from the weather. If you can afford it, insulation under the drywall makes you shop more pleasant for working during extreme weather.

A PRACTICAL PLACE TO WORK

The workplace should also be convenient as well as safe. Your welding projects will be more fun if you can work on them without first having to do a great deal of rearranging. Project storage between work sessions can take up considerable space. Try to avoid using any welding equipment within the house (unless you are doing some job on the house itself). The safety hazards are too great to take chances.

A shop outbuilding is ideal—but great welding

projects are built in garages (car removed) and, in warm climates, a carport.

Electricity and moisture are dangerous, so welding is impractical in an area not reasonably dry. Shelter from wind, rain, and even snow, can be had with temporary windbreaks made with plastic sheeting, a few 2 × 2s, and a staple gun. If you live in a really cold place, do not have the luxury of a separate shop building, and cannot wait to get started welding—you might fashion a snug shelter like those which ice fishermen put up each year. Just remember to allow for continuous ventilation.

A moveable welding bench (refer to Fig. 2-1) that slides on levelling feet is handy for a multipurpose workplace. Cylinder carts, vise and grinder stands, even *power sources* can be made portable. Power sources, or welding machines, supply the special energy needed for electric arc welding. Home welders often build frames and stands with casters for their first projects.

The power source should be located to save extra steps. This might involve wiring changes. Arc welding power sources must be plugged into a suitable source of electrical energy. The supply wiring and wall plug are called the *line connection* and must be protected from overload heating with a circuit breaker or fuse. Most home arc welding equipment uses a 220-volt line supply. These hookups look much like those used for electric ranges or clothes dryers, but the amperage input of a welding machine is greater. Usually, 25 to 40 amps of input current is needed. Figure 3-1 shows the specifications plate on a power source. Table 3-1 lists power requirements of several popular power sources used for home welding.

The input amperage draw of a welding machine is related to its amperage output. As more welding heat energy is used, the draw through the line connection is increased. A 40-amp draw might seem high compared to that of a home appliance like a 25-amp electric clothes dryer, but it is well within the electrical capabilities of most homes.

This leads to power source *size*. A welding machine must produce enough heat energy to easily perform the work you intend to do. The ability to weld plates of a given thickness is largely deter-

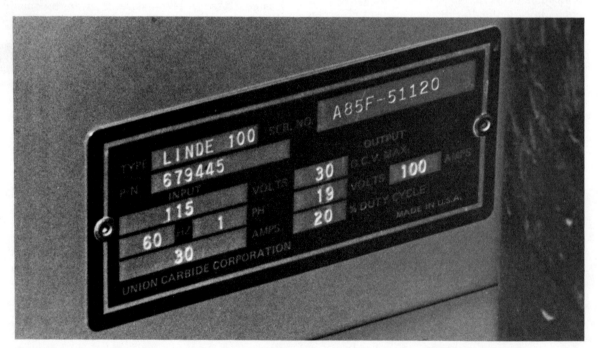

Fig. 3-1. Every electric arc welding power source is marked with its input power requirements in volts and amps. The rated output capacity in amps and duty cycle (time it can work to capacity before needing a rest) are also specified.

20

Table 3-1. Output Ratings, Processes, and Power Requirements for Several Popular Small Power Sources.

MAKE	MODEL	PROCESS	OUTPUT AMPS	INPUT AMPS
Airco	Dip-Pak 200	GMAW	200	46 @ 230 volts
	Mini-Pro 125	GMAW	100	25 @ 115 volts
	Easy Arc ac/dc	SMAW	230 ac 150 dc	45 @ 230 volts
C-K System-atics	MIG 300 SM	GMAW	300	36 @ 230 volts
	FMP 225	GMAW	225	34 @ 230 volts
Cyclo-matic	Powcon 200SMF	GMAW, FCAW, SMAW, and some GTAW	200 or 160	45 @ 230 volts or 33 @ 230 volts
Hobart	Beta Mig 200 and Beta Mig LF	GMAW and FCAW	225	38 @ 230 volts
	Mig-Man	GMAW	105	25 @ 115 volts
Lincoln	ac/dc Arc Welder	SMAW	225 ac 125 dc	50 @ 230 volts
	Idealarc SP-200	GMAW and FCAW	200	46 @ 230 volts
	Idealarc dc-250	GMAW, FCAW, GTAW, and SMAW	250	66 @ 230 volts
Linde	Linde 160	GMAW	160	25 @ 230 volts
	VI-206	GMAW and FCAW	160	40 @ 230 volts
Miller	Miller-matic 200	GMAW	200	38 @ 230 volts

mined by the output (amperage) rating of the power source (Table 3-1). Home projects usually involve metal less than 1/2 inch, most less than 3/8. You may seldom need the maximum current, but a 200-amp power source gives you great flexibility.

Just remember that a 200-(output) amp power source needs to be fed ample input energy from your electrical panel. Wires, fuses (or breakers), and connectors must be sufficiently rated. If this is not possible, consider a smaller unit. It is vexing to stop welding to reset a breaker or blown fuse. See your welding dealer and a competent electrician about line supply questions.

Input plugs for 220-volt equipment come in different styles according to amperage rating. Each plug uses a matching receptacle (Fig. 3-2). Know the amperage capacity of all connectors. It is acceptable to use connectors with a heavier rating than needed, but avoid using lower rated plugs and receptacles.

The best line connection is a separate *fused disconnect* (Fig. 3-3) with a plug-in receptacle for the exclusive use of your power source. It consists of both a heavy-duty switch and a housing for the fuses that protect the circuit (Fig. 3-4).

Having a fused disconnect is not absolutely necessary. Some home craftsmen have begun welding with an extra long primary (plug-in side) cable plugged into the clothes dryer outlet (and maybe dragged through the basement window). Again, know that the amperage draw of any electrical unit is within the capacity of the wiring, plug, and receptacle. If a separate and more permanent arrangement is later made for your power source, extra input cable becomes a primary "extension cord" (Fig. 3-5).

Equipment manufacturers have several pages of practical safety information in their operating booklets. Read and understand this material. Pay special attention to prevention of electrical shock by using a proper ground for the chassis of your power source. Regardless of the electrical power arrangement, follow the electrical code for your locality. Again, it is wise to involve a qualified electrician at least for advice and inspection of your hookup.

If you don't already know, find out what voltage and amperage capability your shop has now. This is practical information to have before you go shopping for equipment. A power source that is too "hungry," could require extensive expansion of the power to your shop. This is not something to find out after your purchase. With the variety of arc welding equipment available, there is probably something that will work well with your present electrical hookups. To handle metal in the 5/16-inch thick range, you need a power source that puts out 150 to 200 amps. To get this type of power, it is necessary to operate from a 220-volt line supply. For light material only, there are some impressive 115-volt welding power sources that work off an ordinary household outlet. More about these in the next chapter.

Steel—the Practical Material

When you begin with welded fabrications, it is best to use the most versatile metal—low carbon, or "mild" steel. This flexible and forgiving substance does not fall apart if you do something slightly wrong. For example, it is difficult to *thermally shock* mild steel with rapid cooling. It is widely available and inexpensive. Oxyacetylene flame cutting is ideal with mild steel. It is welded with any of the home welding processes. Consumables used to weld mild steel are less expensive than for any other metal. Steel fabrication skills also transfer to any other metal you might eventually wish to use.

EXPENSES

There are many different cost factors that affect a given home welding operation. To avoid wallet shock, consider the price tag of the equipment and the ongoing costs of consumables like filler metal, gases, and electricity. Most home craftsmen are pleasantly surprised when comparing the costs of plain carbon steel to other construction materials. For example, a 10-foot length of 1- x -1- x -1/8-inch angle bar weighs about 8 pounds. At $.49 per pound, the angle bar costs only $3.92. This could be used to make a sturdy 28-inch square table top frame. (With shopping, you might find steel as

Fig. 3-2. The electrical receptacle
must match the style of plug on
your power source.

Fig. 3-3. A fused disconnect is both an input power line switch and an overload safety device.

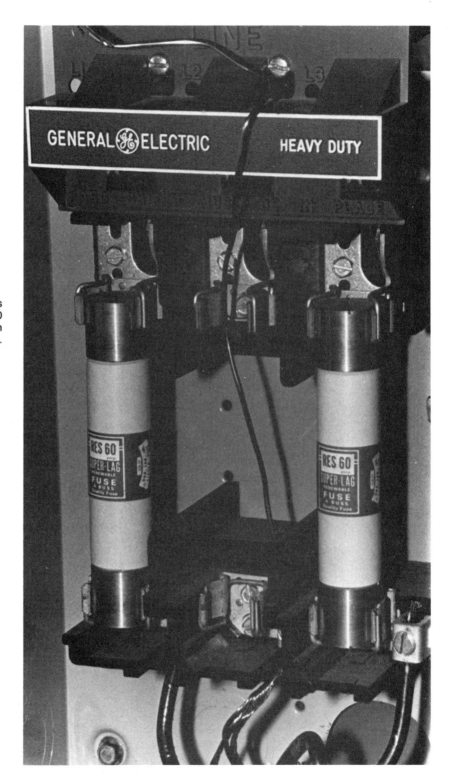

Fig. 3-4. The inside of this fused disconnect contains 60 amp slow-blow fuses that can handle momentary overloads.

Fig. 3-5. A primary (input) power extension cable fitted is very useful. This one enables the power source to be used an extra 25 feet away from the main plug in receptacle.

cheap as $.30 per pound.)

Probably the greatest cost variable is your workplace itself. If you already have an area set aside to fix the car or to build wooden projects, welding activities can probably fit right in with a little fireproofing.

The cost of the compressed gases for cutting and welding must be considered. These are not astronomically high but are often the greatest ongoing expense of home welding. Gas rates, largely determined by transportation and handling costs, vary from region to region. More is said about gas costs in Chapter 7.

Of course, filler metal and electrical power are used during welding. The average home welder uses only a small amount of filler metal. Fifty dollars of filler metal goes a very long way with small home fabrications. Even with a cooling fan, welding power sources use significant power only while the arc is lit.

Energy costs also vary, but even at the highest rates, the cost of adding an arc welder might not even be noticed. The energy to arc weld might cost little more than using a portable electric heater or two. To estimate what it costs to run an arc welder, check your electric bill or call the power company to get the rate per kilowatthour (KWH). (A KWH is 1000 watts consumed for 1 hour.) Working hard, your arc welding equipment can use about 4000 watts of power in an hour or 4 KWH. Multiply your energy rate by 4 to estimate your hourly arc on expense. For example: if your electricity costs $.15 per KWH, 4 times $.15 = $.60 per hour. It might take you weeks to do that much welding.

For most of us, welding equipment qualifies as major purchases and some shopping pays off. Chapter 7 says more about buying equipment.

TIME

With any new enterprise it is important to think about the asset that nearly always seems in short supply—time. Before welding things together you must learn how to weld. This takes time: to observe, experiment, and practice. Time is also used to develop additional welding skills. For a price, certain welding processes and equipment enable you to both learn and work more quickly than others. Welding time is related to welding processes and the associated equipment that then involves expenses.

Once you have learned to weld, how much time will the rest of your schedule allow? Because time gains can be had with equipment selection, a choice might have to be made between spending money or time. For example, wire feed welding is easier and usually faster than stick welding, but the higher initial costs might not be feasible at a particular time. With a very tight schedule, perhaps time is your first priority and the expense of efficient equipment secondary. And again, you might not have either a lot of time or money for home welding. In this case, you must compromise according to the type of project you aim to build.

The time needed to achieve proficiency varies for the different welding processes. Chapter 5 looks at learning time and Chapter 8 discusses advantages and disadvantages of each process.

Your welding activities will be determined by available time and resources. No matter which equipment and process you use, it is important that your welds be strong and your welding enjoyable.

Chapter 4

Equipment You Will Need

Y OU CAN WELD TOGETHER SOME GREAT projects at home without a great deal of equipment. Time and convenience aside, even a sparsely equipped shop can turn out impressive work. A look at what metalcraftsmen in the third world routinely do with only the barest of means can be instructive and encouraging. A lot of sophisticated equipment is simply not needed.

There are three ways to match welding equipment and the job to be done:

● With a complete variety of equipment, you could select a particular joining process from several alternatives.
● With limited equipment, you make it do or forget it.
● When you have no choices, the job can be only done one way.

CUTTING

Material for welded fabrications, like wooden projects, must be cut and formed. An *oxyacetylene cutting torch* is virtually indispensable. It is the equivalent of a woodworker's saw. A cutting torch is part of a *combination oxyacetylene welding outfit* (Fig. 4-1). Another essential time and labor saving cutting tool is a *disc grinder*. Figure 4-2 shows one cutting into a pipe. More is said about cutting in chapter 16.

LAYOUT AND FITUP

Layout is marking material to show where cutting, drilling, forming, finishing and other pre-assembly operations are to be done. *Fitup* (or fitting) is carrying out these operations to produce parts and then to assemble them prior to final welding. Ordinary measuring and shop implements such as a tape measure, marking devices, vises, clamps, pliers, wrenches, hammers, hacksaws, and chisels are layout and fitting tools. Some of these are found in almost every home. Tools are something you may never stop acquiring. Chapter 14 has more about tools, and Chapter 15 says more about layouts.

Fig. 4-1. An oxyacetylene welding and cutting outfit is an essential purchase for the home welding shop.

Fig. 4-2. A disc sander/grinder is a basic metal fabricating tool. It can even be used for cutting, should the need arise.

WELDING

Once metal has been cut and/or properly formed, it is usually welded together with the heat of an electric arc. Your welding activity is largely determined by the power source you choose (Fig. 4-3). This is because most permit arc welding only with a particular type of process. As mentioned in the introduction, home welding processes include SMAW, GMAW, and FCAW. It is necessary that you first consider the process that you intend to use before you make any decisions about power sources. Again, with some hands-on welding experience you will know what process and equipment fits you best.

Certain welding jobs can be done with the heat of an oxyacetylene flame. An oxyacetylene outfit is used for welding and other torch processes as well as for flame cutting.

Process Selection

The particular joining process you use is determined by the welding equipment you have, and conversely, the process you select determines the equipment you must have. For example, oxyacetylene equipment lets you cut, weld, braze, braze weld, and even solder. An arc welding power source is used only for specific welding processes. Some arc cutting is also possible with the home-type power sources.

POWER SOURCE TYPES

The types of power sources that could be used in the home shop are: the transformer, the rectifier, and the motor generator. Furthermore, there are

Fig. 4-3. The power source, also known as a welding machine, provides the proper type of electrical energy for arc welding. This is a single-phase unit suitable for use in a home shop with a 230-volt supply. (Courtesy of Airco.)

two categories of rectifier power sources: constant current (CC) and constant voltage (CV)—also known as constant potential (CP). A fourth type of power source, the inverter, is relatively new to the North American market. Table 4-1 outlines power sources used for home welding.

Transformer Power Sources

In the late 1940s, the home craftsman was mostly limited to arc welding with the basic transformer

Table 4-1. Power Source Types and Nature of Output.

DESIGN TYPE	OUTPUT
Transformer	ac, constant current (CC)
Rectifier (a)	dc (also sometimes ac/dc), constant current (CC)
Rectifier (b)	dc constant voltage (CV/CP)
Motor Generator	dc, ac/dc, or ac; constant current (CC)
Inverter	dc, constant current (CC), or constant current (CC) and constant voltage (CV)

type of power source. Figure 4-4 shows one of the earliest machines available for home use. Transformers are sometimes termed "buzz boxes"—they make a humming electrical type sound during the arc-on time. The small transformer power source has only been promoted as the utility or farm arc welding machine. These machines have been sold through catalog sales, auto parts stores, and even feed and grain stores. They are still made, and they are great little work horses that continue to carry a low price tag of $250 or less (Fig. 4-5).

The transformer power source outputs only *alternating current* (ac) used only with certain SMAW shielded metal (stick) electrodes. For economy, fans are seldom built in. Light construction generally limits these units to a 20-percent *duty cycle*. Duty cycle means the percent of time within a 10-minute time block that a unit can safely be loaded to its rated value. Although these machines permit arc welding with a small initial investment, one is limited to what can be done with only stick electrodes.

Rectifier Power Sources

A slightly more elaborate power source for shielded metal arc welding is the small *rectifier* machine (Figs. 4-6 and 4-7). A rectifier is a device to turn ac into direct current (dc). Rectifier power sources change ac input from the power line into dc output for welding. This means smoother welding without the brief ac arc outages. With refinement of the diode rectifier (Fig. 4-8) in the 1950s and 1960s, dc units became practical and widespread. A few more whistles and bells are often added—such as continuous output control instead of plug-in taps. These make rectifiers cost more than transformer units but also nicer to use. A rectifier unit for home use costs about $300.

Both the transformer and the rectifier power sources described so far produce a *constant current* type of output curve (Fig. 4-9). The droop of the curve shows how the welding current can reach only a certain limit. The welding current needs this limit to prevent the large stick electrode from burning like a fuse if it became stuck to the plate. (Some rectifier machines produce a CV type of output for

wire feed welding, as explained shortly in this chapter.)

Constant current machines are used for the shielded metal arc process and, with certain refinements, for Gas Tungsten Arc Welding (GTAW) as well. Advertising is quick to point this out. GTAW is a way to weld thin aluminum and stainless steel. It is little used by most hobbyists who work mostly in mild steel.

Rectifier add-on units called *dc converters* convert the ac of a transformer power source into dc. These work, but mean more equipment and extra connections. They are mentioned for possible upgrade of an existing power source but not for initial purchase.

Motor Generator Power Sources

Engine driven *motor generator* power sources permit arc welding where no line power exists (Figs. 4-10 to 4-13). They are mostly constant current for stick electrode use, but can be adapted for use with a wire feeder.

With engine noise and fumes, generator units are more suited for rural than city use. Often called "popping johnnys," these units under load sound about like a gas-powered lawnmower.

Inverter Power Sources

The inverter-converter is a compact power source with great promise. It is a design departure from the standard machine that uses 60 Hz (cycles per second) through all of its operating steps. Converters, first used in Europe, have only lately become generally available. They are frequently marketed as a power source and wire feeder package. Some major advantages claimed for the converter power source are:

● Portability—the power source weighs only 70 pounds

● A strong, smooth arc

● Less power consumption than coventional transformer rectifier units

● Multiple process capability for GMAW, FCAW, SMAW, some GTAW, and arc gouging and cutting.

Fig. 4-4. This is a living, working relic from the early 1950s. It is an example of one of the first arc welding power sources built for the small home and farm shop. Because there are no moving parts, they continue to perform decade after decade.

Fig. 4-5. A popular 225-amp transformer type of power source. Outputting only ac, these units are sometimes affectionately called "buzz boxes." (Courtesy of Lincoln Electric.)

Fig. 4-6. A small power source that outputs either ac or dc is very versatile in the home shop. This unit uses a series of plug-in taps for the welding cables. (Courtesy of Hobart Brothers.)

MODEL AC/DC 225/125
INPUT
SINGLE PHASE 60 HERTZ 230 VOLTS 50 AMPS
RATED OUTPUT
225 AMPS AC OR 125 AMPS DC AT 25 VOLTS
79 VOLTS MAX. OCV AT RATED INPUT
TEMPERATURE RISE 115°C MAX.
20% DUTY CYCLE

ELECTRODE
POLARITY

WARNING
READ SAFETY INFORMATION
ON SIDE BEFORE OPERATING.

+
DC
–
AC

CAUTION
DO NOT SWITCH WHEN
WELDING. SWITCH MUST
SNAP INTO EACH OF THE
DIAL POSITIONS.

CODE NUMBER

AMPERES
DO NOT SWITCH WHEN WELDING

SWITCH MUST SNAP IN EACH OF THE DIAL
POSITIONS. SETTING BETWEEN THESE POINTS
MAY CAUSE DAMAGE.

105 120 135
90   150
75   175
60   200
40   225
AC   AC

LINCOLN
ELECTRIC
AC/DC ARC WELDER

THE LINCOLN ELECTRIC COMPANY   CLEVELAND, OHIO   U.S.A.

World's Largest Manufacturer of Arc Welding Products

Fig. 4-7. This small ac/dc power source features a polarity switch at the left. (Courtesy of Lincoln Electric.)

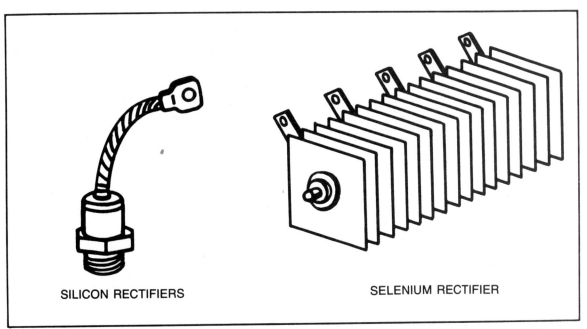

SILICON RECTIFIERS   SELENIUM RECTIFIER

Fig. 4-8. Rectifiers change ac into dc. The advent of the silicon (diode) rectifier made smaller dc units practical. (Courtesy of Alloy Rods.)

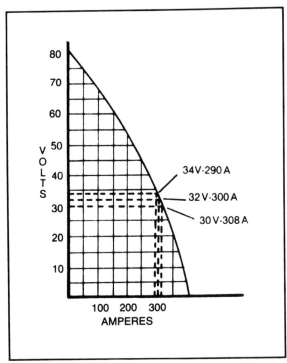

Fig. 4-9. A constant current volt/ampere curve. (Courtesy of Alloy Rods.)

Converter advantages come from sophisticated (delicate, say competitors) control logic circuitry. They output very smooth dc. A conventional power source has a 150-pound transformer; the small converter transformer is 12 pounds! This works because the 60 Hz input is raised to a frequency between 800 Hz and 5000 Hz. This higher frequency is introduced into the small transformer to output as much amperage as a much larger 60 Hz transformer.

The compactness of converters make them attractive to the home craftsman. The arc is also so smooth that it spoils many users into avoiding conventional units. Some forecasters see the inverter-converter design as standard in the near future.

WIRE FEED WELDING

The same power source can be used for all the home variations of wire feed welding to include:

- Gas Metal Arc Welding (GMAW), formerly MIG—used to weld almost anything (Fig. 4-14)

Fig. 4-10. This is a deluxe smaller motor generator power source that outputs either ac or dc and also 110 volts for lights or power tools. It can be used almost anywhere.

Fig. 4-11. This motor generator power source features an ac/dc output and a reliable air-cooled gas engine. (Courtesy of Hobart Brothers.)

● Flux-Cored Arc Welding (FCAW) with and without shielding gas—used mostly to weld mild steel (Fig. 4-15)

Wire feed power sources differ from stick machines in that they have a constant voltage (CV) type of output curve (Fig. 4-16). There is no sudden amperage drop off as with the constant current curve. "Constant potential" (CP) is used interchangeably with constant voltage. Most CV/CP power sources must not be used with stick electrodes, although some are designed for dual service.

Wire Feeder

A continuous wire feeding mechanism is used with

the CV/CP power source. The wire feeder can be part of a self-contained welding unit (Figs. 4-17 through 4-19). With a separate power source and wire feeder, it is best to seek the feeder designed for the power source. Adaptation problems can arise from "bashing" diverse components.

A Brief Wire Feed History

Wire feed welding can seem needlessly complex. A short history might keep things clear.

Gas Metal Arc Welding (GMAW) is wire feed welding using a solid wire electrode and an externally supplied shielding gas. The older term "MIG" is still quite in vogue.

After World War II, a revolutionary type of arc

Fig. 4-12. This compact portable gas powered welding unit outputs only ac but packs up to 4500 watts of 230-volt auxiliary power. (Courtesy of Hobart Brothers.)

Fig. 4-13. This power source is really portable, it goes wherever you drive. The heart of this package made by Resco in Fort Worth consists of an extra heavy-duty alternator that replaces the standard unit on a car or truck engine. The welding cables are being connected to the control box and ground jack.

40

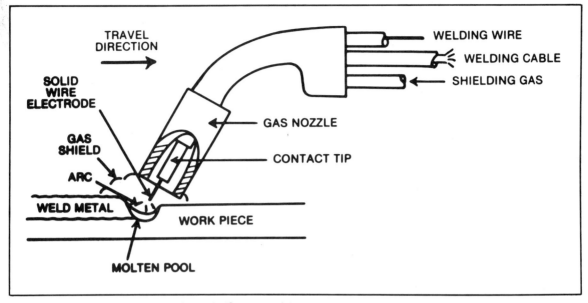

Fig. 4-14. Gas Metal Arc Welding (GMAW). (Courtesy of Alloy Rods.)

welding was introduced that permitted fast welding of heavy aluminum. It was soon called "MIG" which sounded like "TIG." MIG stood for metallic *inert* gas, because an inert gas (helium and/or argon) is required for shielding with aluminum welding. TIG (Tungsten Inert Gas) welding had been around for awhile and was used to weld light aluminum aircraft parts. (Today TIG is called GTAW.) MIG worked with heavier sections and contributed to the increase in the number of aluminum products during the 1950s and 1960s.

An efficient way to weld steel other than with stick electrodes was also sought. By the early 1950s, aluminum MIG technology had been

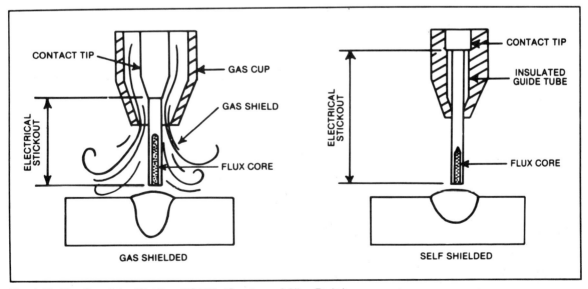

Fig. 4-15. Flux-Cored Arc Welding (FCAW). (Courtesy of Alloy Rods.)

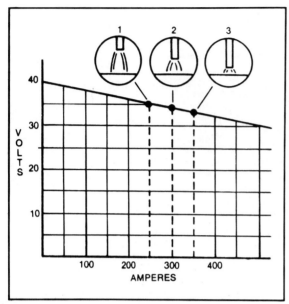

Fig. 4-16. Volt/ampere curve for constant voltage. (Courtesy of Alloy Rods.)

adapted to steel welding using carbon dioxide rather than argon or helium. Carbon dioxide is not an inert gas but the term MIG was still used. Today the preferred term is Gas Metal Arc Welding (GMAW). There are three (or four) variations of GMAW depending on the gas, voltage, and amperage used. These are explained in Chapter 11. GMAW permits high quality, efficient welding of most metals from thin sheet stock to 1-inch plate.

An offshoot of solid wire GMAW is Flux-Cored Arc Welding (FCAW) in which a tubular wire is filled with flux and even alloying ingredients. The flux-cored electrodes are mostly used to weld steel. Some early day proprietary names persist. "Dual Shield" (Alloy Rods Corp) refers to gas shielded FCAW, and "Innershield" (Lincoln Electric Co.) refers to self shielded FCAW. Flux-cored electrodes are made by several manufacturers.

For about 15 years, flux core was confined to welding heavy structures in the flat position. But lately, improvements in these wires has been on the leading edge of commercial welding progress. New small flux-cored electrodes enable quality welding in any position, some with and some without gas. Such small .035-and .045-inch wires have extended

the capabilities of the home welder who deals mostly with lighter sections.

Wire Feed Equipment for the Home Shop

Versatility with metal types and thicknesses is gained with single-phase wire feed welding equipment. As mentioned, some combine a power source and feeder. Other configurations include a separate wire feeder and even a *spool gun* (Fig. 4-20). Chapter 11 covers different equipment options for wire feed equipment.

120-Volt Units

A 220-230 volt wire feed package might offer the greatest versatility, but smaller units are now available that operate from 115 volts. Such units have the great appeal of simply plugging into existing wall outlets (Figs. 4-21 and 4-22).

The 115-volt units are designed for use with auto body and other light applications. The home craftsman for whom a 230-volt hookup is impossible might wish to consider these small units. The overriding fact is that they are intended just for light work. (You cannot get any more energy out of a power source than you put in.) They also have a low *duty cycle time*—the safe "arc-on" time during which a power source can be worked hard without having to cool off. The 115-volt unit works fine if you limit yourself to working in the 3/16- to 1/4-inch range. Many home welding projects do fall within this range.

SHIELDING GAS

Gas-shielded welding requires a gas cylinder, a regulator-flowmeter (Fig. 4-23), and connecting hoses. A small (150-cubic foot) cylinder of argon or argon mix lasts for about 3.5 hours of non-stop welding. Carbon dioxide may cost only 1/5 of an argon mix but produces a comparatively ragged weld. A regulator-flowmeter controls the flow of shielding gas. With the self-shielded flux-cored electrodes, no gas provision is needed.

Self-shielded electrodes represent an inexpensive way to begin wire feed welding. They work well on mild steel even if the metal is not entirely

Fig. 4-17. A popular 200-amp single phase combination power source and wire feeder. (Courtesy of Hobart Brothers.)

Fig. 4-18. This compact combined power source and wire feeder features a Binzel-type quick disconnect gun assembly. (Courtesy of Airco.)

clean, but there is generally some penalty in weld quality. The merits of the wire processes suitable for home use are discussed in Chapter 10.

CLEANING

The term *cleaning* describes any postweld operation to make the welded object useable and attractive. Cleaning includes removal of spatter (splashed metal), slag, smoke, and dust prior to painting or other coating. A final cleaning "once over" can reveal sharp burrs that require filing and/or grinding.

The most elaborate cleaning tool that you will surely appreciate is a heavy-duty disc grinder, capable of 5000 RPM, which can also serve as a sander. These are different from auto body shop sander/polishers that rotate under 2000 RPM. A good disc grinder is expensive, $150 or more, but lasts for years (Fig. 4-24). Avoid flimsy units—they just fall apart. If fitted with a thin reinforced wheel, disc grinders provide a fast way to slice through most metals.

Cleaning equipment also includes the familiar slag hammer and wire brush along with chisels, scrapers, and even liquid cleansing agents. Chapter 19 discusses cleaning procedures in more detail.

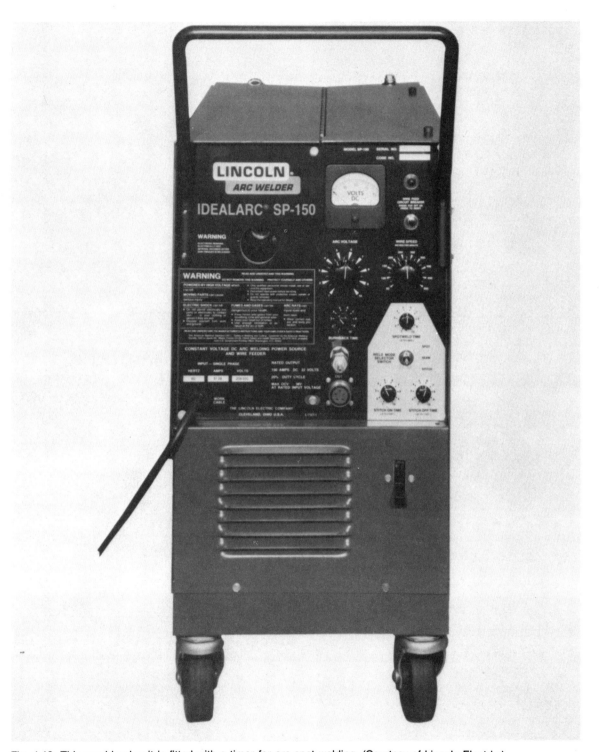

Fig. 4-19. This combined unit is fitted with a timer for arc spot welding. (Courtesy of Lincoln Electric.)

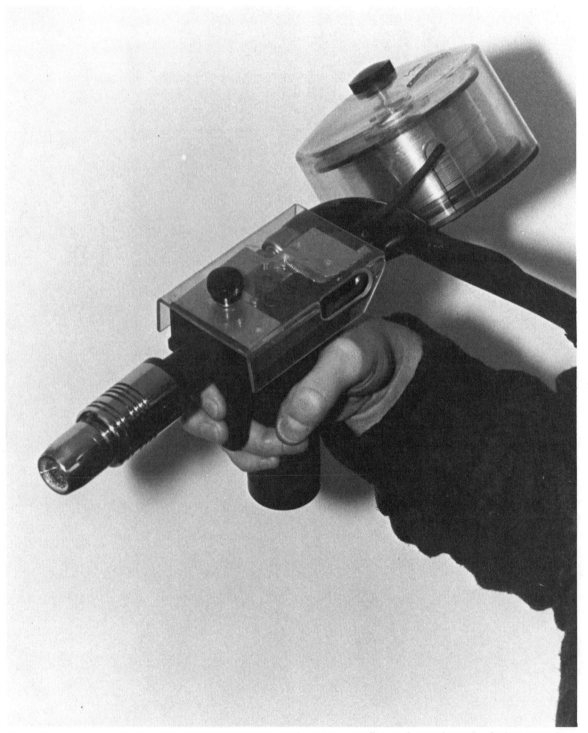

Fig. 4-20. A spool gun has the drive motor in the handle and carries a small 1- or 2-pound supply of wire electrode.

Fig. 4-21. This small wire feeder package that operates from 110-volts is popular with auto body shops.

Fig. 4-22. Equipped with two lifting handles, this 110-volt wire feed package can be carried by two people. (Courtesy of Airco.)

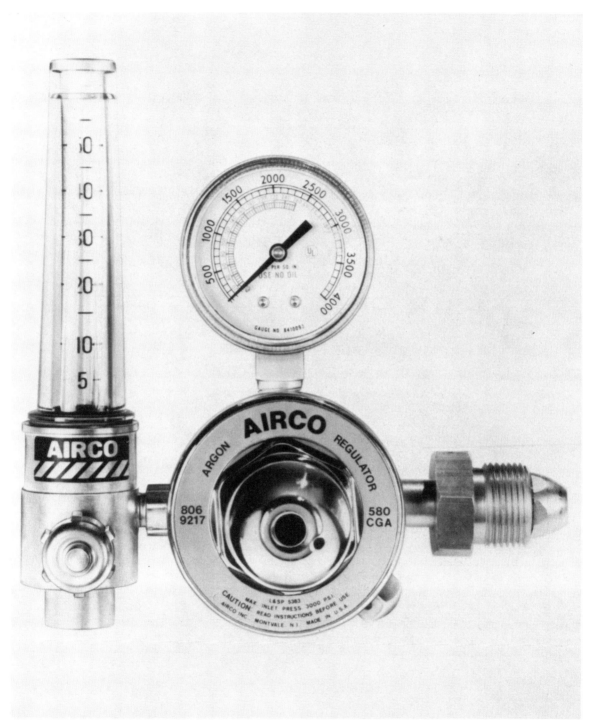

Fig. 4-23. A regulator-flowmeter is used with gas-shielded arc welding processes. This device automatically reduces the high cylinder pressure to about 50 psi. The gas flow rate is set manually. (Courtesy of Airco.)

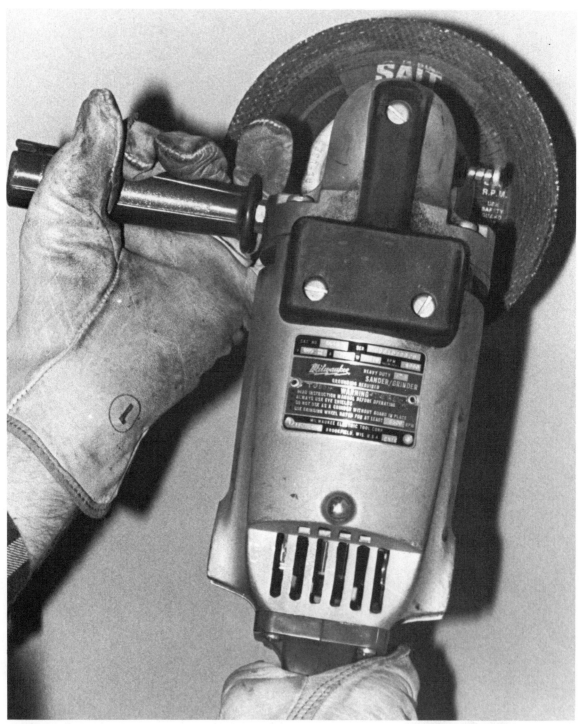

Fig. 4-24. A reputable heavy-duty sander/grinder such as this one is built for years of service. These tools are indispensable for metal fabrication.

50

Chapter 5

Avenues for Learning How to Weld

T HERE IS PROBABLY NO ACCURATE ESTIMATE of the number of welding hobbyists because no one tracks the small sale. However, there is evidence to indicate that the ranks of home welders are increasing. For several years, the best-selling welding equipment has been single-phase wire feed systems. Many of these have gone to auto body and muffler shops, but many have also been sold to do-it-yourselfers. Somehow these non-professionals have learned to use their equipment without the time investment of extensive vocational training.

The owner's manual supplied with a welding machine contains valuable information and specifications, but says little about how to use the equipment in actual fabrication. What are the training avenues for the welding hobbyist?

WHERE TO START

As in the past, there are several ways to learn welding:

- Informally learn from another

- Read and experiment at home
- Take a welding course
- Take a correspondence course
- Use the fumble-and-blunder method

Informally Learn from Another

Manipulative skills are often just picked up by watching and imitating a proficient user of those skills. This is true for woodworking and other do-it-yourself activities, it also holds for welding. Hopefully, the demonstrator is aware of safety procedures and actually has the skill (not to mention the ability to teach it).

A drawback with this approach is the lack of standards for measuring the proficiency gains of the learner. It is tough to know when you have mastered one skill well enough to progress to the next—if the next is even known. If you try to learn welding by this route, get second opinions on your progress. Show your practice welds to a professional welder for evaluation.

Much can be learned by watching others weld,

especially with advance techniques. However, if possible, first learn the basics, especially *safety* basics, in a structured training environment.

Read and Experiment at Home

Hobbyists starting out or seeking to expand and improve their welding could get a textbook and try out techniques in the privacy of their own home shop. If workmanship samples and the text are clear, and you are imaginative, the self-taught approach can work. But you might need more information. If you opt for some structured training at a later time, you will at least have a frame of reference. A problem with self-taught skills is that poor work habits can become ingrained and are hard to overcome later.

Safety training is a part of every welding course. If you teach yourself, be sure you know the safety hazards. Again, there is no substitute for a live safety lesson and class discussion.

Take a Welding Course

It was suggested in Chapter 3 that you explore welding by visiting a welding class. If you are enrolled in a course, you can learn formally and informally from others and still read and experiment at home. A class combines the advantages of the other two methods just outlined with legitimate instruction. A welding course usually gives the most information in the least time.

Full-time vocational welding training is offered by public and private schools. Training for the hobbyist is provided through community colleges or community programs of municipal school districts. Such classes run mostly in the evening for the convenience of daytime workers.

Evening welding classes have a subtle advantage—skills held by class members tend to get spread around. The diversity of such classes is great for learning. A single class can contain a few tradesmen upgrading their skills, a metal sculptor, a fisherman, a farmer, a car buff, a boat buff, a back-to-earth advocate, an amateur scientist, and a few people who just find welding fascinating.

Some will already be skilled. Just rubbing elbows with them will give you at least a few ideas. Even if all your classmates were novices, a class still has the advantage of shared experiences to give depth to learning.

Contact with other class members is not, of course, the only incentive for taking a welding course. Supervision by an experienced instructor makes the most efficient use of your learning time. He or she always passes along a few useful shortcuts and tricks of the trade.

If a welding class is not practical or desirable, try your welding supply house for some hands-on practice. Some offer mini-courses for small groups of customers. It is to their advantage, because you will be likely to buy some supplies from them. Unfortunately, the importance of the small sale is not always appreciated and encouraged by every dealer.

Take a Correspondence Course

Another way to learn welding is through a private correspondence course. In a remote location or with an uncertain schedule, this might be a solution. The overriding problem with self-study is the lack of feedback—the inability to get your questions promptly answered. Also, manipulative welding operations are difficult to convey without first-hand examples. (However, training media such as video cassettes are always improving.)

As with any mail-order operation, investigate the claims of a correspondence course before laying down any hard-earned cash. Ask for a sample of the training package. The Postal Service, your attorney general's office, and the Better Business Bureau should know if there is any question of legitimacy about a particular program. Ask the outfit for the name of a satisfied customer in your area. Such consumer research can only cost a little time. Correspondence training is recommended only as a last resort.

Use the Fumble-and-Blunder Method

The fumble-and-blunder method is not suggested. You, your loved ones, and your environment are to precious to endanger. A license is required for explosives and firearms, yet only cash is needed to obtain potentially destructive welding equipment. Of course, the fact that you are reading this book means that you are interested in high-quality welding in a safe manner. Unfortunately, there are a few foolhardy louts who try to do things by just twisting knobs and dials. Their statistics are one reason insurance rates are so high.

MORE LEARNING

No one uses just one avenue for learning how to weld. Many home welders gained their skills through a combination of informal exposure, reading, experimenting with their own equipment, and taking training. These do not happen in a set sequence, nor do they ever really stop happening.

Welding, like woodworking, is a dynamic activity that seems to grow with time. New tools, new skills, and improved techniques make these activities ever refreshing. Home welders continually discover new things to build along with the techniques needed to build them. Besides books, Appendix C also lists some suggested periodicals that focus mainly on commercial welding but also contain practical information for all welders, including you.

"Learn as you go" is basic to all crafts. As your welding progresses, you will be able to visualize how many structures were fabricated. An exciting aspect of being able to weld is that—almost anywhere and at any time—you can learn something new.

Chapter 6

Spending Time

THIS CHAPTER CONSIDERS THE TIME ASPECTS of welding at home: time to learn; time to shop for equipment, supplies, and materials; travel time; work time; and lost time. With an early idea of how much time is involved with home welding, you can better plan how to include it in the other business of living.

TIME TO LEARN

It is said that learning is fun, but how much time do you want to spend learning how to weld? The less the better? There are ways to speed it up. Use an easy process that requires less practice than others. That's easy—choose an easy process. Yet, as was explained in Chapter 4, the same equipment is not used for all processes. Equipment for the easiest process is not the least expensive, so you might face a dilemma of choosing between dollars or welding ease. Chapter 7 discusses the expenses.

The fastest and easiest way to start building welded projects is to use a wire feed process. This is true whether or not you already know how to arc weld with stick electrodes.

The home craftsman can learn about welding time from industry where time is money. Commercial users of welding must consider time when figuring the bottom line.

The occasional shortage of SMAW welders, along with inherent inefficiencies of that process, has prompted most manufacturers to switch to wire feed processes where possible (Fig. 6-1). Little time is needed to upgrade existing welders; if required, even unskilled workers can be trained to use the wire processes in a mere fraction of the time needed for learning the stick process. The sales of welding consumables shows this trend. In the 1960s almost 80 percent of the filler metal produced went into stick electrodes. By the mid-1980s they accounted for considerably less than half the filler metal produced.

Manufacturers depositing many tons of weld per year get anxious over stick electrode stub losses and the time lost with arc starts and stops. But these concerns have little meaning to the home craftsman who might deposit only 100 pounds of

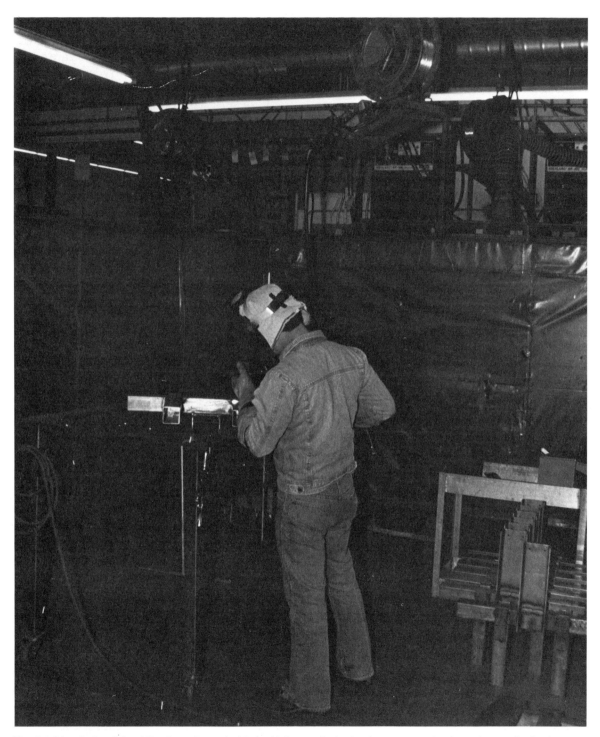

Fig. 6-1. Most industrial welding is performed with the high-speed wire feed processes. As shown here, wire feeders are often suspended to provide an uncluttered work area.

weld in a busy year. So don't worry about time lost changing electrodes. However, you could cut your filler losses in half with wire feed welding. Most manufacturing anxieties over profits do not exist in the home shop, yet training time is something everyone thinks about.

It has been said that there are lies, damn lies, and then statistics. The actual low hour figures in Table 6-1 must fit one of these categories. But, they do allow some time comparisons for gaining minimal welding skills with different processes.

It is apparent from Table 6-1 (taken from the American Welding Society) that shielded metal arc welding is the most difficult to master. While there are similarities between GMAW and FCAW wire feed processes, the figures simply lump these together. The numbers are not realistic, but it is safe to conclude that processes that do not rate any separate learning time probably are not all that tough to learn.

A Semi-Break with Tradition

The beginning home arc welder has the choice of welding steel with SMAW, GMAW, or the FCAW processes. Maybe SMAW has been around for a long time—but it is difficult to be sentimental about a process that uses up a great deal of our valuable recreational time. The wire processes are quick to learn.

Why does oxyacetylene or "gas" welding get so little endorsement as a home welding method? (After all, the figures in Table 6-1 say it only requires 44 hours to master.) Some reasons for this downplay of a genuinely versatile method are: it is expensive to use, involves hazardous gases, requires excessive heat input to all but the lightest of metals, and is incredibly slow. These are a few of the same reasons why industry long ago abandoned gas welding for production.

With such liabilities, there might seem little reason to bother at all with oxyacetylene. Yet, welding is but one oxyacetylene operation. Torch brazing and soldering are essential to maintenance and repair and are used frequently in the home shop. Oxyacetylene is also discussed in Chapters 8, 9, and 21.

By all means, eventually learn to gas weld. Oxyacetylene welding is recognized as the best initial welding experience for training vocational welders. It conveys how to work with molten metal and develops skills needed for more sophisticated processes.

Remember your valuable time. You need some form of arc welding to fabricate steel projects of any size, and the fastest way to get up and running with arc welding is with the use of self-shielded, flux-cored wire electrodes.

There are people who disagree with this analysis. In certain situations, efficiency counts for very little: More about this in Chapter 8.

TIME TO SHOP

Like it or not, it takes time to chase down needed equipment and material. If possible, let your fingers do the walking. Use the Yellow Pages and telephone. With a steel supplier's data book or a products sheet, you can see what is available. Many suppliers give out price sheets (subject to revisions,

Table 6-1. Processes and Minimum Suggested Time for Learning Each.

PROCESS	LEARNING TIME
Oxyacetylene (gas) Welding (OAW), Oxyacetylene (flame) Cutting (OAC), and Torch Brazing	44 hours
Shielded Metal (stick) Arc Welding (SMAW)	176 hours
Gas Metal (MIG) Arc Welding (GMAW)	44 hours
Flux-Cored Arc Welding (FCAW), gas shielded	hours included in GMAW
Flux-Cored Arc Welding (FCAW), self-shielded	hours included in GMAW
Gas Tungsten (TIG) Arc Welding (GTAW)	44 hours

of course). Doing your own estimating can save you time and money.

Shopping helps not only to find the best prices but also to locate alternatives. For example, as long as you are calling, you might ask a steel supplier about using metal with slightly different dimensions. The angle bar that you might have decided to use for your project could actually cost *more* than a size a bit heavier. Prices are partly a matter of supply and demand. By substituting, you might weld up a better project for less money.

Also, check out metal cutting fees. For a few extra dollars, you can sometimes save on hours of cutting time at home.

TRAVEL TIME

A run to the local steel service center, like a trip to the lumber yard, takes away from construction time. This is assuming that you are even able to transport the materials. Delivery of longer or larger metal shapes is often inexpensive and frees you for better uses of your time. You will, however, usually need to allow a day or two for delivery.

Most steel service centers have a "bargain basement" shorts or remnants department. Lower prices of remnants is possible because much of the digging and sorting is left to you. To take advantage of such prices you need time to hunt for what you need.

No matter how well you plan your travel time (or your evasion of it), there is a mysterious law of nature at work to confound you. This law applies to all do-it-yourself activities and goes something like: No one trip to the hardware store, lumber yard, welding supply house, etc., is ever enough.

WORK TIME

At last! You start construction of your project. But do not forget setup time. You might need to drag out equipment, make and test connections, clear the area of flammables, and remove or protect items that could be damaged by smoke or weld spatter.

When finished working, you also need some time to drag up your welding leads, shut off gas cylinders, bleed oxyacetylene lines, and perhaps knock down an earlier set up. It is good practice to take time to sweep and clean your work area before leaving it. This way you will not overlook a smoldering fire.

LOST TIME

There are a few things (besides a power outage) that could shut down your welding in a hurry. You could run out of oxygen or acetylene. In fact, running out of any welding consumable can stop your welding. Besides fuel gases, consumable items include:

- Shielding gas
- Electrode and filler rod
- Flux
- Grinding discs
- Contact tips (Chapter 11)
- Metal
- Hardware

Make a frequent check of your consumables and do not run your gas cylinders bone dry.

It seldom happens, but home welding equipment does sometime malfunction. Some repair shops have loaner equipment. In many areas you can rent welding equipment.

Chapter 7

Spending Money

T HE ENGAGING SUBJECT OF MONEY AND ITS uses might seem like a short story when it concerns our own limited resources (barring lottery windfalls). Our beneficial recreational tools are hardly luxury items, but neither are they cheap. The greatest expense with home welding is nearly always initial equipment purchases, but there are also ongoing costs such as compressed gases, filler materials, steel, and electrical energy.

EQUIPMENT EXPENSES

Choose welding equipment carefully; you will likely live with it for quite awhile. One aim of home welding is to relax from everyday stresses; frustration from inappropriate equipment is the last thing you need. Equipment that is perfect for one home craftsman can irritate another. It is a "buyer beware jungle" out there if you move too fast.

Equipment Package Options

Three very general options for equipping your shop follow. Again, before you buy, try welding with someone else's equipment to better learn what you can and want to do. After a trial period, go find your own equipment.

Option 1: Oxyacetylene and Shielded Metal Arc (Gas and Stick). For less than $500 you could get an oxyacetylene outfit (without cylinders), and a small ac stick transformer power source. This is the economy option, but the equipment can always be used. Outright purchase of oxygen and acetylene cylinders would be an added expense.

Option 2: Oxyacetylene and Wire Feed Arc Methods (Gas and Wire). You might purchase oxyacetylene equipment and a small wire feed unit for arc welding. Figure 7-1 shows a home welder so equipped. This costs between $1200 and $2000. In the short run, wire feed equipment gets you welding fast. In the long run, it also gives you the ability to work with aluminum or bronze.

Option 3: Oxyacetylene, Shielded Metal Arc, and Wire Feed (Gas, Stick, and Wire). You will probably want to have gas, stick, and wire feed equipment eventually. To buy all three at once

Fig. 7-1. This home welder is equipped for building nearly anything with a wire feeder and an oxyacetylene outfit. (He also has a fire extinguisher.)

costs between $1600 and $2800. Power sources that can be used for either wire feed or stick electrodes can save money (Fig. 7-2). Some users of early multi-process machines complained of a rough arc. Spend time testing and comparing potential equipment purchases.

Quality of Equipment

Unfortunately, it is not always easy to tell quality light-duty equipment from flimsy junk. The limitations of reputable light-duty machinery are clearly indicated. It is also forgiving of an occasional overload. Beware of bargain apparatus said to "do everything." Only buy equipment you can try out; you could end up buying replacement equipment later. Equipment manufacturers and dealers plan to stay in business. They would be foolish to sell you less equipment than you will need. In this era of consumer awareness, they stand to lose your repeat business and risk their valuable reputations.

Here are a few questions to ask about your possible equipment purchase:

● What is the duty cycle; what happens if it is exceeded?
● If the equipment acts up, how much time and money will it take to get it quickly repaired?
● Does the dealer also do repairs?
● Are loaners available?
● What are the prospects for welding aluminum and stainless steel in the 1/4-inch-thick range?

COMPRESSED GAS EXPENSES

Compressed gas prices vary according to distance

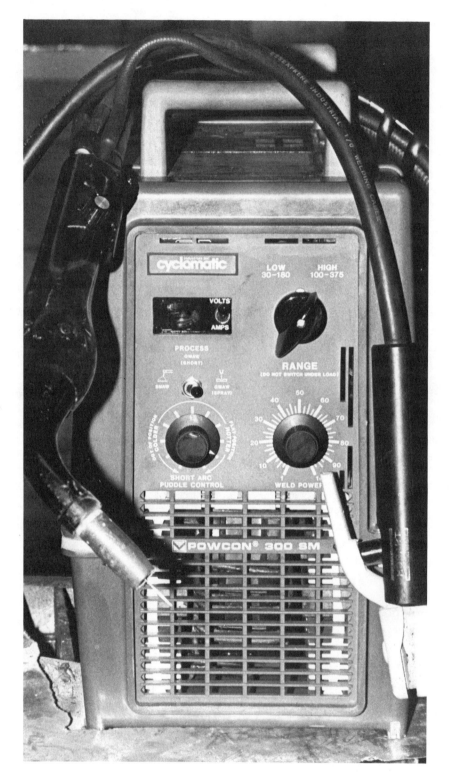

Fig. 7-2. This inverter power source can be used for both SMAW and the wire feed processes.

from the reduction plant and distribution costs. Industrial gases are also priced according to volumes consumed, the small user pays the top price. Still, it is possible to speculate about a typical home shop gas bill.

To flame cut 1/4-inch plate you need a cutting tip with ample gas flow to both preheat and cut. A tip for this purpose uses about 40 cubic feet per hour (cfh) of oxygen and 7 cfh of acetylene. If oxygen costs \$.25 per cubic foot and acetylene costs \$.30, the total gas expense for a non-stop hour of flame cutting would be:

$$40 \text{ times } \$.25 = \$10.00 \text{ for oxygen, and}$$
$$7 \text{ times } \$.30 = \$2.10 \text{ for acetylene}$$
$$\text{Total} = \$12.10$$

An hour of cutting is a lot of cutting; you might not do that much in a couple of months. The cost of the gas in a cylinder is not likely to break the bank.

Most home welders use small 80-cubic foot oxygen cylinders and 75-cubic foot acetylene cylinders. Outright purchase is the cheapest course; they would cost (filled) less than \$300. Refills are mostly on an exchange basis unless you are prepared to wait up to two weeks to get your original cylinders back. (There is no fast way to refill cylinders—especially acetylene). Oxygen refills run under \$15.00 and acetylene under \$20.00. You use at least two cylinders of oxygen to one cylinder of acetylene.

If gas consumption is low, cylinder rental can cost more than the gases they contain. For example, a cylinder deposit fee might be \$100. Then, there is the rental fee of about \$4.00 per month. In many places, a cylinder returned for refill within 30 days is rent free. But, after 30 days, an extra dollar or two, *demurrage fee*, is added. This encourages cylinder turnover with industrial users. You might not use gas that quickly. In a couple of years, these fees could have been used to purchase your own cylinders. You also would not have tied up deposit money.

Shielding gases are used to weld mild steel with most wire feed processes. Carbon dioxide (CO_2) or an argon-CO_2 mix is used. An argon mix is nearly always recommended to smooth out the roughness of single-phase power. At \$40 for a smaller 150-cubic foot cylinder, an argon mix is two or three times more expensive than straight CO_2 but well worth it. A combination regulator and flowmeter is also needed for shielding gas; these run under \$100.

Acetylene is not a practical fuel for a large amount of heating. A lot of heat is used to preheat and heat form. Multiple large acetylene cylinders and a connecting manifold are needed for the high flow rates of large torches (Fig.7-3). More is said in Chapter 9 about handling acetylene. Use a large propane fuel heating torch, like a weed burner, to heat larger castings etc.

FILLER MATERIAL EXPENSES

Shopping can save a few dollars on filler material, but use only the best electrodes, rods, and fluxes. Sometimes, there are electrode bargains with small-lot leftovers returned from a big industrial job. Only beware of broken or opened packages. These could mean water damage, kinked wire, or even oxidation (rust) damage. A spool of wire can be knocked off-center and not run true. Stick electrode coatings can easily be chipped. In light of the small quantities of filler materials used in the home shop, take no chances and go first class.

Filler metal is sold by the pound and quantity means discount prices. However, you would need to buy a supply for several lifetimes to get any appreciable price break. If kept high and dry, electrodes have a long shelf life, but producers do not stick their necks out to say just how long. Flux-cored and coated electrodes can deteriorate within a year of storage. They should be used as soon as possible once the package seal has been broken. It does not pay the home welder to stock up. Most filler rods, on the other hand, seem to last indefinitely if kept dry.

METAL EXPENSES

Most welded material is low-carbon (mild) steel within two varieties: hot rolled steel (HRS) and cold rolled steel (CRS). Hot rolled steel, logically, is rolled at the mill while still hot from the steel mak-

Fig. 7-3. A manifolded system with large cylinders is required for drawing acetylene gas at high flow rates.

ing process. It is economical and stronger, but more expensive, cold rolled steel (CRS) is seldom necessary. CRS is made by first removing millscale from HRS with an acid "pickling" solution. It is then rolled again. Cold rolling imparts a precise dimension and a stronger structure.

Alloy steels, with super corrosion resistance or high impact strength, should be avoided if possible. They cost more, are heat-sensitive, and must be welded with more expensive electrode.

Steel is sold mostly by weight with a premium for unpopular shapes and sizes. To estimate costs, find the weight per foot of structural shapes and multiply that by the number of feet required. Plate costs for a given composition and thickness are found by multiplying the approximate area (square feet, inches, etc.) by the weight per unit of area (pounds per square foot, etc.). With a phone call to get the current price per pound, you have enough information to compute your metal cost.

Most structural shapes come in "20 footers;" small, flexible bars are often cut to 10 feet for handling ease. Take extra steel to avoid cutting fees or to avoid splicing leftovers.

As mentioned before, get a copy of a steel supplier's databook. You might find that new steel is the best buy for structural shapes. Used plate is often a good buy for quantities less than full sheets. As a rule (with definite exceptions), it is more difficult to salvage structural shapes than plates.

New, unblemished steel is sometimes returned to the steel service center, or picked up by a salvage company. It is often resold at a reduced remnant price because it lacks a "pedigree" or *mill report*. A mill report certifies the chemistry of a batch of steel and is necessary in commercial fabricating to ensure meeting code specifications. Most home welding projects can be made with remnant steel.

Used metal with heavy rust, paint, or plating is seldom a bargain. It takes time and effort to remove such surface contaminants. Welding through even a slight residue can produce a bad weld. Also, avoid used alloy steel that is difficult to weld. The fast way to check the suitability of project metal is to weld small samples and make a simple break pretest (Chapter 10).

ELECTRICAL ENERGY EXPENSES

Make certain that your equipment will not overload your electrical service. You will very seldom push any machine to its capacity in the home shop. (Home welding is mostly done on metal less than 3/8 inch thick.) This means that you need not worry about using a machine with a maximum line draw about the same as the recommended capacity of your supply line.

Little energy in kilowatthours (kwh) is drawn by a home-type power source (see "Expenses" in Chapter 3). The greatest energy expense is often not for energy itself, but for the hookup.

Circuit Breakers and Surges

Fuses and circuit breakers are essential devices to protect an overloaded electrical circuit (wiring etc.) from overheating (Fig. 7-4). When the rated amperage of a fuse or breaker is exceeded, it breaks the circuit to stop further current flow. Circuit breakers for household service are easily tripped. Such hair-trigger action does not tolerate momentary startup surges such as occur with many power sources.

An involved solution is to rewire the line circuit for the maximum surge amperage. One popular wire feed unit draws 38 amps but needs a 60-amp breaker to handle the startup surges. (A rule of thumb in industry is to use a breaker 150 percent of the rated draw.) A 60-amp breaker is large for home use; 25 to 45 amps are more common. Replacing a 45- with a 60-amp breaker does not increase protection, but instead means that possibly inadequate supply circuitry can be strained with up to 60 amps if a "real," sustained overload occurred.

Slow-Blow Fuses

There is a way around this problem. A *slow-blow* fuse is a logical substitute for a circuit breaker in a power source circuit. It is carried in a fused disconnect that permits disconnecting the power source without pulling a plug (refer to Fig. 3-4). A slow-blow fuse can handle brief overloads. If the

Fig. 7-4. A circuit breaker and a fuse. The fuse has an advantage in being able to sustain a momentary overload that would trip the breaker.

normal draw of your power source is 38 amps, a 50-amp slow-blow fuse will provide protection and handle the momentary 60-amp startup surge. As always, talk over these matters with your welding equipment dealer and an electrician familiar with welding equipment.

Cooling Fans

It is standard safety practice to turn off any unat-

tended machine; saving energy is a side benefit. The fan in a power source only needs about the same energy as a light bulb, but forces air (and everything in it) across internal components of the machine. After awhile, dust and dirt buildup can cause some undesirable electrical activity.

If you live in a dusty area or generate dust with your work, place a furnace-type filter over the air intake of your power source (Fig. 7-5). Just remem-

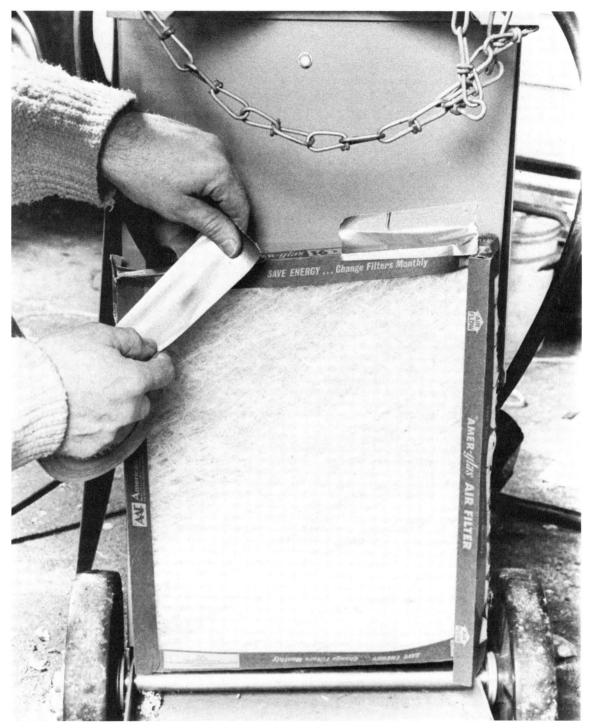

Fig. 7-5. A furnace filter can be used as a temporary means to prevent excessive dust from entering a power source. The flow of cooling air must never be blocked.

ber to replace the filter as soon as it starts to look dirty. A restricted air intake could cause disastrous overheating. Elevating a power source above the shop floor also helps to keep it clean on the inside. A clean power source runs cooler and longer.

TOOL EXPENSES

The tool expense of home welding is small. Other than welding and grinding equipment, there is little difference between hand and power tools used for woodworking or for welding fabrication. For example, wood and metal layout equipment is the same. Some heavy-duty clamps and perhaps a larger hammer or mallet are the only additional items that might be needed.

Chapter 8

Home Welding Processes: Advantages and Disadvantages

SOMETIMES THE HOME CRAFTSMAN SELECTS a project that could be joined by several different welding processes. This chapter gives criteria for deciding which process to use. Such a decision is most often termed (welding) *process selection*. Of course, this is simple if you have equipment for just one process. As with cutting wood, there is probably an optimum choice of saws, but you end up doing the job with the best that you happen to have.

It is worthwhile to examine the nature of each of the home welding processes, see their usual application, and evaluate the advantages and disadvantages.

ACETYLENE

The oxyacetylene flame, an "ancient" welding tool is still used to both cut and join metal. Chapter 9 explains oxyacetylene (or "gas") welding. It is possible to join many different metals with the use of oxyacetylene, but some judgement is required to decide whether gas welding is right for a particular job.

Oxyacetylene Advantages

Oxyacetylene is the most natural way to weld due to the relatively slow pace and excellent visibility of the welding zone. This makes for a comfortable, controlled, and satisfying activity. Metal sculptors often use the torch processes as a tool of imaginative expression.

Small oxyacetylene outfits can be taken just about anywhere. Fire departments use flame cutting outfits for rescue operations. With no need for electrical power, oxyacetylene could be used literally in the middle of a turnip patch.

Oxyacetylene is usually regarded as a way to weld mild steel, but it can also join other metals by other methods. In fact, there is often simply no better way. For example, low temperature brazing and soldering avoids damage to thin or otherwise delicate structures. Torch brazing steel and gray cast iron is a useful repair process for the home shop. Silver brazing copper, bronzes, brasses, and stainless steel is a commonplace process. Even combinations of these metals can be brazed together. It is a bit tricky, but even aluminum can be joined if the sections are not too thin. (This is not recommended for the beginner.)

Many repairs are done with soft soldering using tin and lead solder for filler. Plumbing and other such assemblies are soft-soldered without potentially destructive higher welding or brazing temperatures. With precautions, the hot oxyacetylene flame is used for soft soldering. The cooler air-acetylene flame, however, is easier to use (Fig. 8-1).

Thin sections, even 18-gauge like that of mild steel exhaust tubing, can be welded with oxyacetylene. Not-so-thin sections can be joined to a practical limit in the 1/8-to 3/16-inch range.

An oxyacetylene outfit is comparatively low priced. This is a strong incentive for the continued use of the torch processes.

Overall, oxyacetylene is used more for flame cutting metal than to join it. Flame cutting (or burning) is limited to carbon steels. Because most metal that the home craftsman uses is ordinary carbon steel, this limitation is no problem Cast iron can even be cut using a steel rod to maintain the reaction.

Oxyacetylene Disadvantages

Oxyacetylene is slow because, at 5800 degrees Fahrenheit, the flame is at least 3000 degrees cooler than the electric arc. In addition, the heat from the gas flame is spread out. As a result, the total heat input to the plate is comparatively high. High heat causes distortion and burning of any surface coatings. For these reasons, oxyacetylene should not be used for welding metal much over 1/8 inch except in an emergency.

There is also a safety problem with gas welding heavier plate with the small acetylene cylinders commonly used at home. High acetylene flow rates are needed for the large tips needed for heavier metal. These rates often exceed the 1/7 cylinder volume safety rule (Chapter 9). This rule prevents loss through the torch of the liquid acetone from the acetylene cylinder. Acetone is extremely flammable and destroys hoses and rubber parts in the regulators and torch. With the need for high flow rates, acetylene is also impractical for large heating torches. There are better fuels for this—

propane, for example.

Oxyacetylene equipment is easy to assemble in dangerous combinations. Small cylinders connect right up to oversized torches. A Number 6 tip, for welding 1/8-inch plate, can consume 10 cubic feet per hour. By the 1/7th rule, 7 × 10 or a minimum cylinder size of 70 cubic feet is required. With one of the popular small acetylene cylinders of 80 cubic feet, you would not have much spare capacity. (1/7 × 80 = 11 3/7 maximum allowable draw per hour.) Use of a slightly larger tip would be even worse.

There is nothing uncertain if comparing the high cost of oxyacetylene welding with any arc process. The gas costs expended for an oxyacetylene weld are greater than the costs of arc welding electricity and electrodes. There is also consumption during each flame lightup and adjustment, not to mention the consumption of equipment setup and shutdown.

Summary of Oxyacetylene Advantages

There are strong reasons for the continued use of oxyacetylene into the foreseeable future:

- The naturalness of the torch processes
- Portability
- Electrical hookup is unnecessary
- Ability to join many metals by several methods—welding, brazing, and soldering
- Welds both thin and not-so-thin metal
- Low price of equipment
- Outfit can be used for flame cutting

Summary of Oxyacetylene Disadvantages

The disadvantages of using oxyacetylene include:

- Slow
- Produces distortion
- Requires expensive gases
- Flame consumes breathing oxygen
- Gases and their container are potentially dangerous
- Limited practical range of metal thicknesses from 1/16 to 1/8 inch for small home-type outfits.

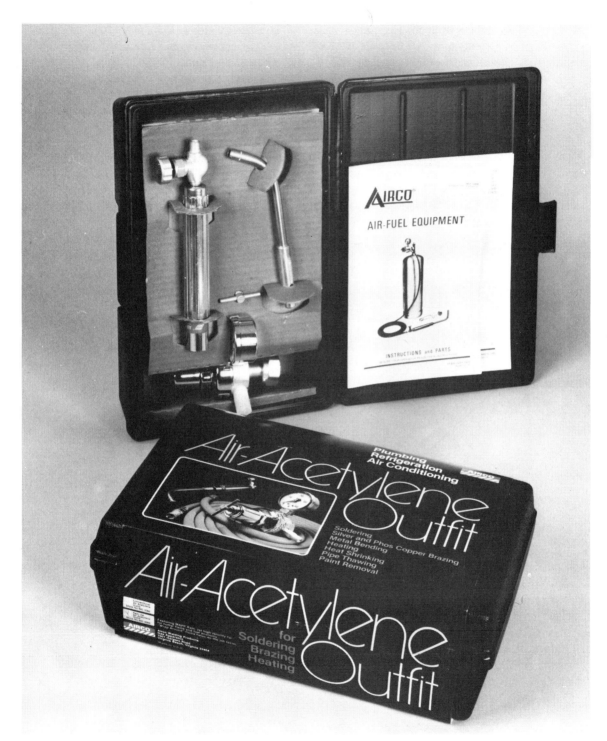

Fig. 8-1. The air-acetylene torch is very useful for soldering applications. Sometimes these are called "Prest-O-Lite" torches. (Courtesy of Airco.)

Oxyacetylene is very much alive and well nearly 100 years after its initial development. It is best as a repair tool for small jobs where electric arc heat might prove excessive. It is used more for cutting and brazing than for gas welding.

SHIELDED METAL ARC WELDING (SMAW)

After World War II, small single-phase arc welding equipment appeared. At that time *arc welding* meant SMAW with coated stick electrodes. Such "buzz boxes," as transformer machines are often called, were heralded as *utility* welding machines for occasional use in a non-manufacturing environment. Small fab shops, machine shops, farm maintenance shops, sheet metal shops, autobody shops, and garages were among the first takers; so were the enterprising home craftsmen.

By the early 1950s, the Lincoln Electric Company was promoting home arc welding with small ac power sources. Today there are thousands of both ac and dc "stick" welding machines in home workshops. There must be some good reasons for this.

SMAW Advantages

Stick electrodes are available in a variety of alloys and sizes for welding a range of metal types and thicknesses. All arc welding processes concentrate heat at the arc and thus avoid distortion from the wide heat dispersion of a torch flame. Figure 8-2 compares an arc "flame" with a torch flame.

If you do mostly maintenance and repair welding, the ease of changing stick electrode types is appreciated. You can literally weld cast iron one minute and stainless steel the next. It is not necessary to buy a large supply of electrodes. Often 5- or 10-pound quantities are available. Certain electrodes are sold by the pound, so it is possible to keep a variety of filler metals without having much money sunk into an inventory.

SMAW equipment is portable and simple. A power source and pair of welding cables constitute the main items. This simplicity makes it quick and easy to set up for stick welding. Equipment simplicity is also reflected in low equipment prices.

The flux coating on stick electrodes permits welding through light rust and other contaminants. Electrode types vary in part due to different flux coatings. Fast-freezing electrode coatings permit welding in all positions.

With practice, many welders are able to join light 16-gauge sheet metal with stick electrodes. At the other extreme, deep-penetrating electrodes permit welding heavier plate.

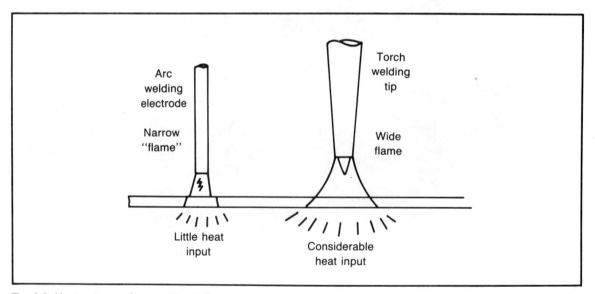

Fig. 8-2. Heat patterns of arc and torch flame.

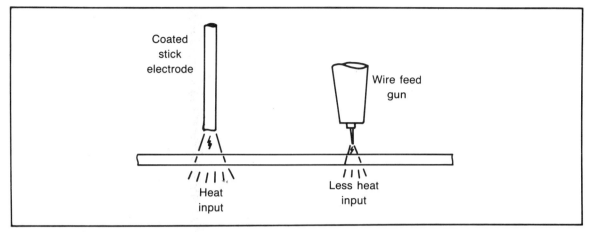

Fig. 8-3. Heat patterns of stick and wire electrodes.

SMAW Disadvantages

SMAW has disadvantages that are significant to an industrial user. Operator in efficiency and short-length weld problems aside, the home welder should still consider certain SMAW drawbacks.

SMAW produces smoke and a lot of spatter, but the biggest disadvantage is the long learning curve. It takes many hours of structured practice to master. There are subtle yet significant differences in welding with various electrodes in different positions. Because SMAW is so demanding, most industrial users are not willing to offer either initial or upgrade training necessary for its use. In fact, most of today's welding is done with more productive processes by operators who might know little about SMAW.

Summary of SMAW Advantages

Among the strong reasons for the SMAW process remaining popular for home welding are:

- Good variety of electrode types
- Ease in changing electrode types
- Simple, low-priced equipment
- Ability to weld through light contaminants
- Welds both thick and reasonably thin

Summary of SMAW Disadvantages

Certain aspects of the SMAW process pose seri-ous limitations for the home craftsman:

- Smoke and spatter
- The long time needed to learn the process

WIRE FEED PROCESSES

The development of wire feed equipment suited to home use is like that of the first home SMAW power sources. Equipment manufacturers set out to produce a light-duty wire feed system for use with single-phase power. Such machines were to bring wire feed welding to smaller commercial shops. Resourceful hobbyists lost little time exploiting these developments. They were using wire feed welding systems in home shops by the mid-1970s.

Advantages of the Wire Feed Processes

All five variations of home wire feed welding share the same advantages. Wire electrodes are smaller than stick electrodes so arc heat is even more concentrated (Fig. 8-3). This means less heat input to the workpiece and less distortion. Small wire electrodes also make it easier to see the welding puddle and position the weld more accurately in tighter joints.

The flux-cored wire produce smoke, but welding with solid wire is practically smoke-free. Less smoke helps arc visibility and means a cleaner workplace. There is generally little spatter associated with wire feed welding. This saves both preweld preparation and postweld cleanup.

Small weld volumes used for home welding diminish the importance of the often-cited high *deposition efficiency*. This refers to whether that expensive electrode ends up as weld rather than as wasted spatter and smoke. With stick electrodes, getting 65 percent of an electrode into the weld is considered pretty good; with wire, 90 percent or more is typical. Even though you might not save hundreds of pounds of electrode with wire welding in your home shop, it is nice to know that you are close to getting what you paid for.

The home craftsman has an occasional need for a long weld such as a tank seam. With each start and stop, there is a possibility of a defect. Weld stops form craters that are prone to crack, starts and restarts are often associated with poor fusion. With continuous wire feed electrodes, there are few reasons to stop welding before reaching the end of the joint. (Sometimes, a long weld is interrupted for repositioning.) This ability to weld non-stop also helps make more attractive welds.

Even with the best intentions, spacing (or fitup) between the member plates of a joint is often not as tight as it should be (Fig. 8-4). Welding such joints can be tricky with the melt-thru tendency at larger "gaps." A big advantage of wire feed welding is the ability to reduce heat input while increasing the amount of weld deposit. This is just what is needed to bridge gaps. The techniques of such control over the welding operation are explained in Chapter 11.

One of the strongest incentives with wire feed processes is that they are so easy to learn and use. Because only a fraction of SMAW learning time is required for wire welding, time is freed for direct project fabrication.

Disadvantages of the Wire Feed Processes

As with most good things, there is a price to be paid for the advantages of wire feed welding. A wire electrode drive mechanism involves more complex and expensive equipment than for SMAW. Some maintenance is also needed from time to time.

Wire electrode is wound on spools, so keeping a variety of small quantities for different types of metals is simply not as practical as it is with SMAW stick electrodes. It also takes a few minutes to change the type of wire electrode.

If a shielding gas is used, there are added expenses for a cylinder, regulator-flowmeter, and, of course, the gas itself. Shielding gases also mean a more cumbersome welding unit. Because the protection of the weld zone depends entirely on the shielding gas, a strong cross breeze can ruin a weld.

Flux-cored wire electrodes usually achieve adequate fusion, but the appearance of a weld made with solid wire can be deceiving. The improper use of solid wire electrodes can cause *cold laps*—poorly fused areas of weld metal deposited mainly on the surface of the parent metal (Fig. 8-5). This happens mainly on heavier plate with a build-up of scale. Grinding or sanding prepares a surface for using solid wires. Of course, it is always important to make test welds.

Summary of Wire Feed Advantages

Wire feed processes have come to the home shop. The main reasons discussed are:

- More concentrated arc, less distortion
- Excellent visibility
- Little smoke and spatter
- High deposition efficiency
- Long non-stop welds possible
- Ability to handle poor fitup
- Easy to learn, easy to use

Summary of Wire Feed Disadvantage

Some drawbacks with the use of the wire processes are:

- Added expense of more complex equipment
- More complex equipment means some maintenance
- Time required to change electrode
- Shielding gas costs and concerns
- Possible tendency for cold lapping

Table 8-1 lists the more popular home welding

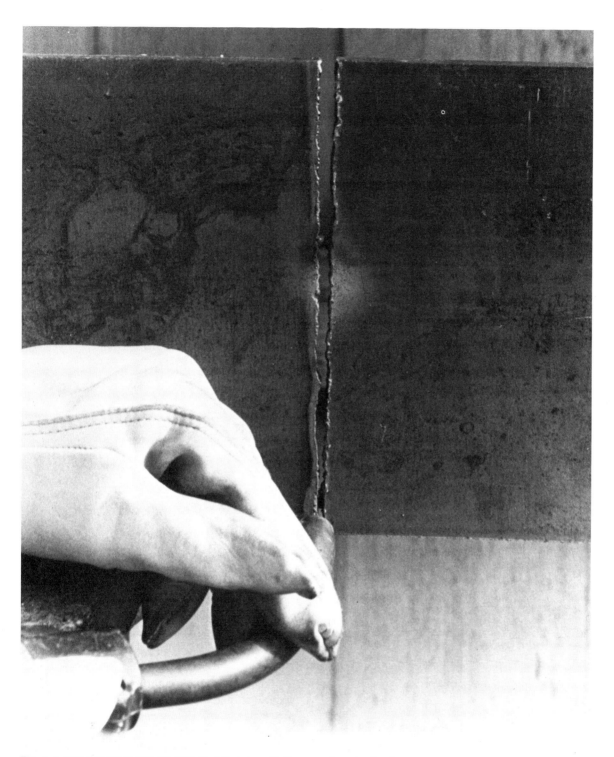

Fig. 8-4. Welding joints with uneven fitup is easy with the use of a wire feed process.

Fig. 8-5. This might be a neat bead but its holding power is worthless because it is cold-lapped. Under load it easily breaks away from joint members.

processes and rates them according to performance qualities.

CONCLUSIONS AND RECOMMENDATIONS

Each welding process has its place in the scheme of the home workshop. Considering the many uses for home welding, it is difficult to remain an advocate of just one process for very long. The process used will finally depend on the intent and purpose of the user. Of course, high-quality (and therefore safe) welds should be the first priority of welders everywhere.

(A = excellent, B = moderate, C = difficult or impractical)			
PERFORMANCE QUALITIES	**OAW**	**SMAW**	**WIRE FEEDS (GMAW AND FCAW)**
Ease of Learning	B	C	A
Lighter Thicknesses	B	B/C	A
Heavier Thicknesses	C	A	A
Poor Fitup	C	C	A
Penetration	B	A	B
Weld Quality	B	A	A
Weld Appearance	A	B	A
Absence of Distortion	C	B	A
Ease of Replacing Filler Metal	A	A	C

Table 8-1. Process Ratings and Selection.

In Chapter 7, three equipment packages were examined; each had an oxyacetylene outfit. Oxyacetylene welding has limitations, but an outfit includes a cutting torch. Flame cutting is essential for steel fabrication in the home shop. As a bonus, oxyacetylene also lets you weld, braze, and solder. You will probably want to do all these. Brazing is used for many repairs.

You also need arc welding to join heavier metal sections. Which form of arc welding? Take a good look at what you want to accomplish with home welding. Stick electrodes have some merits for the repair welder working on a variety of metals. Wire processes offer advantages for almost everything else.

Sometimes, you choose from alternative joining methods. For example, 12- to 14-gauge steel can be brazed, welded with gas, stick electrodes, solid wire, or flux-cored wire. If you have all these, you can select the one that makes the best quality joint with the least amount of damage to the surrounding area. Wire processes are the fastest and "coldest" but can be too fast for certain situations. SMAW is also fast but might produce too much spatter or leave too much weld metal. Yet, the slower pace of oxyacetylene welding or brazing might input excess heat and ruin your project by warping.

Some conclusions can be drawn from Table 8-1. Wire feed welding has advantages, but the equipment costs more. Another possible drawback is the time needed to change wire electrode type. Yet most (if not all) of what you weld on is likely to be mild steel, changing wire types might not be a problem.

Although this book deals with weld fabrication of mild steel, you might eventually want to work with aluminum and light stainless steel. Somewhere, someone could be using SMAW to weld aluminum, but it is impractical except for an emergency repair. Try it and convince yourself. On the other hand, aluminum wire feed welding is practical and growing in popularity.

Stainless steel must be welded with low heat input. SMAW is often used, but welding light gauges is more practical with wire.

If you have a choice, try to select the welding process enabling you to do a quality job in a reasonable time. If in doubt, forget time and convenience and err in favor of quality.

Chapter 9

Working with Oxyacetylene

T HE NEXT THREE CHAPTERS EXPLAIN HOW to weld metal together in the home shop. Be sure to read through all the information before starting to weld. A step-by-step approach is the most efficient way to develop mastery of these processes. The slow and deliberate pace of following these procedures will pick up with practice. Avoid taking any short cut that might eliminate a step involving safety. Reviewing the basic operating procedures from time to time ensures that you will get the most from your equipment.

USES FOR OXYACETYLENE

Oxyacetylene refers to the combination of oxygen gas with acetylene fuel gas. With ignition, this gas mixture burns at nearly 6000 degrees Fahrenheit. Steel melts at about 2800 degrees Fahrenheit; its kindling point for oxygen cutting is about 1600 degrees. So, the oxyacetylene flame can both weld and cut.

Besides oxyacetylene, there are air-acetylene torches that produce less heat—about 2500 degrees

Fahrenheit. Although insufficient to weld steel, these units are ideal for small brazing and soldering jobs like joining copper tubing. A common trade name for air acetylene is Prest-O-Lite.

Other fuel gases can be used with oxygen for flame cutting (or "burning") and heating, yet oxyacetylene alone has the proper flame chemistry to both cut and weld steel. Other metals can also be welded with oxyacetylene; cast iron repairs are quite popular.

Common torch operations include brazing, braze welding, soldering, and heating. Oxyacetylene is used for all these. With such capabilities, oxyacetylene is essential to the home weld shop. Another advantage is its wide availability. Oxyacetylene is used literally worldwide to cut, join, and form metal.

If you are still in the market for equipment, your best buy is a reputable medium-duty outfit. These run $150 to $200 without cylinders. Sometimes good buys include cylinders and a cart as well (Fig. 9-1). For $30 more you can add an air-acetylene torch. The most expensive oxyacetylene

Fig. 9-1. Complete oxyacetylene packages with cylinders and a cart are sometimes offered at special prices.

equipment is "industrial strength" for severe use and is too bulky for the lighter type of job done in the home shop. There might be good used items available but, unless you are expert with this equipment, have it professionally checked out. This service on top of the purchase price might well cost as much as new equipment that at least has some limited warranty. As most mechanisms eventually need servicing, be sure that repair personnel in your area handle your type of equipment.

Thousands of satisfied users can attest to the merits of oxyacetylene. Unfortunately, a sorry few can also attest to its destructive potential if mishandled. Used properly, oxyacetylene is safe. Many experts agree that using oxyacetylene is less hazardous than some more common substances—such as gasoline.

WORKING SAFELY

Following this paragraph are general safety facts about oxyacetylene, developed from years of experience by many users. The procedures for setup and use contain more information. Read these and check off points that are not clear. Take no chances. If, after reading the material, you still have questions, ask an expert like a gas supplier or call your local welding school.

● Do not transport or store gas cylinders in a closed space such as an automobile. They have been known to leak and explode.

● Oxygen is not to be confused with "air." Oxygen vigorously accelerates combustion of nearly anything. Never use it in place of compressed air. Never use a fuel gas hose for oxygen.

● Permit no oil to come in contact with *any* of the gas equipment. Oil and oxygen can explode.

● Avoid using compressed air to blow out gas hoses; some oil from the air compressor is generally passed into the supplied air. Use 5 to 10 psi of regulated oxygen.

● Acetylene is extremely unstable at pressures over 15 psi (Fig. 9-2). Use only 5 psi for most work. Acetylene cylinders in use must stand upright. To prevent dangerous loss of the liquid acetone (used as a carrier for dissolved acetylene)

within, an acetylene cylinder that has been tipped on its side must stand upright for at least 2 hours before using. Do not use copper lines to carry acetylene, they will deteriorate. Learn to recognize the telltale odor of acetylene.

● To prevent loss of acetone, never use a heating tip that draws over 1/7 the capacity of the acetylene cylinder.

● Always secure gas cylinders to prevent their falling (chains are best). Move no cylinder without a properly installed safety cap. Never allow a welding arc to be struck on a cylinder.

● Consider cylinders empty when the pressures reach 50 psi. Be sure to securely close empty cylinders and replace caps.

● Never attempt to fill one cylinder from another.

● Do not use any flame around chlorinated solvent vapors or refrigerants like Freon; these form deadly phosgene gas.

● Avoid cutting or welding on a container which has held flammable or explosive materials. Do not apply heat to an unvented container.

● Clear your work area of combustibles and have a fire extinguisher handy.

● Light the torch only with a proper spark lighter (striker).

● Work only in well ventilated areas. The oxyacetylene flame uses up breathing oxygen from the surrounding air.

● Do not allow a torch flame to directly contact concrete. This causes "spalling" or explosive cratering of the concrete with the resulting danger of high-speed particles.

● Reverse flow check valves provide some protection against crossover and are generally recommended. These are not designed to stop a *flashback*.

● In the event of a flashback (the torch squeals because the flame has retreated inside the tip or even farther back), *immediately* shut off the torch oxygen valve and then the fuel valve. Find and correct the flashback cause before proceeding.

● Never look at the flame without proper eye protection (welding goggles or glasses). Use at least a Number 5 density filter lens. Wear unfrayed

Fig. 9-2. Acetylene is unstable at pressures above 15 psi.

clothing. Avoid wearing synthetic fabrics and eliminate cuffs. Eight-inch work boots are ideal. Keep sleeves and collars buttoned, and wear gloves. Handle hot work only with pliers or tongs.

• Protect gas hoses from sparks, hot slag, and restrictions.

• Do not modify or attempt to repair any gas equipment yourself. Discard damaged or worn-out hoses.

• When leaving your equipment unattended, shut off the gases at the cylinders and drain the lines.

Fig. 9-3. The valve safety cap protects the valve stem. It is removed only after the cylinder is chained.

● Take no chances. If in doubt, stop and think. Seek expert advice.

● Do not be discouraged by all these cautions—they are based upon common sense and soon become second nature.

EQUIPMENT SETUP

Setup, or assembly of oxyacetylene equipment, is quite straightforward. Be sure you understand the general safety facts and the literature for your equipment. Take extra care with cleanliness and force no connections. Setup includes:

● Securing and "cracking" the cylinders
● Attaching regulators
● Leak testing the cylinders and regulator
● Opening the cylinders for use
● Attaching the hoses
● Blowing out the hoses
● Attaching the torch
● Adjusting regulator pressures
● Purging the lines

Compressed Gas Cylinders

Secure gas cylinders in an upright position with safety chains, then remove the safety caps (Fig. 9-3). The safety cap protects the cylinder valve. If the valve were broken off, the cylinder could be jet-propelled through the air. Very dangerous.

Wear eye protection when opening cylinders. Stand to the side of the outlet and slightly open or "crack" each valve one at a time to remove dust, bugs, etc. Then close the valve (Fig. 9-4). Check the threaded outlet connections for absolute cleanliness. If necessary, use only a clean, oil-free cloth to remove any dirt or other material. (CAUTION. Shop towels are oil-treated.)

Regulators

A pressure regulator reduces high cylinder pressure to a lower adjusted value for the hoses and torch (Fig. 9-5). It maintains the adjusted pressure as the cylinder empties with use.

Attach each regulator to its cylinder. Fuel gas fittings always use left-hand threads to avoid con-

Fig. 9-4. "Cracking" cylinders clears the connections of any foreign matter. Always stand to one side.

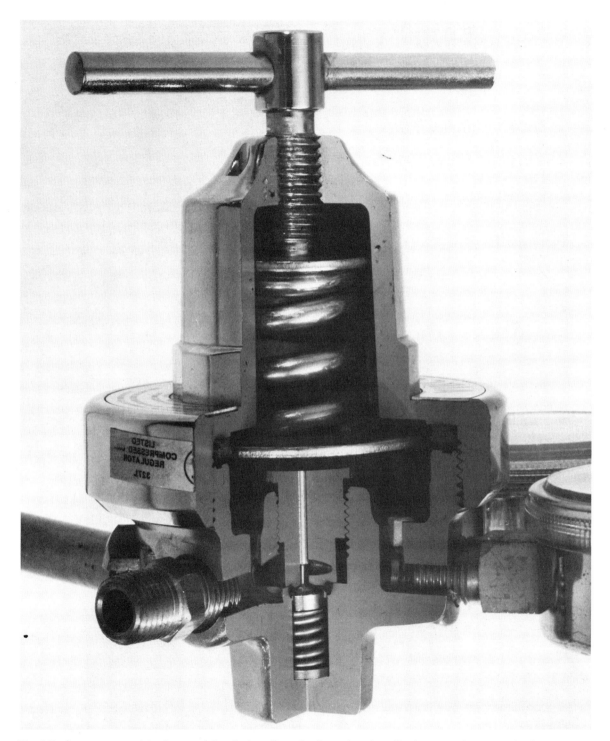

Fig. 9-5. A pressure regulator is a precision device. Screwing inward on the adjusting screw increases the line pressure to the torch. (Courtesy of Airco.)

fusing them with oxygen fittings. Support the regulator with one hand and tighten the attaching nut with the other (Fig. 9-6). Note the use of a clean properly fitting wrench. Do not overtighten any gas fitting.

Cylinder and Regulator Leak Test

With regulators attached, the cylinder connections and regulators themselves can be leak tested. Before opening the cylinders, back out the regulator adjusting screws until they are loose in their threads. This prevents injury should a defective regulator allow a sudden inrush of gas. (Adjusting screws not backed off have been propelled out of defective regulators.) Loosened screws prevent gas flow to the outlet hose connections.

Leak test the cylinder valves and regulators as follows:

● Stand to the side. Slowly open the oxygen cylinder valve 1/4 turn. Then close it again immediately.

● The acetylene cylinder valve is tested in like manner.

● See if the cylinder pressure gauges on the regulators hold steady readings; a pressure drop indicates a leak.

● If indicated, test more for leaks with leak detector fluid or Ivory soap suds at suspected connections. If you find a leak at the cylinder valve to be more than a loose connection, stop everything and refer the problem to your gas supplier. Take a leaking regulator to a qualified repair shop.

Opening the Cylinders

After the leak test, the cylinders are opened for use. Again, make sure that the regulator screws are loose in their threads. Open the oxygen cylinder valve slowly at first. Then continue to open it faster as far as possible to backseat the valve to prevent leaking at the valve stem. Opening the cylinder valves gradually at first avoids a sudden inrush of high-pressure gas into the regulators. Such an inrush is accompanied by extremely high temperatures called the *heat of recompression*. This heat

Fig. 9-6. Attaching the regulator to the cylinder.

could easily melt a regulator and cause injury.

The acetylene cylinder valve is (slowly) opened only 1/2 to one full turn—no further! This permits you to shut off the fuel gas with a quick twist of the wrist in an emergency. If your type of acetylene cylinder valve requires a key or "bottle wrench," keep it on the valve stem when you are working.

Hoses and Torch

The hoses are next attached to the regulators. It is wise to use reverse flow check valves (Fig. 9-7). These are usually sold separately from the outfits, although some valves are part of the torch or regulator. Open (screw in) one regulator adjusting screw to clear a hose of talc (a shipping preservative), dirt, bugs, etc. that could clog a torch. Then close (back out) the adjusting screw. Repeat this operation to clear the other hose.

Attach the torch handle to the hoses. Again, you ought to place reverse flow check valves somewhere in the line.

Attachments to Torch

Most oxyacetylene outfits contain a torch body that accepts either welding tips or a cutting attachment. Attach the correct type and size of item to the torch body (Fig. 9-8). Alas, tip sizes differ with manufacturers but each uses a series of digits. Use the recommended tip for your make of equipment. For welding on 14- to 16-gauge, a Number 1 or 2 tip should be about right.

Do not use a "rosebud" heating tip unless you are sure that its gas flow rate in cubic feet per hour (cfh) is less than 1/7 the capacity of your acetylene cylinder. For those occasional big heating jobs, use propane with a special propane outfit. Rental fees are usually quite reasonable. Propane has no flow problems and, unlike acetylene, there is no need to combine cylinders.

Regulator Pressures—Initial Setting

With the equipment properly assembled, set the line pressures for oxygen and acetylene according to the tip size. This is done with the regulator ad-justing screws. Turn inward to increase line pressure, back the screw out to reduce it.

If you know them, use the recommended settings for your equipment; if not, the following general pressures will get things started:

- For welding, brazing, etc.—5 psi acetylene, 5 psi oxygen
- For flame cutting—5 psi acetylene, 25-40 psi oxygen.

These are "working" pressures set while the flame is lit and adjusted to a *neutral* condition as explained later under "Flame Adjustments." Higher oxygen pressure is used for flame cutting because an additional stream of pure oxygen makes the cut.

One reason to keep pressures within specified limits is to avoid a dangerous *crossover* or explosive blend of backed-up gases inside the torch or hoses. Another is to avoid overly sensitive torch valves that make flame adjustment difficult.

Initial Purging of the Lines

Purge the lines before lighting the torch. Open the acetylene torch valve about one-half turn for one second, then close it. Repeat for the oxygen. With these quick steps you can be sure that only acetylene is in the acetylene line and only oxygen is in the oxygen line. It's sad, but many users are unaware of this simple way to avoid a dangerous gas crossover.

LIGHTING THE TORCH

Torch lighting and adjusting is easy. Hold it in your right hand (if right-handed), and the striker in your left (Fig. 9-9). Use only a sparklighter or a striker, never matches or a cigar lighter. Positioning the torch valves under the handle allows you to gracefully reach them with your left hand while still holding the striker. Hang onto the striker while lighting and adjusting the flame because the flame can blow out.

Use welding goggles with a Number 5 lens, open the acetylene torch valve about 1/4 turn, and light the flame immediately. Acetylene burning

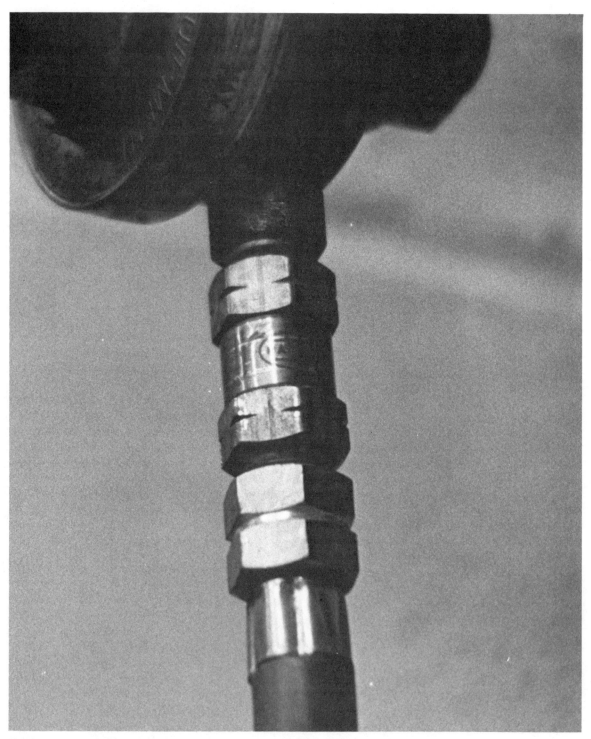

Fig. 9-7. A reverse flow check valve, claimed to prevent crossover of the gases.

Fig. 9-8. Attaching the torch handle to the hoses.

alone produces a brilliant white flame (Fig. 9-10). Add some oxygen to see what effect this has on the flame. Now extinguish the flame by turning off the acetylene torch valve first and then the oxygen torch valve. Next, learn how to light and adjust the flame so that you can do some work.

Flame Adjustments

It is generally agreed that there are three distinct flame types. Most torch work is done with the *neutral flame* in which equal parts of oxygen and acetylene are burned. This flame is "neutral" in the sense that it neither adds nor subtracts from the base metal chemistry. It is just a source of "clean" heat.

After setting the regulators and trial lighting the torch, you adjust the flame to neutral. For a neutral flame, open the torch acetylene valve and light

Fig. 9-9. The correct way to light a torch.

the torch. (Remember your welding goggles.) If needed, increase the acetylene flow enough to avoid smoke. Now gradually add oxygen until you can see the three distinct parts of the flame—the cone, feather, and envelope (Fig. 9-11). Carefully adding more oxygen causes the feather to fade entirely into the flame cone. At this point, the flame is balanced—that is, neutral as shown in Fig. 9-12. Take care not to exceed the amount of oxygen needed to just "burn off" the feather. Extinguish the flame by first turning off the torch acetylene valve and then the torch oxygen valve.

Some real pros leave just a trace of a feather to ensure that no excess oxygen is used. This is termed a *reducing flame*. Regulator settings can "drift" and the flame change enough to damage the work. It is easy to watch for the small acetylene feather as you work. If you can see the small feather, you know that you are not running with too much oxygen. The feather left should be less than 1/6 the length of the flame cone. To adjust for a neutral flame:

● Light the acetylene alone

Fig. 9-10. Acetylene burning alone.

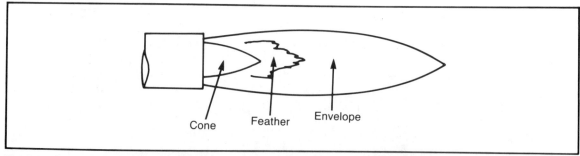

Cone   Feather   Envelope

Fig. 9-11. Parts of the oxyacetylene flame.

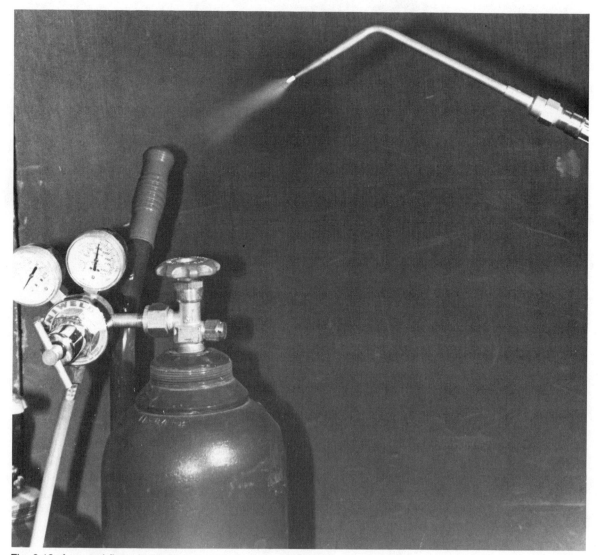

Fig. 9-12. A neutral flame has just enough oxygen to burn off the acetylene feather.

- Add oxygen until the 3 flame parts are distinct
- Add more oxygen until the feather fades into the cone

Final Acetylene Regulator Setting

Many manufacturers recommended a final acetylene regulator setting be made while the torch is lit. Use pressure values specified for your equipment or use the following general method.

Open the torch acetylene valve about 1/2 turn and light the torch. Use enough flow to avoid sooty smoke. Increase (screw in) the regulator adjustment to first cause the flame to separate from the tip and then reduce (screw out) the adjustment until the flame rejoins the tip. Add enough oxygen for a neutral flame.

This final or "fine" acetylene regulator adjustment ensures that the working pressures and gas flow will not be so high as to easily blow out the flame, nor so low as to risk tip *starvation*. Starvation is dangerous—the flame (which normally burns "away from" the tip) retreats to inside the tip. Starvation overheats the tip causing *backfiring* or, a far more serious, *flashback*.

The oxygen regulator adjustment is not as sensitive as the acetylene. Most jobs (besides cutting), use an oxygen pressure about equal to the acetylene setting. In no case should the oxygen pressure exceed the acetylene setting by more than 1 1/2 times. The recommended oxygen setting is usually alright.

Other Flames

An *oxidizing flame* (excess oxygen) has few uses; avoid it. A *carburizing flame* has an excess of acetylene; avoid it too. A slightly carburizing, or reducing, flame as mentioned is mostly used to monitor the flame adjustment. A more pronounced reducing flame is used when applying hardfacing material (Chapter 21).

Relighting

You will need to repeatedly light and extinguish the torch. Remember to first turn off the acetylene, and then the oxygen. You might be tempted or even actively encouraged to light the torch after first turning on a little oxygen and then the acetylene. This might seem to be a shortcut to instantly produce a neutral flame. However, the latest studies show that such "popping" starts drive carbon from the fuel to inside the tip. Relight the torch with the procedure just covered.

EQUIPMENT
SHUT-DOWN AND DISASSEMBLY

If you are finished working for an hour or so but will remain in the work area:

- Be sure both torch valves are completely closed
- Roll up the hoses
- Hang up the torch to prevent jarring the valves

If you are finished for a longer period or if you must leave the work area:

- Close both cylinders completely
- For one gas line—bleed the line at the torch (Fig. 9-13), close the torch valve, and back out the regulator adjusting screw
- For the remaining gas line—bleed the line at the torch, close the torch valve, and back out the regulator adjusting screw.

Bleeding the lines separately in this manner also prevents a dangerous *crossover* in the lines. When working, reverse flow check valves on the torch should also prevent a crossover. Discourage others from using the equipment by removing the regulator adjusting screws.

If the cylinders might be disturbed, remove the regulators and replace the safety caps. Store regulators, hoses, and torch in a clean dry place; seal the hose ends. Keep stored cylinders from falling over.

RUNNING A BEAD

Obtain some uncoated and unpainted mild steel scraps about 2 × 8 inches and no less than 16 gauge (about 1/16 inch) thick. Practice welding on bare

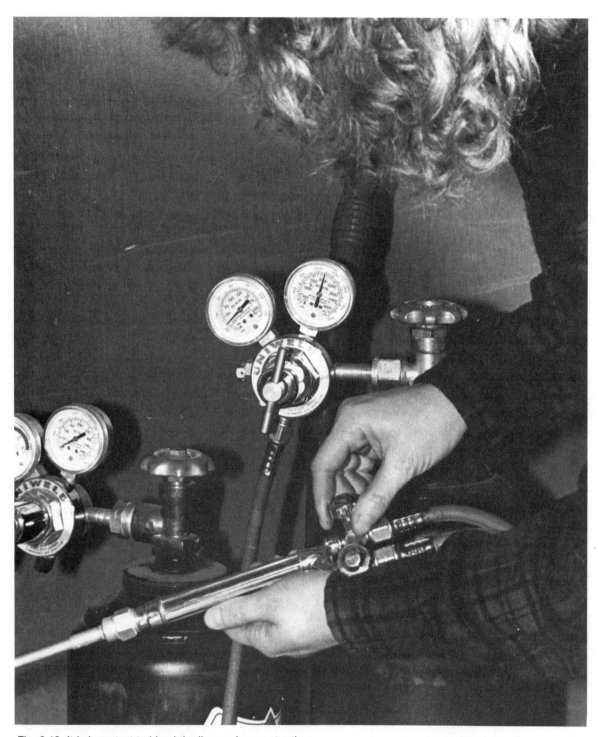

Fig. 9-13. It is important to bleed the lines only one at a time.

steel avoids toxic fumes. Your welding keeps you busy enough without having to hold your breath or lean to the side to avoid toxic smoke.

The practice plates should be placed on a minimum-contact support or applied heat can just soak away. Light and adjust the torch for a neutral flame. Point the flame directly at the plate (Fig. 9-14) and move it in tight circles until the metal begins to melt or puddle. As the puddle forms, point the flame forward along the joint and "push" the puddle uniformly across the plate—the flame moving *behind* the puddle. Practice "running a puddle" until you are reasonably comfortable doing so.

Try to position yourself so that you have total control of the welding operation. Arms should be close to the body with the wrist and/or palms resting on the same support as the plate. This stability permits you to easily move the torch in a uniform forward and back, crescent weave, or oval torch pattern.

A somewhat improvised but effective surface for supporting practice plates could be a table top

Fig. 9-14. Heating a small area of the plate prior to running a bead.

Hibachi type of barbecue (Fig. 9-15). If you try this, be sure that the grills are reasonably clean and that no combustibles remain in the charcoal area. A more permanent table with a fire brick top could be a future project.

The flame cone is held about 1/8 inch from the plate—close enough to heat only a concentrated area, yet not so close as to allow the cone to con-

tact the surface. Avoid holding the torch too far away because heat "fans out" as was shown in Fig. 8-2. Dipping the flame into the puddle causes gases to enter into the weld metal. This causes minute bubbles to form in the weld leading to a porous and generally weakened structure.

To speed things up, it is alright to rapidly cool practice welds by quenching in a water bucket. Use

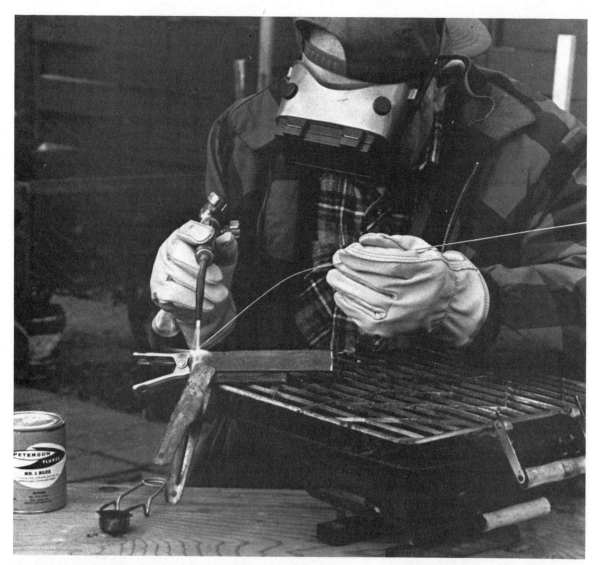

Fig. 9-15. Why not? This resourceful home welder got started welding on a barbeque top. (He built a table after gaining some welding experience.)

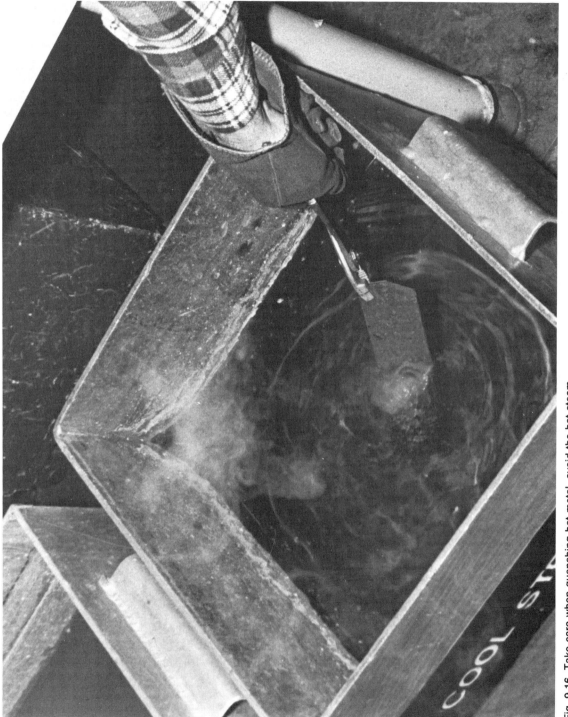

Fig. 9-16. Take care when quenching hot metal, avoid the hot steam.

pliers for the hot work and avoid scalding your hand from fast-rising steam or boiling water (Fig. 9-16). Some steels become hard and brittle if quenched. So, avoid rapid cooling all but practice welds.

Inspect and evaluate each welding effort immediately. In this case, the bead should look fairly even on the *face* (front) side and show a uniform heat effect on the *root* (back) side. Torch angle(s), stand-off distance from the plate, and travel speed all affect the heat input to the plate. If you have trouble duplicating the examples, change just one part of your technique at a time. Practice until you can avoid "skipping" spots and melting through the plate (Figs. 9-17 and 9-18).

USING FILLER ROD

Next try adding 1/16-inch filler rod (type RG-45) to the puddle. Most gas welding is done using filler. Filler rod is added as the torch is moved back (Fig. 9-19). Keeping the rod angle quite low to the plate prevents a large blob from forming on the end of the rod. The trick is to "dab" filler into the leading edge of the puddle rather than to "drip" it onto the plate. Avoid overheating your rod hand by progressively heating and bending the rod (Fig. 9-20). This allows you to keep your hand out of the flame path. If you find the rod sticking in the puddle, slow down your travel speed and add the rod more slowly.

A rather obscure but important fact is that the oxyacetylene flame envelope is an indirect source of shielding gas. This shield protects the vulnerable liquid weld metal from atmospheric contamination, try to keep the filler rod inside the flame envelope. Also avoid any extra removal of the flame entirely from the plate. These practices will help in making strong welds.

WELDING JOINTS

The home welder may torch weld each of the common weld joints—corner, lap, **T**, and butt. Each is discussed.

Welding a Corner Joint

Clamp and tack together a few pieces of 14- or

16-gauge steel as shown in Figs. 9-21 and 9-22. This is an open corner joint or a "corner-to-corner" fit. Welders often try to assemble corners like this. A good weld is made by simply filling the corner with a properly bonded fillet weld.

Place tack welds every 2 inches along the joint. Always tack light-gauge metal by starting at one end of the joint and then progressing to the other end. The metal heated needs freedom to both expand and contract. Attempting to place a tack weld between other tacks causes a wrinkle or buckle along the joint.

The welding technique is basic. Start a puddle and then run it the length of the joint. A tightly fit corner joint can even be fused without adding filler rod. However, adding filler provides a fuller bead profile to bridge across minor (or even major) gaps. A common error involves running a puddle across the joint without adequate fusion to one or both surfaces of the member plates. You can check for fusion by looking at the penetration at the far side of the joint (Fig. 9-23).

After cooling, test the strength of your practice weld by flattening the joint (Fig. 9-24). Look for good bond, or *fusion*, at the edge surfaces. The weld must not "peel" easily from the plates, cracks should be in the center of the joint (Fig. 9-25).

Welding a Lap Joint

Clamp and tack up a few pieces of metal in stepped fashion (Figs. 9-26 and 9-27). These lap joints consist of an easy-to-melt edge and a more difficult-to-melt surface. To melt the surface and not lose the edge, direct heat towards the surface. Do this by torch aim and by holding the torch close to the work. Form the puddle on the surface member, add filler and push or wash the puddle onto the edge member. The edge is kept cooler by adding filler at the edge side of the puddle (Fig. 9-28).

Test the strength of your lap joint fillet weld by prying the joint apart in a vise. Fusion with these lighter thicknesses can be seen by the heat effect on the back side of the joint.

Welding a T Joint

Tack up some T joints (Figs. 9-29 and 9-30). These,

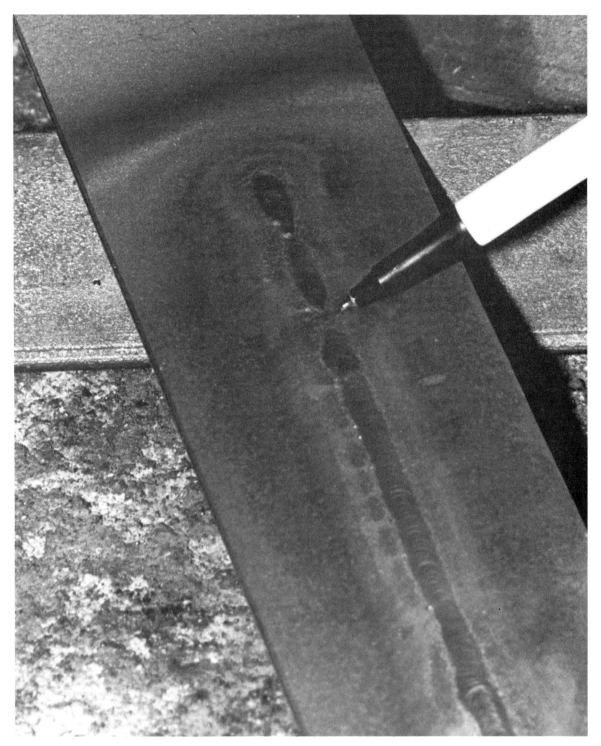

Fig. 9-17. Try to avoid skipping ahead, keep the torch behind the puddle.

Fig. 9-18. The heat line on the backside of the practice plate should be continuous and free from melt-thru.

Fig. 9-19. When the torch is forward, the filler is back.

Fig. 9-20. Bending the filler rod enables you to add it outside of the heat path.

Fig. 9-21. Clamping practice strips into a corner joint is easy with the use of an angle bar support.

Fig. 9-22. A properly tacked corner joint ready for oxyacetylene welding.

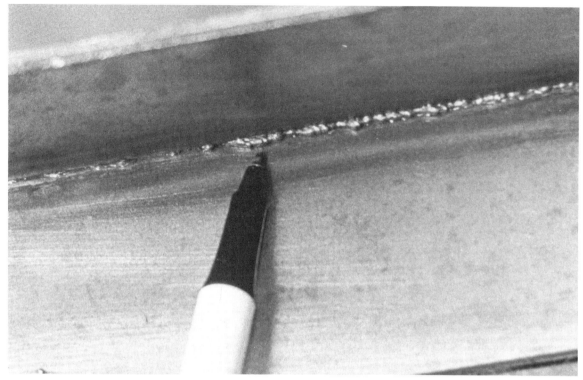

Fig. 9-23. Penetration on the inside of the joint.

like laps, take a fillet weld and involve a surface and an edge. The edge is less likely to be melted away if the joint assembly is tight with the edge directly against the surface piece. With a tight fitup the heat carrying ability of the assembled joint serves as a heat sink to disperse torch energy from the edge.

As with the lap joint, torch energy is directed to the flat surface. Weld metal is worked from the puddle on the flat surface onto the standing member. Avoid overheating the edge by adding filler rod towards the edge side of the joint. Holding the torch too far into the joint overheats the tip, causing backfiring. Keep the torch travel angle low enough to allow some of its energy to escape into the air (Fig. 9-31).

Test the completed T joint by hammering over a sample welded on just one side. Figure 9-32 shows how a good-quality weld joint takes the bend without cracking or tearing.

Welding a Butt Joint

A butt joint in 14-gauge steel is best welded with a small Number 0 or 1 tip. The vulnerable edges demand extra care to achieve adequate penetration without overheating. The best way to do this is to maintain a "keyhole" while progressing along the joint (Fig. 9-33). This usually takes a bit of practice.

Tack welding a butt joint before the main or *production* welding is very important. Spacing between the plates is called the *root opening*. With light metal this should not exceed the metal thickness. For heavier plate (over 1/8 inch thick) try to keep the root opening equal to the appropriate filler rod diameter. Maintain uniformity with the root opening (or gap) and with the *fairness* of the plates. Figure 17-9 illustrates plates fair and out-of-fair. Slight deviation from a proper fitup has a great effect upon weld appearance and penetration.

Tack plates together as shown in Fig. 9-34 with the gap wider at one end to allow heat expansion

Fig. 9-24. Testing a welded corner joint.

and contraction at the joint edges. Place tacks, progressing from one end of the joint to the other, about 2 inches apart (Fig. 9-35).

Begin welding at the tighter end of the joint. Form a puddle and move the torch in a uniform fashion. Add filler rod only:

- as the torch is moved away
- to the rear of the puddle or at an overheated area
- fast enough to maintain a slight bead on top of the plate

Keep the plates hot enough to avoid cold lapping. While welding, a slight sinking of the puddle or presence of a keyhole indicate good bead penetration. A welded butt joint is tested by bending. Check your technique with a short weld sample (Fig. 9-36).

Avoid making long welds on butt joints with oxyacetylene. On all but the heaviest sections (which ought to be arc welded anyway) the dispersed heat and slow travel speeds of this process produce excessive distortion. Normal shrinkage of the cooling edges also causes distortion. If you must

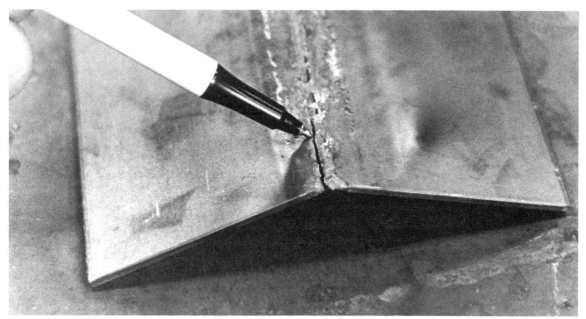

Fig. 9-25. This joint did not have adequate penetration and filler to withstand flattening; yet the cracking took place inside the weld, indicating good fusion to the plates.

Fig. 9-26. A practice lap joint with 14-gauge material.

Fig. 9-27. Tacking should begin at one end and progress towards the other.

Fig. 9-28. Adding filler at the edge side of the lap joint prevents it from melting away.

Fig. 9-29. One way to assemble a **T** joint for tack welding. Note also the minimum-contact platforms with the angle bars.

Fig. 9-30. Another way to assemble a **T** joint.

Fig. 9-31. Some of the torch energy is diverted to the air by lowering the torch angle.

Fig. 9-32. The weld should take the bend without tearing away.

weld a long joint, weld alternately at different areas to avoid accumulating heat.

OXYACETYLENE TROUBLESHOOTING

Several possible problem areas with the use of oxy-acetylene include:

- Loss of gas when equipment is inactive
- Hard-to-turn threads on fittings
- Loss or gain in gas pressure while welding
- Smoke or soot when lighting torch

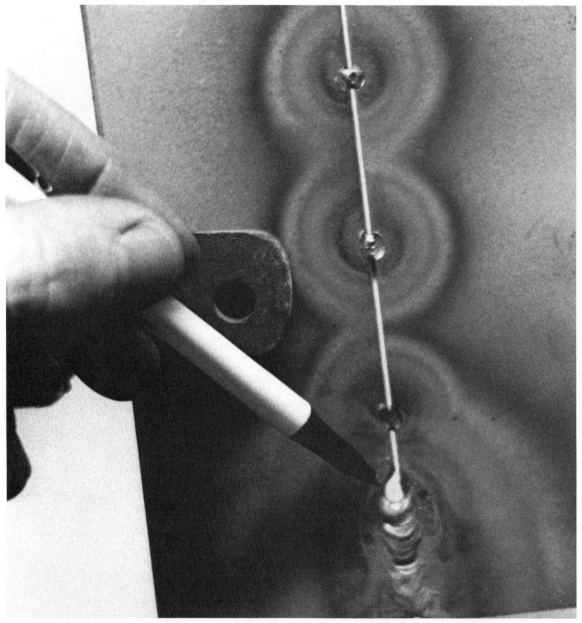

Fig. 9-33. A keyhole on a butt joint.

- Overly delicate torch valves
- Backfiring
- Ragged flame appearance
- Porous or brittle welds

- Welds not fused to base metal

Table 9-1 lists these troubles, possible causes, and corrective steps.

BRAZING, BRAZE WELDING, AND SOLDERING

The oxyacetylene flame has so far been shown as a means by which steel parts are first melted (fused) and then welded together. This versatile heat source also joins steel and some other metals without melting. These lower heat processes are brazing, braze welding, and soldering—whereby metal is joined by *adhesion*, or bonding with a low melting temperature filler alloy. These handy processes are solutions where welding temperatures are impractical. A few applications are: thin-wall or light gauge fabrications, cast iron repairs, art metal, and electrical and plumbing connections. Home welders

Fig. 9-34. The wider space at the far end of the gap will close up with each successive tack weld.

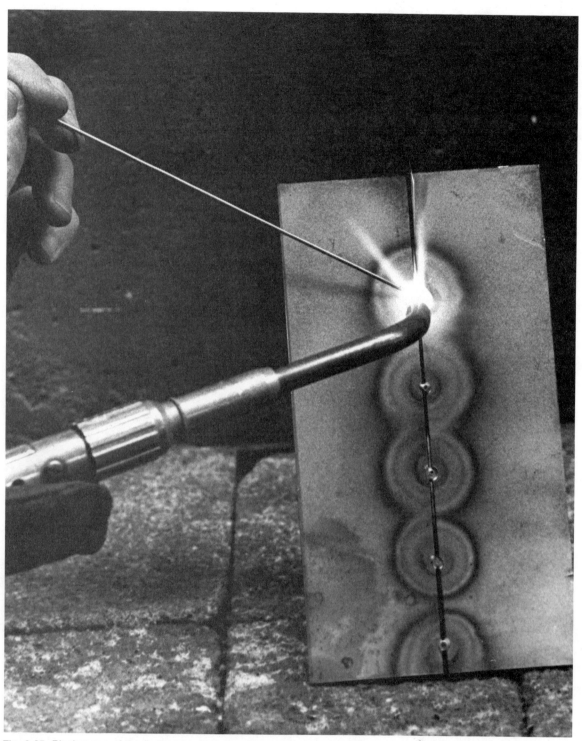

Fig. 9-35. Placing one of the final tacks. Note how the gap has closed up.

Fig. 9-36. The joint should not crack before it is completely flattened.

often use brazing, braze welding, and soldering—especially for repair jobs.

There are some important differences among these "colder" torch processes. Table 9-2 lists temperatures and some applications for these processes.

Both brazing and braze welding require temperatures over 840 degrees Fahrenheit. Most home applications are with mild steel and temperatures of 1600 to 1800 degrees Fahrenheit. Both processes use the same filler rod and cleaning agent or flux. Soldering (or soft soldering), done at temperatures under 840 degrees Fahrenheit, requires a different sort of flux. Soldered joints must be tightly fit and are generally a form of lap joint.

Some Pros and Cons

There are costs and rewards associated with brazing, braze welding, and soldering. Compared to fusion welding, joints are not as strong, often are a different color, and weaken greatly at high temperatures. (Do not braze together a fireplace poker or exhaust manifold.) On the other hand, because less heat is required (nearly 1000 degrees less), braze welding saves time and energy and reduces distortion and stress. Typically, braze welds also look smoother than steel welds. High temperature fusion welding can cause the loss of delicate parts.

The "no melt" processes have other drawbacks. Low melt fillers are easy to overheat and boil or vaporize—producing toxic fumes and a weak joint. The fluxes produce harsh fumes even with proper use. If, at some future time, you should want to repair a brazed joint with fusion welding, the brazing alloy must first be entirely removed to avoid forming a hard, brittle weld.

Flames, Fuels, and Equipment

Oxyacetylene, air acetylene, propane, and natural (city) gas flames are all used to torch braze, braze weld, and solder. Sometimes it is impossible to do very close work with oxyacetylene. The energy is concentrated at the flame cone. Other fuel gases spread the heat more throughout the flame. Such flames are softer and therefore cause less overheating. Brazing with other fuels often involves a torch

Table 9-1. Oxyacetylene Troubleshooting.

PROBLEM	PROBABLE CAUSE	REMEDY
1. loss of gas when equipment is inactive	a. cylinder valve(s) left open b. leaking cylinder c. defective cylinder valve	a. close cylinder valves securely when finished working b. test cylinder with soap bubbles; if leaking move to open air and call supplier c. same as (b) above
2. hard-to-turn threads	a. foreign matter on threads b. threads damaged	a. remove foreign matter with gentle wire brushing; do not force fittings b. repair w/thread file if minor, otherwise call supplier
3. loss or gain in gas pressure while welding	a. large change in cylinder temperature	a. relocate cylinder to area of even temperature
4. smoke or soot when lighting torch	a. inadequate flow of acetylene b. insufficient acetylene line pressure	a. open acetylene torch valve a full 1/2 turn before lighting b. set acetylene to 5 psi
5. overly delicate torch valves	a. line pressure too high b. torch valve packing nut(s) loose	a. set to specified values b. tighten packing nut(s) slightly
6. backfiring	a. tip too close to hot work b. "starved" tip/insufficient gas volumes	a. keep tip proper distance from plate b. maintain sharp cone; use smaller tips for more delicate work
7. ragged flame appearance	a. tip dirty b. tip damaged	a. clean with sparing use of tip cleaners b. file end and clean orifice; or, replace tip
8. porous or brittle welds	a. flame too oxidizing b. flame too carburizing c. flame cone touching puddle d. foreign matter on plate	a. use neutral flame b. use neutral flame c. keep flame cone out of puddle d. remove all paint, rust, etc. before welding
9. welds not fused to base metal	a. tip too small b. travel speed too fast c. use of excess filler d. filler applied before puddle formed	a. use recommended size tip b. travel slow enough to permit complete fusion c. do not "flood" the weld zone d. add filler to the puddle; avoid dripping filler onto plate

Table 9-2. Temperatures and Applications for Torch Processes.

PROCESS	TEMPERATURE (DEGREES FAHRENHEIT)	USUAL APPLICATION
Brazing (with low-fuming bronze: RCuZn-C filler)	1625	joining steel, cast iron, brass, and bronze with tight joints; capillary action necessary
Braze Welding (w/RCuZn-C filler)	1625	same as above except larger filler deposits used without need for capillary action
Silver Brazing ("Silver" or "Hard" Soldering)	under 1200	for joining ferrous, nonferrous, and dissimilar metals and alloys with close joint clearances
Soldering (soft)	under 500	for joining coppers, brasses, and ferrous metals where high strength is not required; joints should be tightly fit and free from high temps and vibration

that mixes air, rather than pure oxygen alone, with the fuel gas.

An oxyacetylene torch handle can be adapted for air acetylene use, but a more convenient arrangement is the specially designed air acetylene torch and regulator shown in Fig. 8-1. These items are designed for use on a Prestolite (POL) style cylinder, so a "POL" acetylene adapter (Fig. 9-37) is needed.

Comparing Brazing and Braze Welding

The main distinction between brazing and braze welding lies in the joint design. Brazing involves a tight lap-type joint into which a thin deposit of filler alloy is drawn by *capillary* action. (Capillary action refers to the physical process by which sponges, towels, and lamp wicks soak up liquids.) Ideal fitup for a brazed joint is a .003-inch gap between the members. On many braze joints, the drawn-in filler is nearly undetectable. The usual filler metals are bronze alloys and silver brazing alloys, with the latter also called hard (or "silver") solder.

Brazing is often an ideal solution for main-

tenance and repair problems both at home and in industry. Dissimilar metal combinations with carbon steel, alloy steel, stainless steel, copper, brass, and bronze can be joined with brazing. While most brazing (and soldering) of manufactured articles such as heat exchangers is performed in sophisticated furnaces or fixtures, a great deal of work can be done using the ordinary oxyfuel torch.

Braze (or bronze) welding is done in much the same way as steel fusion welding except the base metal is not melted. A distinct fillet or reinforcing bead is common. Braze welding is done with all of the "open" weld joints—corner, lap, and T. Filler is deposited onto properly heated joint members and does not depend upon capillary force. Heavy castings are often braze welded because lower heat is less intrusive and cheaper.

Brasses, Bronzes, and Fillers

The most popular filler rod for brazing and braze welding is often called low-fuming bronze. *Bronze* loosely describes an alloy of copper with tin, while brass is a copper and zinc blend. There are not always clear distinctions between these metal groups.

Fig. 9-37. A "POL" adapter for connecting an air acetylene torch to a standard acetylene cylinder fitting.

Copper-based or *cuprometals* have many alloys with uses ranging from bearing materials to electrical components. Only a few are useful in the home shop. The most often-used copper alloy base metals and filler metals are outlined in Table 9-3.

For filler rod matchup problems there are "broad spectrum" fillers to cover many situations. Check with your welding equipment supplier if you have such a problem.

Fluxes for Brazing and Soldering

Fluxes are cleaning agents to remove surface oxides that keep filler alloys from properly flowing and adhering to the base metal. Fluxes also prevent formation of additional oxides during the joining operation and assist the capillary flow of brazing and soldering alloys. There are liquid, paste, and granulated fluxes. These all form a liquid when sufficiently heated which flows just before the filler alloy is applied. Melting of the flux indicates that the right temperature has been reached to apply the filler. There are many fluxes, but the home craftsman can do almost any job with four basic types.

For brazing and braze welding with low-fuming bronze, use a borax flux. Silver braze with fluoride flux. For soft soldering use acid flux if rinsing is possible, and a rosin flux where rinsing is impractical—like electrical work. To braze or solder an unusual metal, your dealer has special purpose flux and fillers. Remember to use:

- Borax flux for brazing with low-fuming bronze
  - Fluoride flux for silver brazing
  - Acid flux for soft soldering with post rinse
  - Rosin flux for soft soldering electrical work

HOW TO BRAZE AND BRAZE WELD

Braze welding techniques are similar to gas welding with filler rod. Obtain 16-gauge to 1/8-inch steel scraps, 1/16-inch low-fuming bronze filler rod, and brazing flux.

Tinning and Buildup

Before attempting to join anything, first experiment by applying bronze filler rod onto a flat steel surface. This is called overlaying and has uses besides mere exercise. Bronze overlays provide bearing surfaces, corrosion resistance, and build-up for worn surfaces. Also, some dramatic effects consist of bronze laid over steel (Fig. 9-38). When making heavy bronze deposits, filler is applied to the work in layers. The first layer is the most critical. Coating with the first thin layer is termed *tinning* and the plate is said to be "tinned;" flux is used. Add layers of filler require little flux. Once tinned, more build up of braze welded deposits is like fusion welding.

To tin, first remove surface scale by sanding or light grinding. The surface must be clean for the flux to work and the bronze to adhere. Minimum-contact plate supports save on heat.

Use a neutral or slightly reducing flame and keep the torch constantly moving. Heat the end of the filler enough to allow flux to cling as it is dipped into the flux can. This is repeated as the coated end

Table 9-3. Copper Alloy Filler Metals.

COMMON NAME	AWS CLASSIFICATION	APPLICATION
Low Fuming Bronze	RCuZn-C	Braze welding steel, cast iron, brass, and bronze; the most popular filler
Nickel Silver	RCuZn-D	Higher temperature and higher strength than Low Fuming Bronze, white color sometimes good for matching base metal color
Manganese Bronze	RCuZn-B	For a hard, wear-resistant deposit
Silicon Copper (Silicon Bronze)	RCuSi-A	Deposit is almost pure copper, good electrical conductivity

Fig. 9-38. Some interesting effects are obtained with bronze applied to steel.

of the rod melts into the joint. Heat the plate with the flame envelope, moving the torch in a rhythmic fashion until grains of flux dropped on the plate melt. This indicates the right temperature for applying the brazing alloy. Experiment to see how filler flows on the clean and heated areas only. A smokey flareup of the bronze means you are overheating the metal and burning out the zinc component. Figure 9-39 shows an unmistakable pale yellow burned zinc dust left on the plate. Overheating makes weak and porous joints.

Brazing Procedures

After practicing tinning, make some practice lap joints (Fig. 9-40). Like fusion welding light metal, tack every 2 to 3 inches starting from one end. Also heat and add filler from just one end of the joint to avoid a bubble forming and becoming trapped somewhere in the center. Hot gas must escape. Metal oozing from the joint opposite the applica-

tion side indicates complete penetration. In order to braze:

- Provide ventilation, use welding goggles.
- Use a reducing flame.
- Heat end of filler and coat with flux.
- Assemble and clamp the work.
- Work from only one end of the joint.
- Heat work to a dull red with a moving flame.
- Add filler with a quick back and forth wiping motion, repeat flux coating the rod as required.
- Continue to play the flame on the joint, heating the upper surface to encourage the capillary draw of filler.
- Use filler sparingly.

When the brazing is completed, cool the plates and test for joint strength as you would for fusion welds. The plates should resist tearing evenly along the joint.

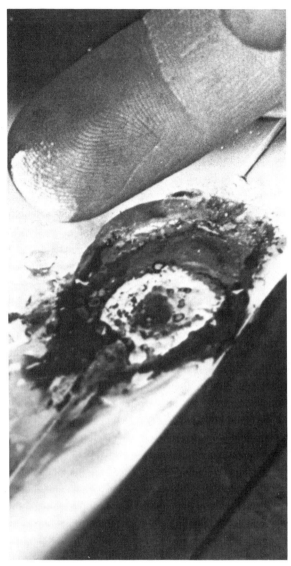

Fig. 9-39. A zinc dust residue from an overheated low-fuming bronze deposit.

Braze Welding Procedures

Braze welding, or "bronze" welding, is a low-heat process whereby the base metal is unmelted and filler is not distributed by capillary action. The filler rod is low-fuming bronze as for brazing. Enough is used to form a definite fillet or bead. Borax flux is used. Home welders use braze welding to join parts where stresses are not severe, such as on decorative ironwork.

Butt joints in steel are generally not braze-welded; the joint surface area is small and the strength of the filler alloy alone is low. Butt joints are usually fusion-welded. To braze weld:

- Assemble 1/8-inch (11 gauge) to 3/16-inch steel strips about 2 inches × 8 inches into a T joint and tack as for brazing.
- Use a neutral or slightly reducing flame.
- Directing more heat toward the surface than the edge, heat the joint until it is dull red.
- Apply the filler coated with flux with a wiping motion in line with the joint, do not overheat.
- If required, make additional passes until the fillet measures one to one and one-half times the material thickness.
- Braze weld one side only and test the T joint by hammering; the bond should be uniform throughout.

If braze welding heavier plates (which for some reason cannot be fusion-welded) make several distinct passes; avoid flooding the joint with an overlapped mass. This ensures a good bond at the root of the joint.

Silver Brazing Procedures

Silver brazing, often called "silver soldering," differs from brazing with low-fuming bronze in both the flux and filler used. Only small amounts of approximately 50 percent silver brazing alloy are required. The process is more heat-sensitive than working with low-fuming bronze and should be practiced before being performed on an important assembly.

Silver brazing is often used to join assemblies of mild and stainless steel, also copper and nickel. This ability to join such dissimilar metals makes it especially useful both to industry and at home. Silver brazing bears up well on joints subjected to moderately high temperatures and vibration. Electrical heating elements are often joined by silver brazing.

Silver brazing flux is a paste and is applied carefully or the brazing alloy will stick to any surface upon which the flux has acted. Flux should spread out evenly on a slightly warmed surface. If it

119

Fig. 9-40. Brazing practice lap joints.

gathers in clumps like water on a waxed surface, the base metal requires better cleaning (Fig. 9-41). To silver braze:

- Remove any scale, dirt, grease, etc.
- If required, dilute the flux until it is about the same consistency as melted ice cream.
- Apply a small amount of flux to the joint faces and assemble the joint.
- Using a neutral flame, "flash" the flame over the joint to dry the flux.
- Continue to heat the joint until the flux remelts.
- Add the silver brazing alloy sparingly.
- Continue to flash the flame over the joint until the alloy has been drawn along the entire joint;

add filler only if required. There is no advantage in a buildup of filler.

Silver brazing flux is soluble in hot water. Flux residues should be removed to prevent their continued corrosive activity.

"Sil-fos" is a "poor man's silver solder"—a substitute for high-percentage silver brazing alloys. These phosphorous alloys were patented years ago by Handy and Harman, silver brazing authorities. Such bargain alloys are about 5 percent phosphorous and little silver. (A 2- to 6-percent silver alloy is the most versatile.) Phosphorous works as a flux but only for copper alloys. Sil-fos works well on copper-to-copper tube joints without flux. Keep sil-fos well identified.

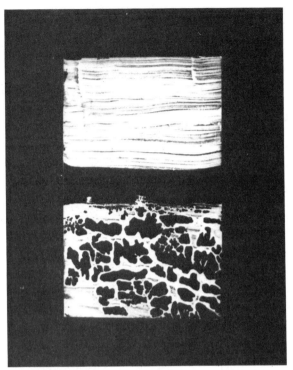

Fig. 9-41. Silver brazing flux should spread out evenly on a slightly warmed plate (top). If it collects into islands, the work is not clean enough. (Courtesy of Handy and Harman.)

Silver brazing for food service items should be done with a non-cadmium bearing alloy. Your welding supplier has a variety of safe silver brazing alloys in small packages.

HOW TO SOLDER

Soldering is the lowest heat metal joining process. It is a good way to join or seal sheet metal and other thin sections. Solder is widely used to seal fluid systems.

Most soldering is done with a tin and lead filler. Often called "soft soldering," it is used to seal or fasten lightly loaded assemblies. Soldering can join combinations of steel, brass, bronze, and copper—these are most of the commonly-used metals. Familiar applications are car radiators and electrical connections. Recall that so-called "silver soldering" is actually higher temperature brazing. Never mix these solders or fluxes.

Solder joints get their strength by adhering to base metal surfaces. Like glue joints, they are strong only with ample base metal surface area and only a thin layer of the bonding agent.

Heat for Soldering

Soldering is done with various heat sources including:

• Torches—oxyacetylene, air acetylene, air propane, air natural gas. These torches provide very fast heat in a wide pattern and are especially suitable for larger assemblies. Figure 9-42 depicts a very popular air-propane style torch.
• Irons or coppers—heat-transferring surfaces that are separately heated by an oven or torch, or integrally heated by a flame or electrical resistance coils (Fig. 9-43). Coppers afford excellent control and are made in several sizes and shapes.
• Guns and Pencils—miniature electrically-heated surfaces (Fig. 9-44). These are mostly limited to precise spot soldering.

It is absolutely essential that a solder joint be clean and at the proper temperature. The hot surface of the heat source must also be clean and tinned with a thin layer of solder. This prevents the formation of oxides and transfers heat to the metal.

Solder Filler Metal

Most soldering is done with "50-50" solder: 50 percent each of tin and lead. The ratio controls the melting temperature. See Table 9-4 for melting temperatures of different solders. Having two solders with different melting points is useful where two joints are made near each other and you wish to avoid undoing the first one as you apply heat to solder the second.

Solder packages have several forms. The most familiar is wire, solid core or hollow, filled with flux. Flux is either corrosive acid or a less caustic rosin—the familiar acid-and rosin-core solders. Recall that electrical connections that cannot be rinsed are soldered only with rosin flux. Solder is available in a combined alloy-flux paste for maximum convenience.

Fig. 9-42. Soft soldering has long been used for sealing sheet metal fittings.

Procedures

The following procedures are for soldering a mild steel lap joint using both a torch and a soldering copper.

First obtain these materials:

- 18- or 20-gauge sheet metal strips measuring approximately 2 inches × 4 inches
- 50-50 solder, either solid or acid core
- Soldering acid and acid brush
- Two (2) spring clamps

Table 9-4. Soft Solders, Melting Temperatures, and Applications.

SOLDER (TIN TO LEAD)	MELTING TEMPERATURE (DEGREES FAHRENHEIT)	APPLICATIONS
40/60 (acid core)	460	General Purpose other than electrical
40/60 (rosin core)	460	Electrical Connections
50/50	420	General Purpose
40/60	460	Stainless Steel with special flux

Fig. 9-43. An electric soldering copper provides a great deal of control.

Fig. 9-44. Soldering pencils are used for delicate work.

Torch Soldering. Remove millscale, rust, grease, or paint from the metal. Steel wool is effective for lighter accumulations; use sandpaper or emery cloth to remove heavier buildups. Follow all precautions about safe use of the flux. In order to solder with a torch:

● Clamp the strips as shown for welding in Fig. 9-31.

● Brush acid flux on the lap joint faces of both strips.

● With oxyacetylene, use a small tip (00), light the torch, and adjust to a reducing flame; for other fuels adjust until the flame cone is sharp.

● Slightly preheat the metal by flashing the flame across the joint and immediately follow with another light application of the acid flux.

● With the torch in one hand and solder in the other, heat the metal and apply the solder along the joint; keep the torch moving at all times to avoid excessive warping.

● Allow the solder to "freeze" completely before moving the joint—it will change from shiny to slightly dull.

Soldering with an Iron or Copper. The copper must be properly tinned at all times and hot enough to keep the tip shiny. It is overheated if the solder coat on the tip burns away quickly. Electric coppers without thermostats need plugging and unplugging to maintain the proper temperature. Never cool an electric copper by dipping in water.

● As with torch soldering, apply flux, assemble, and clamp the lap joint.

● Apply the full face of the copper to the outside of the joint and add solder without quite touching the copper—solder is drawn into the joint by capillary action.

Soft soldering has many applications. Two common uses are sealing copper tubing joints and securing electrical connections. Procedures for these follow:

Soldering Tubing Joints. Residential plumbing is often copper tubing and connections. Such systems are joined by soft solder and a propane torch.

Tubing joints are round lap or socket joints. Use steel wool to clean the joint surfaces. Apply rosin paste flux, heat the heavy part, and flow solder into the joint. Continue to flash the flame on the joint until a slight fillet forms all around. Avoid overheating! Besides possibly burning the filler, solder is heavy and might just run out of the joint.

The plumbing system should be drained of water before soldering and vented at some point. If water is in the system a propane or city gas torch might not produce enough heat. (Torch energy is used up changing water into steam.) In such a case,

Fig. 9-45. A heat sink (the alligator clip just above the soldering pencil tip) is used to protect the heat-sensitive transistor during a soldering operation.

use oxyacetylene to solder. An old-timer's trick for dealing with a "wet" system is to stuff white bread into the tubing a few inches from the joint. This acts as a sponge long enough for soldering to be done and later flushes out. Just avoid whole-grain breads that could clog the system—the plumbing system, that is.

Soldering Electrical Connections. Light-load electrical connections like electronic parts may be held together only with solder. Even light-duty mechanical connections and splices can be solder reinforced. Just recognize that soft solder is not a good electrical conductor and that resistance heating can develop in the joint. If the electrical connection is subject to heating or excessive vibration, it is best to silver braze it following the earlier procedure.

With a soldering pencil, the tip must be securely fastened to the body of the heating unit. A soldering gun tip is the highest resistance of an electrical circuit and needs tight connections. File the tip to remove oxides and tin with heat from the unit.

To solder an electrical connection, first apply rosin flux to the joint. With rosin-cored solder, you might not need extra flux. Then place the hot tip alongside the joint until flux boils. Add solder sparingly at the connection, not the tip. Do not overheat the work—use heat sinks (Fig. 9-45) if soldering semiconductors. Allow solder to freeze before moving the joint. Remove excess solder and/or flux with a clean dry cloth.

To summarize the steps in soldering an electrical joint:

- Apply rosin flux.
- Apply heat.
- When the flux boils, add solder.
- Allow solder to solidify before moving.
- Remove excess.

This concludes the coverage of brazing, braze welding, and soldering. Although not welding in the stricter sense of fusing the base metal, these handy processes should be a part of every welder's "process repertoire."

Chapter 10

Welding with Shielded Metal Arc Welding

T HE NEXT TWO CHAPTERS DESCRIBE ARC welding processes most suitable for home use, including shielded metal arc welding and the wire feed processes with both solid and tubular electrodes. A basic understanding of how it works is useful.

Welding with an electric arc is faster than with an oxyacetylene flame, but this speed places additional demands on the welder. Manipulative arc welding skills are at their best only if the welder is correctly positioned before, during, and even at the end of a welding operation. Arc welding should not be a struggle with an uncertain outcome; it should be performed with confidence and even a certain grace.

PRINCIPLES OF ARC WELDING

All arc welding methods use heat energy derived from electrical energy. The concentrated heat of an electric arc at nearly 10,000 degrees Fahrenheit is even hotter than the oxyacetylene flame and, therefore, more than sufficient to fuse steel. The complete subject of circuit electricity is fascinating yet complex. Fortunately, it is not necessary to know what electrical energy is all about to use it in a welding circuit—but a basic understanding helps. For clarity in the following explanation, *Voltage* is defined as electrical force or pressure; *Amperage* is the flow of an electrical current and is the main control of welding "heat."

The Welding Circuit and Polarity

Welding energy is delivered from the power source to the point of arc welding at the plate. Figure 10-1 compares a simple electrical circuit to a closed water system—a helpful analogy. The energy flows through a large welding cable, connectors, and either an electrode holder or a torch (gun) containing an *electrode*. The electrode is a sort of "take off and/or landing" place for electrical energy, where the break in the circuit (and greatest heat) is located. Like a water system, the welding energy path must form a complete closed *circuit* from the power source, through the arc and back to the

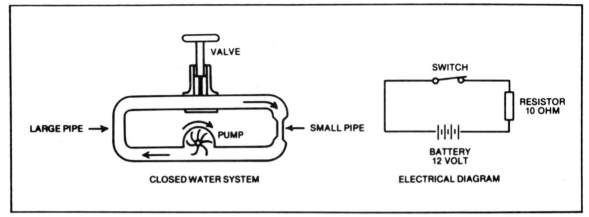

Fig. 10-1. A simple electrical circuit, in many respects, resembles a closed water system with elements of pressure, flow, and restriction. (Courtesy of Alloy Rods.)

power source. This is done with another (ground) cable also connected to the power source. Figure 10-2 shows the basic welding circuit for Shielded Metal Arc Welding (SMAW) with coated stick electrodes. In a dc (direct current) circuit, current flow is generally considered to be from negative (−) to positive (+). A dc current always flows the same direction depending upon the way the cables are connected. An ac welding current, on the other hand, goes back and forth in the cables—usually 60 times per second.

Polarity describes the direction in which the current flows across the arc. It is important because it determines just how a particular electrode will behave at the arc. The flux coatings on each type of stick electrode and the granulated fluxes inside tubular electrodes provide effective shielding, penetration, arc force, and other functions only if the specified polarity is used. Exactly why this is so remains a mystery to most, but provides abundant material for theoretical discussions.

The specified direction of dc current flow is often from the power source to the work (ground) connection, across the arc onto the electrode, and back to the power source. This is referred to as direct current electrode positive (DCEP) (Fig. 10-3). DCEP was formerly known as direct current, "reversed polarity" (DCRP). Even today, the older term is widely used. Figure 10-4 shows a modern power source with polarity still marked "straight" and "reverse." Most stick electrode coatings and

the solid wire electrodes are designed for use with DCEP.

If the welding cables were switched positive to negative, the current flow would be from the power source to the (−) electrode, across the arc, onto the (+) work, and back to the power source. This arrangement is called direct current electrode negative (DCEN) or "straight polarity" (Fig. 10-3). A couple of coatings are tailored for use with DCEN as are most self-shielded flux-cored wire electrodes.

The *transformer* type of power source produces an alternating current (ac) output. Here the polarity alternates from straight to reverse 60 times a second. Figure 10-5 shows one cycle. Such polarity is just called "ac." (The business about turns in the figure refers to rotation of an electrical generator.) In the home shop, ac is used only with certain coated stick electrodes and does not work at all with wire electrodes.

The Welding Arc and Voltage

As just mentioned, electrical energy flows as a current in a closed path called a circuit. Breaking this path causes a spark or *arc* to form across the gap (Fig. 10-6). The arc conducts energy (current) across the break. The arc continues as long as the pressure or voltage behind the current flow is strong enough to overcome the resistance of the circuit (Fig. 10-7).

A circuit, connected to a voltage yet having a break too large for current flow, is called an *open*

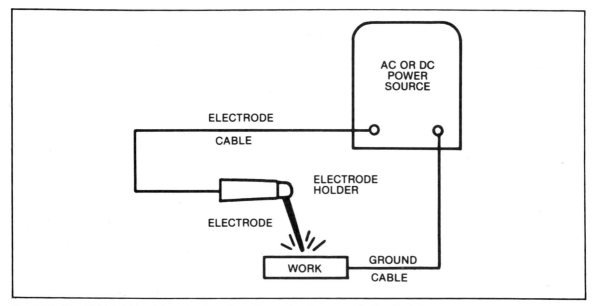

Fig. 10-2. The Shielded Metal Arc Welding (SMAW) circuit. (Courtesy of Alloy Rods.)

circuit. With the power source "on" and the electrode not yet in contact with the plate, the welding circuit is "open." Voltage during this open circuit condition is called *open circuit voltage* (OCV). No current flows during the OCV period. Again, this is the situation before the arc has been struck on the plate. Once the electrode touches the grounded plate, current flows and the voltage drops to a welding voltage level (Fig. 10-8). Most power sources for home use have no meters so the voltage-amperage rela-

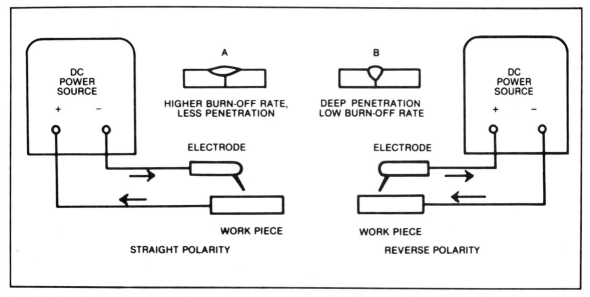

Fig. 10-3. Polarity determines which direction the current flows in the welding circuit. It has a great effect on the condition of the weld deposit. DCEN or straight polarity is shown to the left and DCEP or reverse polarity is shown at the right. (Courtesy of Alloy Rods.)

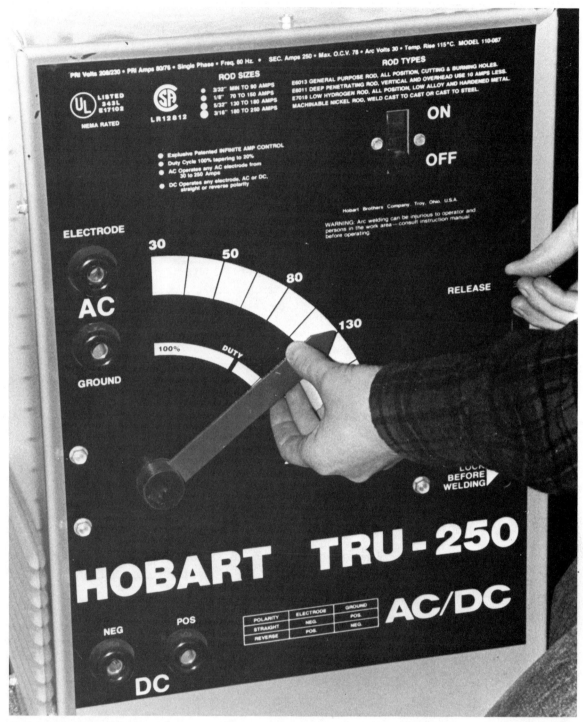

Fig. 10-4. This is the face of a power source that produces both ac and dc welding power. Notice the polarity table near the bottom.

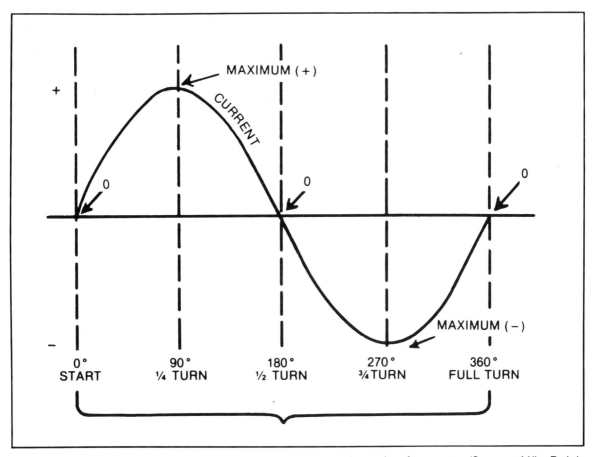

Fig. 10-5. One cycle of alternating current. The degrees and turns refer to the rotation of a generator. (Courtesy of Alloy Rods.)

tionship is seen by its effect upon the plate.

Arc length is the size of the gap or break in the circuit. The greater the arc length, the greater is the arc voltage (and vice versa). Arc voltages for home welding range from 17 to about 32 volts; arc lengths are about 1/8 inch.

To recap, the break in the circuit (arc) takes place between a consumable electrode and the work itself. The electrode is located in a torch (an electrode holder or gun depending upon the specific welding method used). It is connected to the insulated "or hot" side of the arc circuit and the work is connected to the ground side. Besides heat, a consumable electrode also provides filler.

With stick electrodes, you control the arc length manually by holding the arcing electrode slightly closer to or farther from the plate. This volt-

age change in turn controls the amount of current (amps or "heat"). It also determines "arc fan" (Fig. 10-9). Wire processes are less demanding because the welder sets voltage before starting to weld. Welding with wire is easy; a slight change in the gun-to-plate distance has little effect on the plate. This is discussed in Chapter 11.

OTHER ARC WELDING METHODS

The welding industry uses some other arc welding processes. Some require elaborate equipment for joining very heavy or very thin metal. A few home welders who work with light stainless steel or aluminum use the Gas Tungsten Arc Welding (GTAW) or TIG process, in which the electrode is not consumed (Fig. 10-10). With GTAW, filler is added separately like oxyacetylene welding.

Fig. 10-6. The electric arc at 10,000 degrees is more than adequate to fuse most metals. SMAW with a coated electrode is shown here.

Line Energy and Welding Energy

Line power usually comes to your home shop at 230 volts. In its trip from the power plant, voltage was stepped up or down for transmission efficiency (Fig. 10-11). 230 volts needs to be transformed to a lower (and safer) value for arc welding. The *transformer*, a major component of a power source does this (Fig. 10-12). A simple transformer consists of a primary (input) coil and a secondary (output) coil (Fig. 10-13). Note that the primary side is high voltage

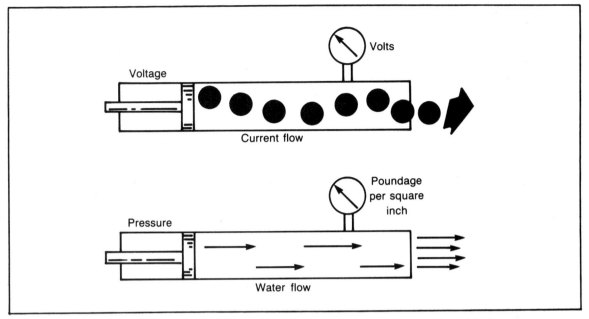

Fig. 10-7. Pressure is needed to make water flow, voltage is needed to cause a current (amps) to flow.

and the secondary output is lower. Figure 10-14 shows a simplified welding transformer in which the primary plugs into the 230-volt line, and the secondary side goes to the welding circuit. A transformer does not generate energy (here expressed as watts); it changes it. In the process, all energy is accounted for as either useable or lost as heat.

Note that while a transformer steps down input voltage it raises the current (amps) at its output. This is fortunate. High voltage and low amperage go into the transformer; low voltage and high amperage (heat) come out—exactly what is needed for arc welding. Can anything ever be quite that simple? The inside of a power source is actually stuffed with electrical and electronic components that provide control of the transformer output. How a power source works is interesting; there are some excellent books on the subject. Keep

Fig. 10-8. The high open circuit voltage (OCV) drops rapidly as the arc is struck. The voltage fluctuates somewhat during welding until the arc is broken when it again returns to the OCV level.

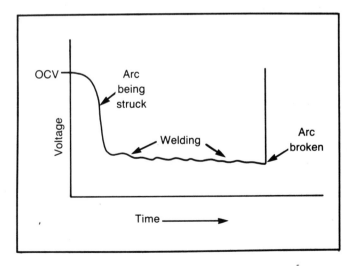

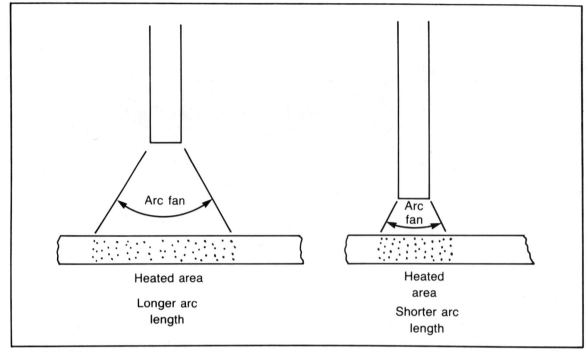

Fig. 10-9. The arc length affects the amount of base metal heated.

in mind, however, that only two types of people can tinker with the inside of a power source: a trained and qualified technician—and a fool.

PROPER USE OF A WELDING HOOD

Besides the obvious bright light of the arc, there are also infra-red and ultra-violet rays at wavelengths above and below that of the visible light. Eyes and skin must be shielded from these burning rays. A welding hood (Fig. 10-15) does a good job in providing protection and is the most common form of shield. Use at least a Number 10 density filter lens for arc welding. Place clear plastic protective plates over the filter lens. (Always keep a few extra clear plates on hand.)

There is an easily formed bad habit with the use of a hood. Try to avoid reaching up to lower your hood; it can be lowered with a nod of the head. Notice from Fig. 10-15 that there are occasions when both hands are required for the work, so learn to arc weld using both hands. With both hands dedicated to the work, you have excellent control for locating the weld.

The Flip Visor and Slag Removal

The flip visor with the filter lens (Fig. 10-16) raises and lowers to allow the hood to double as a face shield. (Note that there is a stationary clear lens in the hood window.) The flip visor should not be dropped when striking an arc at the beginning of most welds. Doing so occupies one of your hands and, for best control, both hands are needed for the work. Precise welds are made only if your hood can be dropped simply by nodding your head. As Fig. 10-17 shows, the headgear has adjustments for:

- The size of your head
- The height at which you will wear it on your head
- The tension used to hold the hood up
- The distance that the hood will drop

Spend a few minutes examining the means for these adjustments. Your hood must work right and be comfortable; otherwise it will detract from your welding. If you have questions, your welding supplier should be able to help.

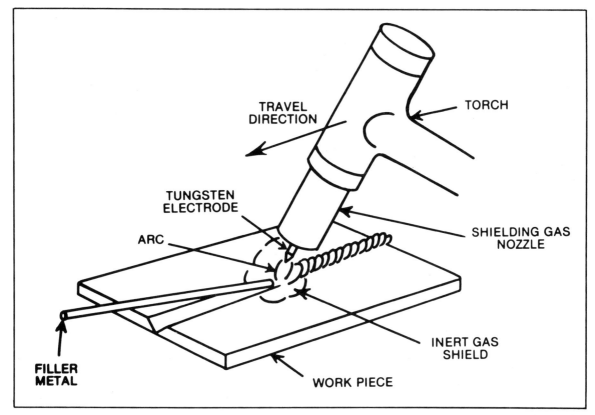

Fig. 10-10. Gas Tungsten Arc Welding (GTAW) is rarely performed in the home shop. (Courtesy of Alloy Rods.)

Accomplished welders use a standardized routine. The flip visor (with filter) is lifted as welding is finished. This leaves a clear lens and the hood itself for eye and face protection when chipping *slag* and wire brushing the weld (Fig. 10-18).

Slag is the crust that forms over the top of the

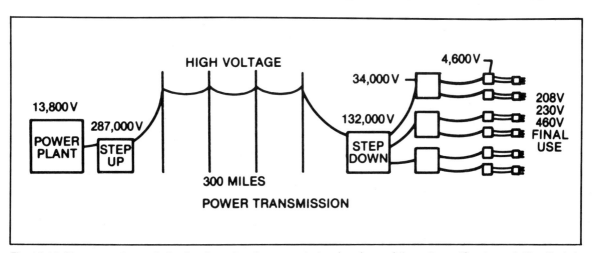

Fig. 10-11. Line power transmission involves stepping up and stepping down of the voltage. (Courtesy of Alloy Rods.)

135

Fig. 10-12. The transformer is the heart of the power source. It steps down the high input voltage to a value suitable for welding. The large bands are secondary output wire windings capable of carrying heavy welding currents.

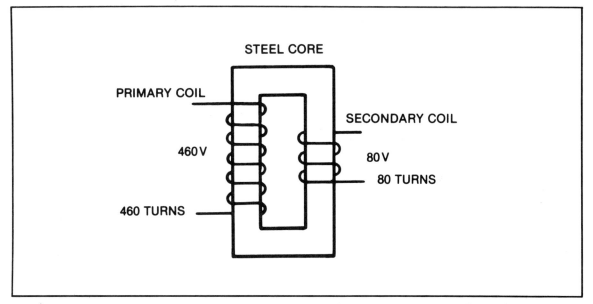

Fig. 10-13. A basic transformer. The ratio of turns controls the ratio of voltage transformation. (Courtesy of Alloy Rods.)

weld (Fig. 10-19). It protects the weld by slowing its cooling rate and by excluding the atmosphere. Slag also contains the solid impurities washed by the arc from the weld zone. As soon as the weld is cleaned, the hood is raised and the visor lowered in one operation. With more welding, the hood is lowered as far as possible to still permit viewing the plate. The electrode is then positioned just off the plate, and the hood lowered by a nod without losing the electrode position.

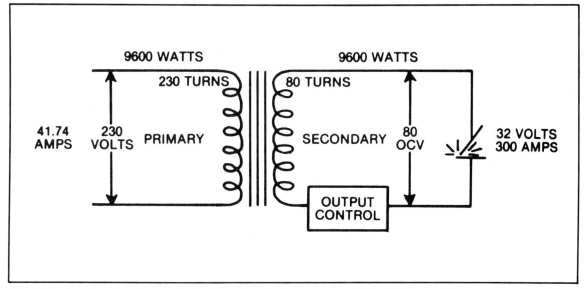

Fig. 10-14. A welding transformer steps voltage down but amperage up. Notice that the wattage energy on the primary (input) side is equal to the wattage on the secondary (output) side, at least on paper. Actually, some energy is lost as heat. (Courtesy of Alloy Rods.)

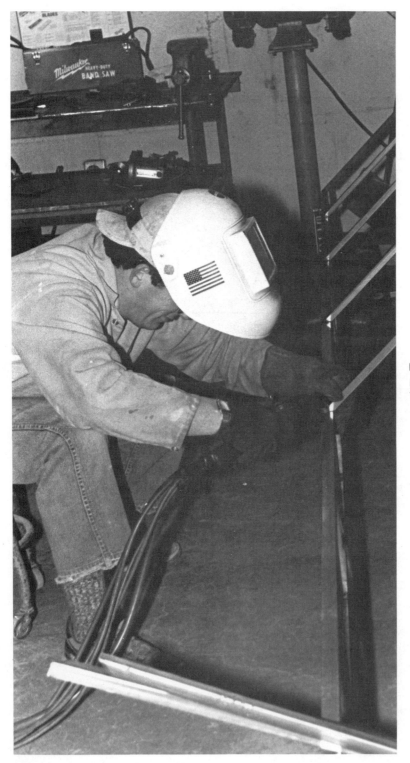

Fig. 10-15. Use a welding hood for protection, and use both hands for the job.

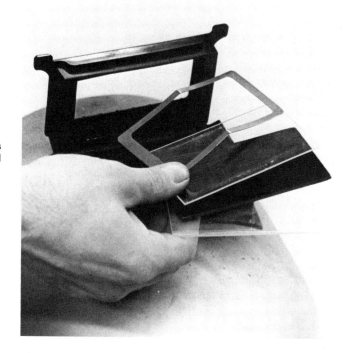

Fig. 10-16. A flip visor on a welding hood contains both a dark filter lens (number 10 is usual) and one or two clear plastic protective plates.

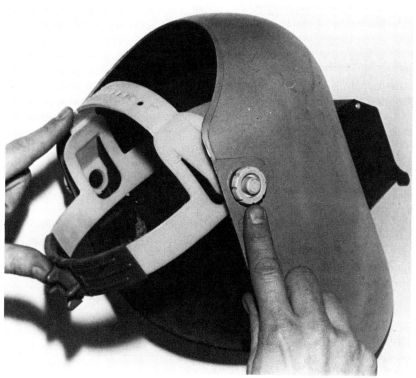

Fig. 10-17. The head gear of a hood has four adjustments.

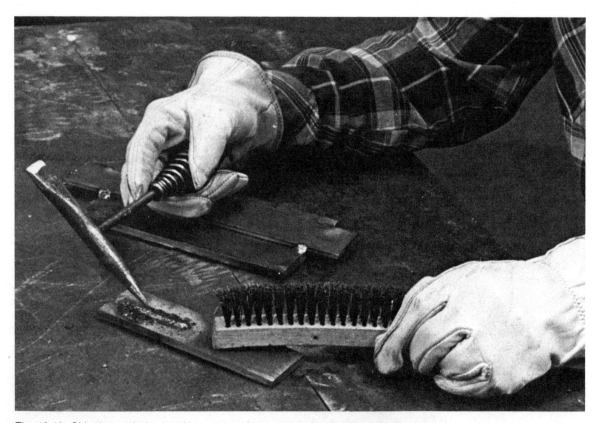

Fig. 10-18. Chipping and wire brushing are used to remove slag from a weld deposit.

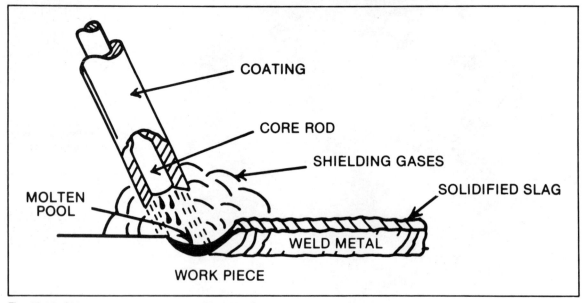

Fig. 10-19. Shielded Metal Arc Welding (SMAW). (Courtesy of Alloy Rods.)

Use of the Hands

The momentary darkness just before the arc is struck, the arc start, and the welding period itself must be controlled. An *arc strike* (Fig. 10-20) is a potentially serious defect resulting from a shakey and misplaced arc start.

The importance of proper body positioning to good welding cannot be over-emphasized. Many welding problems can be traced to operator frustration rooted in improper body positioning. The best control is gained by gripping the torch with only one hand while the other hand guides the torch hand. ("Torch" here means either a stick electrode holder or a wire feed gun.) Examine Figs. 10-21 and 10-22 to see how the welding operation is smoothly guided through the joint by the fingers of the "guide hand."

Hand-Held Shields

It is interesting to note that many welders outside of North America do not normally wear a welding hood. They use hand-held shields (Fig. 10-23) when-

Fig. 10-20. An arc strike is a potentially serious defect.

Fig. 10-21. Position of hands at the start of a pass.

ever possible. The incidence of respiratory problems for Western Hemisphere welders is also higher than for the rest of the world. It is easy to suspect that proximity to the welding smoke is a contributing factor. You may not be ready to trade your hood for a hand shield, but place yourself upwind out of the welding smoke and fumes.

Other Protective Gear

The arc welding described in the following paragraphs produces sparks, slag, and rays that cause painful burns and sunburns. Protect yourself with:

● 8-inch leather boots, free of stitching on toes

● Clothing made from natural fibers—wool is best

● No cuffs or open pockets

● At least a Number 10 density filter lens, up to 14 if required

For welding in any position where reaching up is required use:

● Welders gloves with gauntlets

● A leather or other type protective jacket or sleeves

● Safety glasses

● Ear plugs

● A cap

Avoid a painful lesson. Do not weld while wearing athletic or tennis shoes, cloth gloves, short pants, short sleeves, or synthetic fibers. In addition, do not risk a burn from a book of matches or a lighter in your pocket becoming ignited by sparks.

WELDING AT HOME WITH SMAW

Shielded metal arc welding using coated stick electrodes has been used by home welders since the late 1940s. This would not have been the case if it were unsafe or impractical.

Arc Welding Safely

There are safety concerns with electricity, arcs, and sparks. Reasonable precautions should be taken.

Chapter 3 discussed some facts about workplace safety and practicality for yourself and others. Read and ponder the safety information supplied with equipment instructions.

The presence of moisture around electricity can be extremely dangerous. Avoid using a damaged electrode holder or wet gloves that could place your body within an electrical circuit. Be sure that your power source is properly grounded. Never use defective plugs or wires. When leaving your equipment unattended, disconnect and perhaps even lock the power.

Setting up Shielded Metal Arc Welding Equipment

The equipment used for SMAW is simple. The

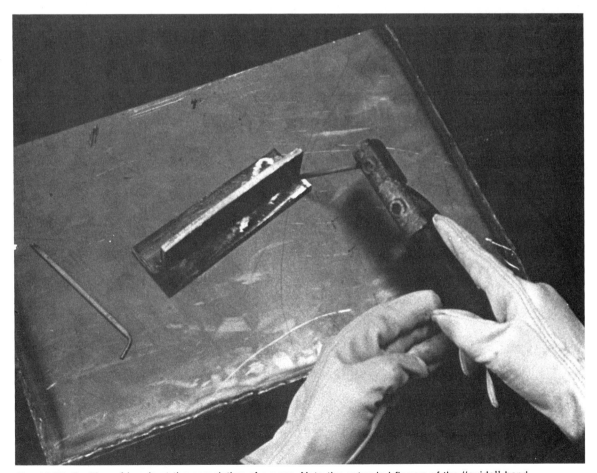

Fig. 10-22. Position of hands at the completion of a pass. Note the extended fingers of the "guide" hand.

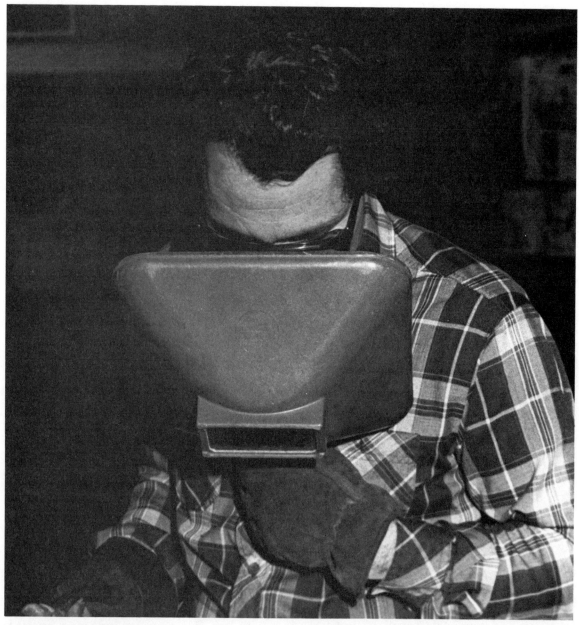

Fig. 10-23. A hand-held shield.

power source must be plugged in and welding cables properly located. Stick electrodes, a slag hammer, a wire brush, and little else are required (Fig. 10-24). Before plugging in a power source, check that it is "off." This prevents unplanned arcing.

The power source should be located where it can receive a good supply of cooling air. If the primary (input) cable crosses a walkway, protect it with a simple wooden ramp. Also route the welding cables safely.

The work (ground) connection is often overlooked. Be certain that it is secure and attached only to a clean surface. When first arc welding at home, it is best to use a clean steel plate for a working surface. Attach the ground clamp to this plate. This also grounds any parts to be welded located on the plate. *Spatter* or splash from the molten weld pool eventually accumulates on this plate and should be removed with a grinder.

Clear the Ground

A good ground connection is needed in spite of whatever finish or substance might be on the metal. If necessary, remove paint, rust, or any other electrical obstacle from project parts by chipping or grinding.

To weld on a vehicle equipped with a charging circuit such as an automobile, or even a trailer attached to the car, disconnect both battery cables.

First remove the grounded battery cable and then the insulated cable. This eliminates possibly "zapping" sensitive and expensive electronic components. Also trace through the welding circuit ground path to prevent current travelling through bearing surfaces. Expensive ball and roller type bearings get instantly pitted if current arcs through them (Fig. 10-25).

Practice Welding with SMAW

To practice weld with SMAW, use reasonably clean mild steel either 3/16 or 1/4 inch thick. If you are in a hurry, buy metal already sheared to size. You can, however, flame cut this plate yourself. Refer to oxyacetylene setup in Chapter 9, and see Chapter 16 for the proper use of a cutting torch.

Six-inch or larger squares of plate should be

Fig. 10-24. SMAW does not require elaborate tools.

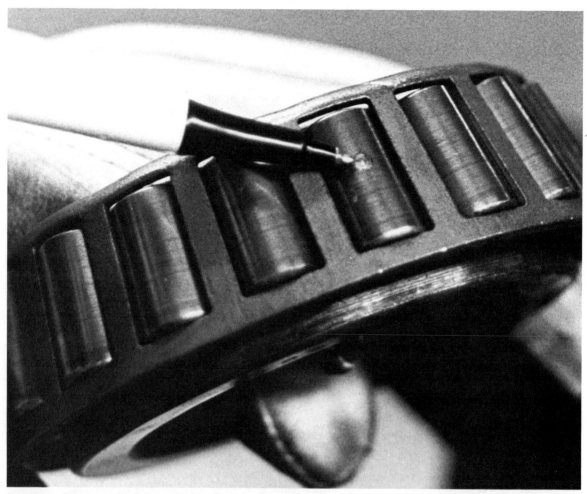

Fig. 10-25. A roller bearing pitted from arc damage. Place ground connections to avoid including a bearing in the welding circuit.

adequate for learning how to strike the arc and run beads. Once you can run a decent bead with starts and stops, you will be able to weld something together. Practice joints are readily made with strips about 2 inches wide and 6 inches long. The thickness depends on the diameter of electrode you use. Plate that is 3/16 or 1/4 inch is ideal with 3/32-inch diameter stick electrodes. SMAW can be used in any position, but first practice the flat and horizontal positions. Most home welding jobs are small enough to allow moving the work into these positions. The vertical and overhead positions are discussed later in this chapter.

Stick Electrodes

Stick electrodes come in several diameters to suit different metal thicknesses. Because stick-type power sources for home are limited to about 180 amps output, 5/32 inch is the largest electrode you will normally use. The next smaller size, 1/8 inch, is the most popular diameter for industrial work. The 3/32-inch diameter stick electrode is the most popular for the lighter home shop welding.

Each electrode producer has a different name for essentially the same type of electrode. Fortunately the American Welding Society has a classification system adhered to by electrode manufacturers. Thus, each part of the designation E6011 has significance:

- E means electrode.
- 60 means 60,000 pounds per square inch tensile strength of the deposited weld metal—this exceeds the strength of the steel with which you will be working.
- The first 1 means this electrode can be used in all positions—flat, horizontal, vertical, and overhead. A 2 would mean that the electrode was intended for flat and horizontal use only: E7024, for example.
- The second 1 means that the coating is mainly cellulose which produces a deep penetrating arc. The coating also contains potassium, which permits welding with ac.

What distinguishes an electrode type is the flux coating. While general "flux recipes" exist, electrode manufacturers keep the exact composition of the coatings pretty much to themselves. There can be significant differences among operating characteristics of the same type electrode made by different companies. These include: striking ease, spatter, penetration, ease of slag removal, and bead appearance (smoothness). The coatings on stick electrodes are to:

- Shield the weld metal
- Stabilize the arc
- Add alloys to the weld pool
- Concentrate the arc stream
- Act as a flux and form a slag
- Permit welding in different positions
- Control the "freeze" or solidification rate
- Electrically insulate the core wire

The stick electrode most reliable for weld integrity is type E6011. In storage it only needs to be kept reasonably high and dry. It should never be placed in a *rod oven* (Fig. 10-26). Ovens are needed to store the low hydrogen stick electrodes that are not generally suitable for use in the home shop. E6011 also works well with either ac or dc.

Electrodes for ac use have *ionizers* such as potassium in their coatings. Ionizers are elements that produce an atmosphere for easy arc re-ignition. Ac has 120 arc outages per second as the wave passes through zero, so ionizers are essential for stick welding with ac. E6011, E6013, and E7024 are electrodes that can be used with ac (or dc).

Striking an Arc and Running a Bead

In order to successfully join metal with stick electrodes, it is first necessary to master striking an arc and running a bead across a plate. Practice these skills with 1/4-inch plate, 1/8-inch E6011, and a power source set to about 100 amps. Draw several soapstone lines across the plate. Mark the lines with a center punch mark every inch or so.

REMINDER:
    Wear proper clothing for welding and practice the correct use of the welding hood as outlined earlier. Use pliers to handle hot metal. When quenching hot plates, remember to keep water away from your power source and wiring, also keep your gloves dry.

With the power source "off," practice striking an arc and running a short bead for 1 or 2 inches. Insert the electrode into the holder and position yourself as was shown in Fig. 10-21. Note the hand positioning and how to lower your hood. When possible, try to weld across your line of vision. For right handers, this means travelling from left to right.

Two angles determine the position of the electrode—the *electrode angle* and the *travel angle* (Figs. 10-27 and 10-28). Changing these angles affects the location of the weld deposit. While running a simple bead, try to keep the electrode angle at 0 degrees, nearly perpendicular to the plate; electrode travel angle should be about 20 degrees in the direction of travel.

The arc is easily struck if the electrode is scratched like a big match upon the plate. For a second or so after contact with the plate is made, the electrode must be held 1/8 to 3/16 inch off the plate. This enables the arc to establish itself. See Figs. 10-29 to 10-31 for the starting technique.

Fig. 10-26. A special rod oven is needed for the low hydrogen electrodes that are seldom used in the home shop.

Fig. 10-27. The electrode (side-to-side) angle is shown here.

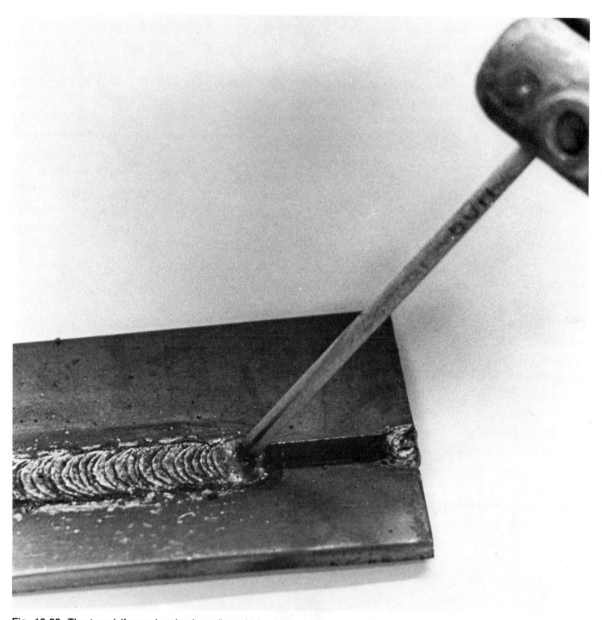

Fig. 10-28. The travel (forward or backward) angle is shown here.

The E6011 electrodes are designed to be *oscillated*. This means that you should move the electrode in a series of regular back-and-forth steps across the plate. Each oscillation moves the electrode in a path about 1/4 inch long.

The arc is broken by lifting the electrode away from the plate. Try to slant the electrode in the opposite travel direction just as you pull away; this helps fill the *crater*. Practice starting until it feels fairly natural. Then turn the power on and try the real thing. Try to maintain an arc length 1/16 to 1/8 inch. Watch the metal being deposited and vary the two electrode angles to place metal exactly where you want it. This is not a natural thing to

150

Fig. 10-29. When striking an arc, first touch the plate lightly with the electrode slightly ahead of where you wish to weld.

Fig. 10-30. Next, lift the electrode slightly until the arc becomes firmly established.

Fig. 10-31. Finally, move the electrode down and back to where you wish to weld and run the bead.

do, because your eyes are attracted to the brightest area of the view through your filter lens—the arc. But watching the arc does not show what is deposited on the plate. It takes some effort to learn to look behind the arc.

Setting the Amperage

Amperage settings on power sources are only approximations. The amperage is correct if the arc is easily struck and is forceful enough to dig slightly into the plate. If the electrode sticks to the plate, keep your hood down and try to bend it free. If that doesn't release it, open the electrode holder. If the electrode repeatedly gets stuck to the plate, increase the current setting 5 to 10 amps and try it again. If the electrode flares up with a great deal of spatter and the arc digs excessively, lower the current setting 5 to 10 amps.

NOTE: Unless you have it in writing that it is alright to change amperage on your power source while welding, adjust the current setting only when the arc is broken. Changing the setting under load could cause a damaging arc inside the unit or worse: serious personal injury. Inform anyone working with you of this precaution.

Welding on overheated metal is unrealistic and frustrating. Either have several plates to work on, or make an arrangement to cool hot plates. At the completion of each bead, remove slag, wire brush the weld, and examine it. Look for a deposit uniform in ripple pattern, bead width, and bead height. Figure 10-32 shows a correctly formed bead. Note the amount of spatter and heat effect (warpage) upon your practice plate.

After you can run a 2-inch-long bead, experiment with both higher and lower amperage settings, different angles, long and short arc lengths, and different travel speeds. These are your controls over the welding operation. Amps, angles, arc lengths, and travel speed are collectively called welding *variables*. Information such as this is very important in commercial welding where strict procedures are followed as part of a welding code. Welding for yourself is not burdened with such for-

Fig. 10-32. Run only short beads and examine each.

Fig. 10-33. Changing the amperage setting, arc length, and travel speed produces different results: (A) Normal conditions, normal bead. (B) Insufficient amps. (C) Excessive amps. (D) Arc length too short. (E) Arc length too long. (F) Slow travel speed. (G) Excessive travel speed.

malities, but starting with the recommended polarity and amperage setting for an electrode reduces some trial and error. Figure 10-33 shows the effects of changing several variables.

Filling the Crater

Depressions or craters form at the end of your weld beads. These must be filled to avoid weak areas and/or cracks. Fill (*close off*) a crater by moving the electrode in 3 or 4 tight circles before breaking the arc. This puts down extra weld metal to fill the crater. With E6011, it is alright to restrike right away and deposit more filler without removing the slag first.

The tendency to form a crater is exaggerated where a weld runs right to the edge of a plate (Fig. 10-34). This is due in part to the accumulation of heat. Also, heat at a plate edge has nowhere to go. Approaching the end of a plate, move the electrode rapidly forward about 3/4 inch, reverse the travel angle, and travel the other way back into the bead. (It might be wise to practice welding backhanded from your usual travel direction before you try this.) This technique not only prevents the plate edge from being gnawed away by the arc but also places the crater back within the bead where it is not likely to weaken the joint.

Starts and Stops

Sometimes it is necessary to stop a weld, then re-

Fig. 10-34. The edge of a plate that has been gnawed away by improper electrode manipulation.

154

Fig. 10-35. Adjacent and overlapping beads are sometimes used for restoring worn surfaces or depositing overlays.

strike and continue with the same bead. Practice such starts and stops by restriking about 1 inch ahead of the last crater, moving up and back, and continuing back through the cratered area. Refer again to Figs. 10-29 to 10-31. Restriking ahead of the last stop gains time for the arc to get well established before again depositing weld metal. This avoids a sputtering arc and defective cold weld deposit at the point of pickup. In time, you will be able to blend your starts and stops.

Building a Pad/Surfacing

The next exercise is to run beads adjacent to one another. They overlap to form a solid *pad* or layer of weld deposit (Fig. 10-35). This is a buildup technique for restoring worn surfaces, applying a wear and/or corrosion-resistant surface, or to create a special surface effect such as bronze over steel. Besides salvaging otherwise worn-out items, an expensive overlay metal can be applied to an ordinary and less expensive base material.

Use steel at least 1/4 inch thick for this exercise. The first bead is run for 4 or 5 inches, slagged off, and brushed as before. The second and additional beads are placed ever closer to the welder (Figs. 10-36 and 10-37). Use enough electrode angle for each succeeding bead to obtain the correct overlap.

Practice with building a short pad until it is reasonably uniform. *Avoid overheating the plate*; if necessary, cool your work between passes.

155

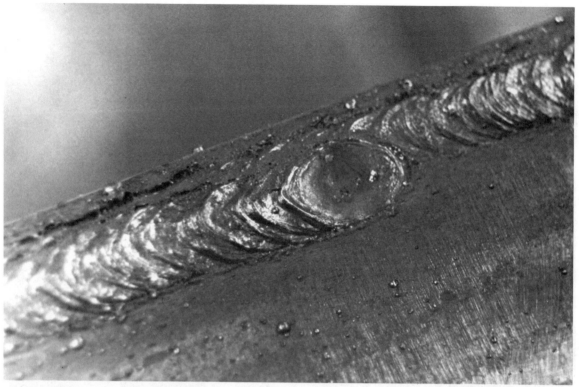

Fig. 10-36. Each successive bead is deposited on the near side. Be sure to use sufficient electrode angle to obtain overlapping beads.

Removal of Slag and Spatter

Notice how easy slag from E6011 is to remove; "hammering" with your slag hammer is not needed. In fact, be careful to avoid leaving any blemishes on the plate from your slagging operation. It looks bad and it weakens the welded structure.

Contending with tiny specks of weld spatter is a significant part of the welding supply business. There are several preweld anti-spatter compounds that work well. Use these on your fancier jobs. With a compound, spatter can be wiped or brushed away. Much of the time you will remove spatter by scraping with your slag hammer. For more precision, use a small chisel and hammer. Use eye protection when slagging, brushing, or removing spatter and try to avoid excessive grinding of spatter (or anything else for that matter). Grinder marks do little to help the appearance of your work and by interrupting the surface of the metal can even cause weakening of the structure.

Welding a T Joint with SMAW

The most common weld joint is the T joint which uses a fillet weld. See Appendix A for types of weld joints. The most common position is the horizontal (Fig. 10-38). Welding a T joint uses the skills you gained with the buildup exercise. You also have a few additional concerns: a tendency for your welds to undercut the vertical plate and to overlap the flat plate. These defect conditions are illustrated in Figs. 10-39 and 10-40. For a T joint, increase amperage 5 to 10 amps over that used to weld on the flat surface.

Tack weld two pieces of steel into a T joint and position horizontally. A tack (weld) is small but must be of a high quality if it is to be incorporated into the final welded structure (Fig. 10-41).

Fig. 10-37. A built-up pad showing the placement of the craters away from the edge of the plate.

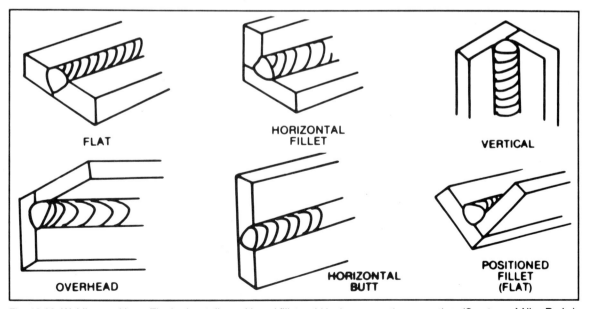

Fig. 10-38. Welding positions. The horizontally positioned fillet weld is done more than any other. (Courtesy of Alloy Rods.)

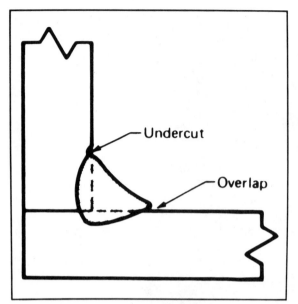

Fig. 10-39. Undercut and overlap are common defects. (Courtesy of American Welding Society.)

Weld the T joint by first striking at a spot slightly down the joint from the starting point of the weld deposit and then bringing the electrode back to the starting point. This is the same technique as used in starting and stopping with a continuous bead. Another starting method, used at the end of a plate, is to strike the arc and hold the electrode slightly off the end of the joint until the arc is strongly established.

Use a uniform oscillating pattern and watch the deposited metal—not the arc. With each oscillation, push some metal from the puddle on the (flat) surface plate onto the vertical member. Close off weld craters as before. Slag and brush your weld and examine it for possible undercut or overlap.

Weld several more T joints; experiment with changing the electrode angle, travel angle, and then travel speed. Watch the effect these changes have on the depositing weld metal. Take time with these exercises until you feel comfortable with T joints. Try to keep the leg dimensions (Fig. 10-42) equal.

Intermittent (Skip) Welds

Intermittent, or skip, welds (Fig. 10-43) are explained in Appendix A. These are used where a continuous seal is not required or the heat effect of a continuous weld would cause excessive distor-

Fig. 10-40. Undercut is caused mainly from either an excessive arc length or an excessive travel speed.

Fig. 10-41. Position tack welds with care.

tion. While welding T joints, mark off increments along a practice joint with soapstone. Try placing tacks at points where the skip weld increments will end. This not only hides the tack but also provides a little extra metal for filling craters.

Welding a Lap Joint with SMAW

A lap joint, like a T joint, requires a fillet weld. Welding differs, however, in that there is less heat carrying capacity with the edge of the outside member laying flat against the broader surface (Fig.

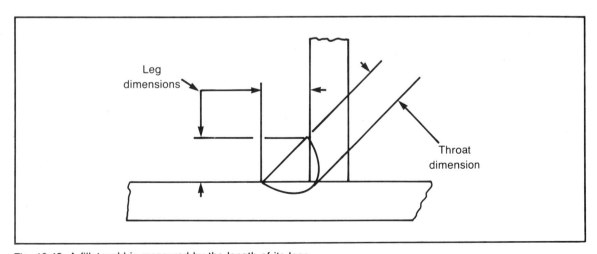

Leg dimensions

Throat dimension

Fig. 10-42. A fillet weld is measured by the length of its legs.

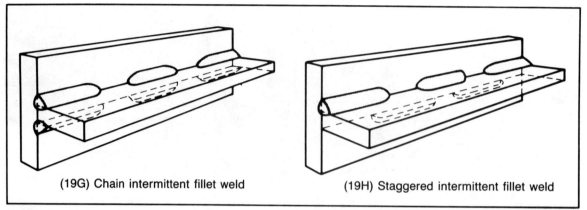

| (19G) Chain intermittent fillet weld | (19H) Staggered intermittent fillet weld |

Fig. 10-43. Intermittent or skip welds provide adequate bond yet avoid excessive heat buildup to any one area. (Courtesy of American Welding Society.)

10-44). The electrode must be directed more towards the surface member than the edge member with the effect of washing metal from the surface onto the edge with the arc force. Again, watch the metal behind the arc and not the arc itself. Figure 10-45 shows a lap joint welded with SMAW.

If welding a lap joint produces excessive burning away of the corner of the edge member—use more than one pass, travel faster, deposit smaller beads, and use slightly less amperage. In many applications there is no real need to fill a lap joint anyway. Also practice making unequal leg fillets (Fig. 10-46).

Welding an Outside Corner Joint with SMAW

Corner joints can be fitup in several different ways, such as fully or partially lapped. Welding is easiest if it is fit together corner-to-corner as shown in Figs. 10-47 and 10-48. If the corner joint is filled with weld, the thickness and structural strength of the corner is uniform throughout.

When welding a corner joint arc energy should be directed from the outside towards the areas of substantial plate mass to avoid burning through the joint (Fig. 10-49). It is impractical to weld from the inside because the arc easily burns through the plate corners. The electrode angle to the plate must

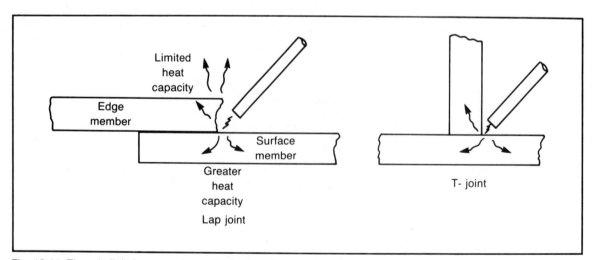

Fig. 10-44. There is little heat carrying ability in the edge member of a lap joint.

160

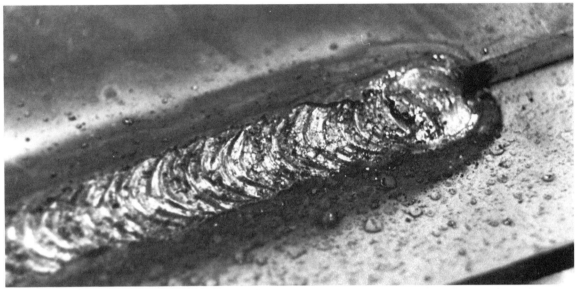

Fig. 10-45. A completed SMAW lap joint with a full fillet.

change as you oscillate and weave the electrode. Be sure to pause at the sides of the joint to deposit enough filler. Figure 10-50 shows a properly welded outside corner joint.

Welding a Butt Joint with SMAW

There are many kinds of butt joints if you count all the possible ways of preparing the member plates. Figure 10-51 shows "prepared" butt joints. Professional welders may agonize to master welding butt joints, which occur often in fluid systems and larger commercially welded fabrications.

Welders often must pass a performance test for a welding code. These involve X-ray and/or standardized destructive testing of their sample welds (Figs. 10-52 and 10-53). The home craftsman can avoid being overly concerned about the variety of butt joints and all the possible weld positions. In the home shop, you can often avoid welding butt joints altogether; and the ones you must weld can usually be done in the flat. The next several paragraphs describe welding butt joints only in a generalized way. Welding instruction is recommended if you need to weld some critical butt joints.

A butt joint for welding with the SMAW

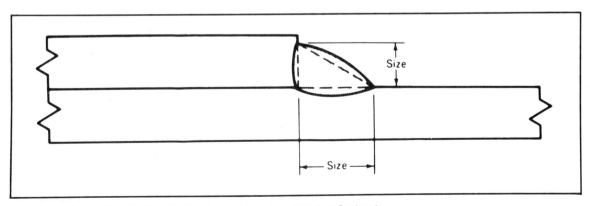

Fig. 10-46. Unequal leg fillet weld. (Courtesy of American Welding Society.)

Fig. 10-47. Tack welding with a full open or corner-to-corner fit. Notice the use of temporary spacer plate.

process often requires edge preparation with a V or U-bevel or other "prep" only if the plate is over 1/4 inch thick. Preparing the joint gives access to its bottom for complete root penetration (Fig. 10-54) and maximum strength. Welding a prepared joint often involves more than one pass. The first or *root* pass is deposited in the thinner area of the joint and so is similar to welding light-gauge material. Low amperage and a fairly high travel speed keep this root pass small. Subsequent passes can be welded more slowly and with higher amperage.

With material 1/4 inch or less, a square butt

joint with a root opening (gap) equal to the electrode diameter is usually alright for a full-penetration weld. The easiest position is vertical up, especially if the welding is from only one side. The *keyhole* technique ensures complete penetration (Fig. 10-55).

Fig. 10-48. A corner joint ready for welding.

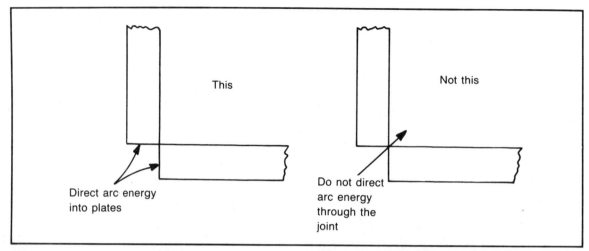

This

Not this

Direct arc energy into plates

Do not direct arc energy through the joint

Fig. 10-49. To avoid burn-thru, the electrode positioning is important when welding a corner joint.

Fig. 10-50. A properly welded outside corner joint will be filled but not overfilled.

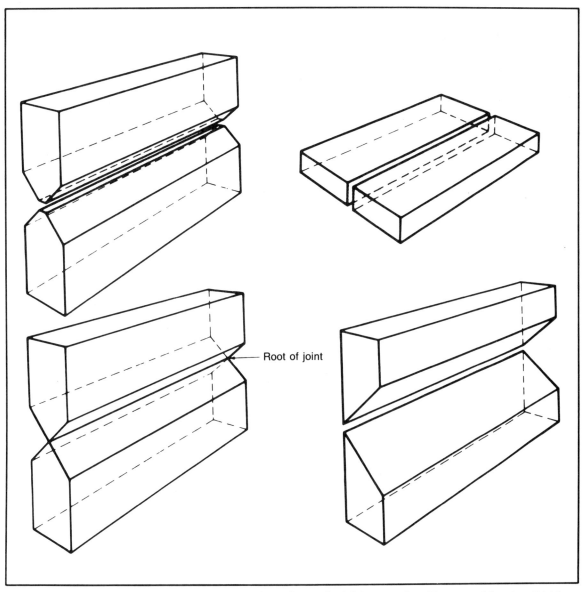

Fig. 10-51. There are a variety of different butt joints depending on the joint preparation. (Courtesy of American Welding Society.)

A butt joint in material less than 1/8 inch (11 gauge) is more difficult to weld than thicker sections because there is so little mass to carry heat away from the weld pool. Also the heat effect of shrinkage along the joint edges can cause severe warpage and destroy the usefulness of the object. If you must weld lighter metal with SMAW, use 3/32-inch electrodes, a rapid travel speed, a downhill direction when possible, and try to clamp the joint to prevent distortion.

To weld a butt joint with a wide root opening (gap), a backup bar (Fig. 10-56) can be used to eliminate burn-thru. If it is the same as the parent metal, it becomes part of the structure that might or might

165

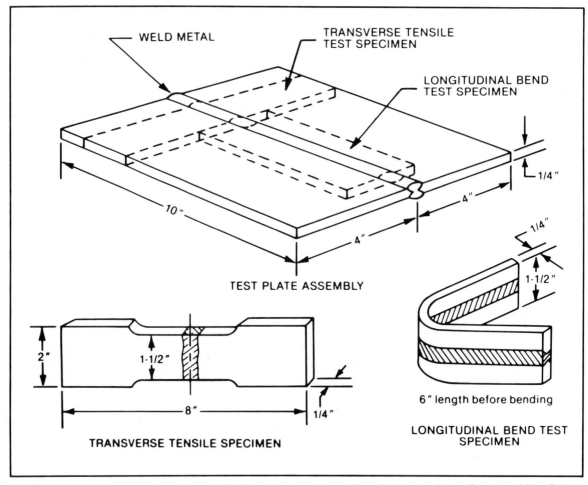

WELD METAL

TRANSVERSE TENSILE TEST SPECIMEN

LONGITUDINAL BEND TEST SPECIMEN

1/4"

10"

4"

4"

TEST PLATE ASSEMBLY

2"

1-1/2"

8"

1/4"

TRANSVERSE TENSILE SPECIMEN

1/4"

1-1/2"

6" length before bending

LONGITUDINAL BEND TEST SPECIMEN

Fig. 10-52. Details of transverse tension and guided bend tests used to certify professional welders. (Courtesy of Alloy Rods.)

not pose a problem. Sometimes a thick copper or aluminum bar is placed behind a steel butt joint (Fig. 10-57). The steel weld does not adhere to copper or aluminum so this provides an effective temporary backup. Ceramic backup tapes are also used to weld a butt joint from one side.

Arc Spot Welding

The term "spot welding" is usually associated with Resistance Spot Welding (RSW) light-gauge lap joints. RSW requires specialized equipment not normally found in a home shop (Fig. 10-58). Arc welding material thicknesses less than 1/16 inch (16 gauge) is not always practical because of the tendency to burn through. However, it is possible to arc *spot* weld lap joints of quite thin metal, like 24 gauge, to heavier plate without vaporizing the lighter plate.

Thick-to-thin assemblies are joined with arc spot welding. The arc is struck on the thinner member, melts through, and deposits a bead upon the heavier member. The bead acts like a rivet attached to the heavier plate. The completed product looks similar to a *plug weld* (except no prior holes are made in the plate) (Fig. 10-59). While this is mostly an automated industrial process, it can be done quite well at home with stick electrodes. For the procedure outlined below, use 2-inch wide scrap

166

strips of 24-gauge steel and at least 1/4-inch plate. To join thick to thin by SMAW arc spot welding:

- Use E6011, 3/32-diameter electrodes
- Set (and test) the welding amperage—it must be just hot enough to permit the arc to dig into the heavier plate without the electrode sticking
- Wire brush the mating plate surfaces
- Clamp the plates tightly together—this is essential

- Strike the arc upon the (top) thin plate and immediately plunge the electrode through to the heavier plate—keep the electrode square to the plate
- Move the electrode rapidly in tight circles until the growing bead spills over onto the thinner plate

A variation of the arc spot weld involves placing a flat washer on top of light sheet metal. This,

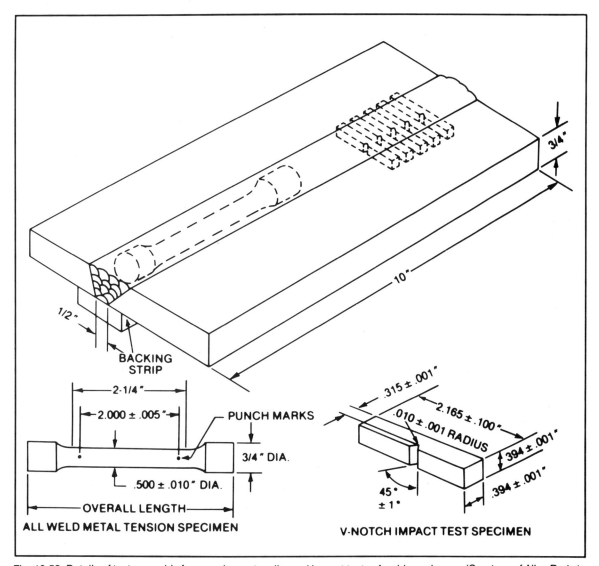

Fig. 10-53. Details of test assembly for soundness, tensile, and impact tests of weld specimens. (Courtesy of Alloy Rods.)

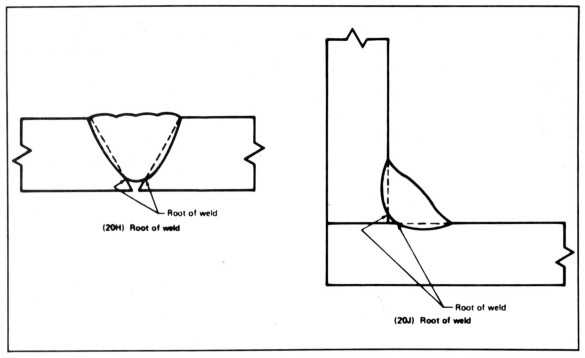

Fig. 10-54. This shows insufficient penetration of the weld root. (Courtesy of American Welding Society.)

(20H) Root of weld

(20J) Root of weld

in turn, is held tightly against the heavier plate. The arc is struck inside the washer, burns through the thin plate, and forms a bead on the heavier plate. The weld is continued to fill the hole in the washer.

OTHER WELDING POSITIONS

You have been encouraged to weld with the work in the flat position for the buildup exercise and butt joint, and horizontal for the T and lap joint. There are several reasons for this. First of all, it takes time to learn to weld well in all positions. Second, most of your projects will be small or light enough to permit locating to the flat position. Third, the best work is usually done in the flat position. Even commercial welders try to weld everything flat. There are, of course, cases in which a large or fixed workpiece is "there and staying there," and must be welded out-of-position.

A few pointers will be given about learning how to weld in any position. The best way to learn all-position welding is under the watchful eye of an experienced welder and/or teacher.

Gravity

Gravity affects the molten weld pool of a T joint in the horizontal, vertical, and overhead positions. This necessitates lower amps, smaller beads, and using the arc to locate molten metal. Like pushing leaves with a garden hose, the jet action of the arc pushes the puddle; the angle of the jet to the plate is important. The necessity for electrode oscillation, whereby molten metal is alternately deposited and left to freeze, is more evident vertically and overhead.

Down and Up

Welding a horizontal T joint showed the tendency to undercut the vertical plate. Heat naturally rises, so welding downhill on a vertical plate minimizes accumulated heat input and produces minimum penetration. This is especially effective to avoid burn-thru and distortion with lighter metal. Conversely, welding uphill accumulates heat for maximum penetration of heavier plate.

168

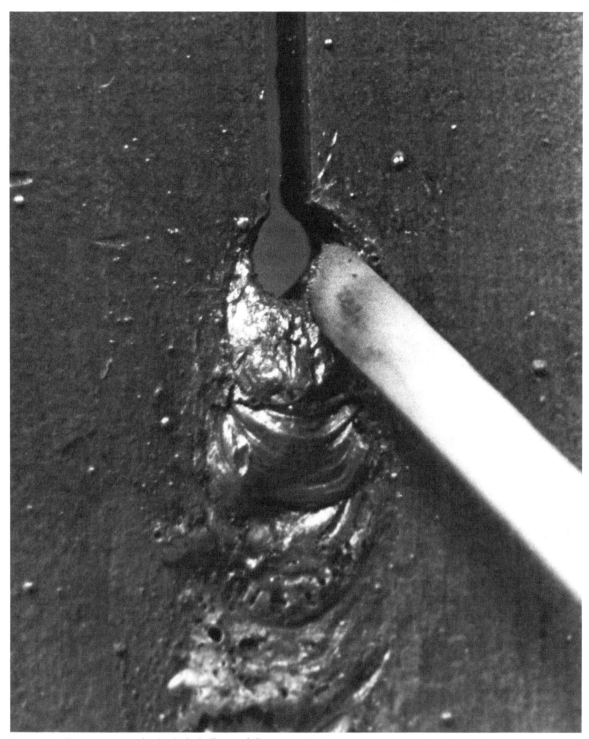

Fig. 10-55. The presence of a keyhole indicates full penetration through to the backside of the joint.

Fig. 10-56. A backup bar.

Fig. 10-57. A heavy aluminum backup bar beneath a butt joint.

Fig. 10-58. Resistance spot welding is often used to join sheet metal fabrications.

Fig. 10-59. An arc spot weld involves burning through a thin member to the heavier plate beneath. The fitup must be tight.

Horizontal Buildup

Run a bead across a vertical plate (at least 1/4 inch thick) with E6011. This is welding in the horizontal position (and direction). Use a "swinging" elec-trode motion (Fig. 10-60). A lower electrode angle keeps molten metal from falling. Also, a tight arc length allows the arc force to push the molten mass slightly upward, or back onto the plate. Prevent un-

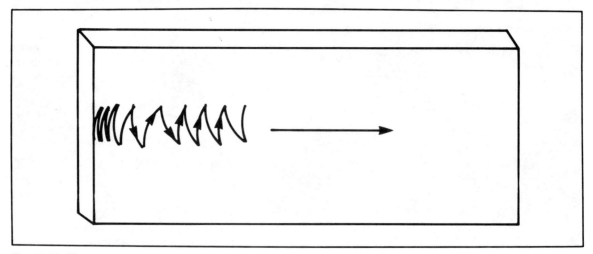

Fig. 10-60. The path of the electrode when depositing a bead in the horizontal position.

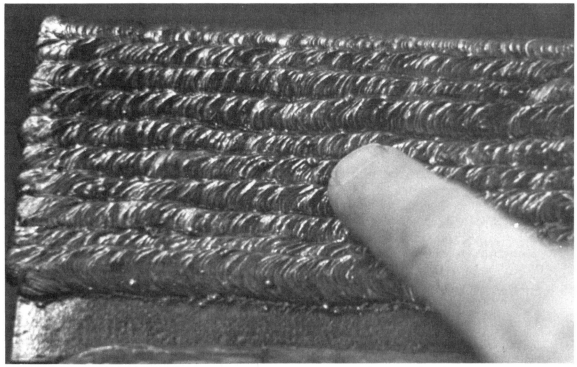

Fig. 10-61. No matter what position is used, a buildup should present a surface of uniform height, free from low areas and voids between the beads.

dercut by pausing at the extreme upward point of the oscillation just long enough to add sufficient filler. Maintaining a tight arc also avoids overheating surrounding metal with the outward arc fan. Place another bead just above and partly overlapping the first. Continue adding beads and compare your results with Fig. 10-61.

Vertical Buildup

Running beads uphill or downhill on a vertically positioned plate is done by maintaining a close arc, depositing some filler, and then quickly moving the electrode away to allow the puddle to freeze. As with horizontal welding, the electrode (angle) must always be directed upward so that the puddle is held up onto the plate. There are several possible ways of "whipping" the electrode for making a vertical deposit. Probably the easiest way is to simply use an upside-down T motion.

Overhead Welding

Welding overhead is basically upside-down flat welding. *Surface tension* keeps molten weld from dripping off an overhead surface. Surface tension works like a thin skin of weld to hold up the rest of the deposit. Of course, gravity prevails over surface tension if the weld deposit is allowed to grow too large. Welding overhead is done with low amperage and a fast travel speed; a tight arc is essential.

Multiple Pass Welds

Welding heavier metal might require more than one pass. Each is important, but the first pass is the most critical. A crack in the root pass can migrate through the rest of the weld to the surface. Avoid welding over a cracked weld. It might be necessary to first remove the crack by grinding or chipping.

The use of multiple passes avoids overheating

a T or a lap joint in moderately thick material. These both use fillet welds, and the right pass sequence is important. The first pass is slightly convex in profile to avoid cracking. A second partly overlapping bead is placed in front of the first, a third above and behind the first (Fig. 10-62). Undercutting is avoided if the weld is allowed to cool before running the final bead.

Other Stick Electrodes For Mild Steel

Some of the differences in electrode characteristics can be very subtle, often too subtle to interest the home welder. Rest easy and be sure of clean, solid welds by using E6011 for most of your SMAW jobs. Unfortunately, over the years a number of small SMAW power sources have been sold with a supply of E6013. There is nothing wrong with this electrode, but it is often misused.

E6013

E6013 electrodes contain *rutile*—naturally occurring titanium dioxide. Unlike cellulose in E6011 flux, rutile produces light penetration. Most industrial users consider E6013 to be a shallow penetration rod limited to welding sheet metal.

Easy arc striking, a gentle arc, little spatter, and beautiful beads characterize E6013 as a rela-

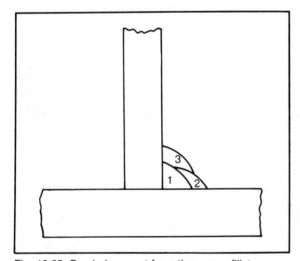

Fig. 10-62. Bend placement for a three-pass fillet.

tively pleasant electrode compared to E6011. It might, therefore, seem like no contest; but E6013 should not be used to weld heavier plate where structural strength is required. It has neither sufficient penetration nor *scavenging*. If an impurity such as a bit of slag remains on the plate, running over it with E6013 produces an inclusion defect. It is far better to learn to live with the old reliable E6011. E6013 is a valuable tool for lighter gauges. It has sufficient potassium compounds to permit welding with ac and, like E6011, has no special storage requirements. Avoid rod ovens, these electrodes require a 3- to 5-percent moisture content.

E7024

During the Second World War, there was a severe need for more defense production welding. Many housewives went to work in shipyards and defense plants. "Rosie the Riveter" (Fig. 10-63) was a popular characterization of these workers. The inexperienced were trained in a few hours for limited stick welding of flat and horizontal fillets. The electrode they used, often dubbed "drag rod," contains iron powder in the coating for high deposition rates (in pounds of weld per hour). It is available today as E7024 and sometimes called "jet rod." Like Rosie, you too can use E7024 for similar flat and horizontal applications in the home shop

The heavy flux of E7024 produces a slag even heavier than E6013 take care to avoid inclusions. It is run hot—approximately 155 amps for the 1/8-inch diameter. Both the electrode and travel angles require attention to push the heavy slag back from the weld pool. E7024 welds are slow to freeze and have time to smooth out, so there are few ripples. The digit 2 means the electrode is used flat and horizontal only. If appearance is a prime concern and joint accessibility is good, E7024 is a good choice. It runs well on ac or DCRP and has no special storage requirements.

E7018 AND OTHER LOW-HYDROGEN ELECTRODES

Don't use low-hydrogen electrodes unless you have

Fig. 10-63. "Rosie."

a special purpose vented rod oven for baking and storing them and you have a special need for using them. The E7018 (E8018, E9018, etc.) electrodes are designed to weld low-alloy steels prone to hydrogen cracking. These electrodes cause more grief than they should because of misuse. *Low-hydrogen* means low water content in the flux coating—0.3 to 0.5 percent. Hydrogen is the ingredient of water (vapor) that causes underbead cracking—a sneaky, delayed-action cracking from the inside out (Fig. 10-64). These electrodes are *hygroscopic*, they absorb moisture from the sur-

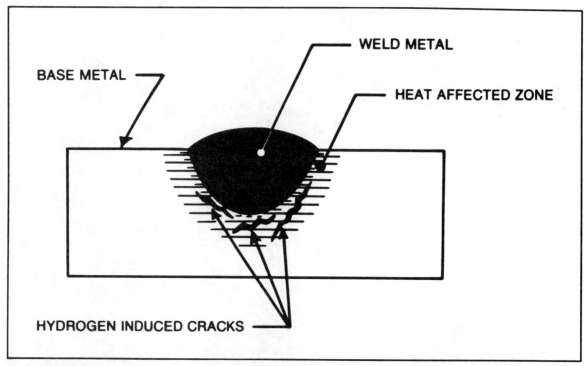

BASE METAL

WELD METAL

HEAT AFFECTED ZONE

HYDROGEN INDUCED CRACKS

Fig. 10-64. Underbead cracking from entrapped hydrogen is insidious, sometimes occurring many hours after the welding is completed. (Courtesy of Alloy Rods.)

rounding when their hermetically sealed container is opened.

Low-hydrogen electrodes need to be in an oven at 300 degrees Fahrenheit when the can or box is opened. If out of the oven for an hour, they must be reconditioned at 700 degrees Fahrenheit for 1 hour. Do not handle them with damp hands or gloves. Does this sound like a lot of trouble? It is, and you don't need it. The main problem with these electrodes is that they are seldom properly dried. As the coatings absorb hydrogen, without an oven, they actually end up adding hydrogen to the weld. Even if beads are not as pretty, the home welder is safer with trusty E6011.

Stick Electrodes for Other Metals

The preceding paragraphs have pointed out that a home weld shop equipped with only E6011 and E7024 electrodes can handle almost all mild steel jobs. Additionally, E6013 for sheet metal and E7018 low-hydrogen type for critical welds with the

proper safeguards, round off the mild steel electrode inventory.

Besides the four mild steel varieties discussed so far, there are many stick electrodes for other metals. Some of these are:

- Alloy steel
- Stainless steel
- Hardfacing deposits
- Bearing metals such as bronzes
- Nickel electrodes for cast iron
- Aluminum (as a last resort)

Each electrode type involves slightly different techniques. Specific instructions for each should be obtained and followed.

Cutting Electrodes

With high amperage, E6011 will cut steel (sort of), but it also deposits some filler. Special cutting electrodes deposit no metal—not the same as carbon

Table 10-1 SMAW Troubleshooting.

PROBLEM	PROBABLE CAUSE	REMEDY
1. Arc weak, difficult to strike	a. Insufficient Amperage	a. Increase amp setting
	b. Faulty connection	b. Check and secure all connections including work (ground) clamp
2. Electrode sticks to the plate	a. Insufficient amperage	a. Increase amp setting
	b. Improper technique	b. Review how to strike the arc
3. Lack of fusion	a. Insufficient amperage	a. Increase amp setting
	b. Travel speed too high	b. Reduce travel speed
4. Burn-thru	a. Excessive amperage	a. Reduce amp setting
	b. Arc length too short	b. Maintain 1/16-in. arc length
	c. Travel speed too slow	c. Increase travel speed
	d. Root opening too wide	d. Reduce root opening, use a backup material
5. Inclusions	a. Insufficient amperage	a. Increase amp setting
	b. Excessive arc length	b. Maintain 1/16-in. arc length
	c. Uneven oscillations and/or travel speed	c. Move electrode uniformly
	d. Dirty plate	d. Remove rust, grease, paint, etc.
6. Porosity	a. Dirty plate	a. Remove rust, grease, paint, etc.
	b. Excessive amperage	b. Lower amp setting
	c. Excessive arc length	c. Maintain 1/16-in. arc length
7. Undercut	a. Excessive arc length	a. Maintain 1/16-in. arc length
	b. Improper electrode angle	b. Direct electrode more into area of undercut
	c. Travel speed too high	c. Reduce travel speed
	d. Excessive amperage	d. Lower amp setting
8. Overlap	a. Improper electrode angle	a. Lower electrode angle
	b. Travel speed too slow	b. Increase travel speed
9. Cracking	a. Bead too small or too concave	a. Reduce travel speed
	b. Failure to fill craters	b. Circle electrode at end of bead, restrike to fill as required.

PROBLEM	PROBABLE CAUSE	REMEDY
	c. Wet or dirty plate	c. Dry or clean plate as needed
	d. Wet or dirty electrode	d. Use only dry and clean electrodes
10. Excess spatter	a. Excessive amperage (fine sized spatter)	a. Lower amp setting
	b. Excessive arc length (large sized spatter)	b. Maintain 1/16-in. arc length
11. Rough appearance	a. Oscillations spaced too far apart	a. Use more oscillations per inch of travel
	b. Improper travel angle	b. Reduce travel angle
12. Arc blow	a. Work (ground) clamp improperly located	a. Move clamp to different place relative to weld
	b. Direct current	b. Use ac if possible
13. Fingernailing (of flux)	a. Flux coating cracked or chipped	a. Use undamaged electrode
	b. Flux coating not concentric with rod	b. Exchange for quality electrode

cutting electrodes used with compressed air (see Chapter 16). Two popular cutting electrodes are "Chamfer Arc" and "Chamfertrode."

Cutting electrodes work but produce incredible smoke and leave a rough kerf or groove. They also need more amperage than that available with home-type power sources (250 amps for the 1/8-inch diameter). To use, amperage is set to maximum and the electrode laid over nearly parallel to the plate. A sawing motion digs out and pushes the metal to be removed off the plate.

SMAW TROUBLESHOOTING

Something less than perfection is always to be expected. Yet home welders should be sure that welds are of sufficient quality for their intended purpose. It is a good idea to analyze test welds by cutting them apart and breaking or bending the smaller sections. Look for: fusion, inclusions, porosity, undercut, overlap, cracks, spatter, and overall appearance.

There are at least 13 possible problem areas with the use of SMAW. These include:

1. Arc weak, difficult to strike
2. Electrode sticking to the plate
3. Lack of fusion
4. Burn-thru
5. Inclusions
6. Porosity
7. Undercut
8. Overlap
9. Cracking
10. Excess spatter
11. Rough appearance
12. Arc blow
13. Fingernailing

Table 10-1 lists these SMAW problems, probable causes, and remedies.

Other Troubleshooting Suggestions

Keep a handful of electrodes from a batch that you know is alright. These can be used for comparing any problem electrodes. Welds can be porous or crack if the base metal has too much sulphur or phosphorous. So save a few scraps of steel for comparisons.

Should you have serious problems like cracking welds, first contact the electrode sales people. They usually try to solve problems with their products. If you are still stumped and the project is a major one, see a professional welding engineer.

The Future of SMAW

Commercial use of stick electrodes has declined in recent years because of some inherent inefficiencies. Yet, like oxyacetylene welding, SMAW continues to be a vital maintenance and repair tool. For the home craftsman, SMAW is an inexpensive way to begin arc welding.

This concludes the specific discussion of SMAW. The next chapter delves into wire feed processes for the home shop. These give the do-it-yourselfer even more exciting potentials.

Chapter 11

Welding with Wire

H OW EASY CAN ARC WELDING BE? JUST POSI-
tion the gun, nod to drop the hood, pull the trigger and Presto! you are doing it. No sputtering arc strike, no sticking electrodes, no need to stop until you want to. This is wire feed welding and home craftsmen everywhere are discovering its creative advantages. Wire feed welding helps you build projects faster and often better.

This chapter on home welding could not have been written just a few years ago. Most equipment and consumables discussed here are recent developments. Although increasing, the home welding market is not seen as large by manufacturers of welding equipment and consumables. As a home welder you are the indirect beneficiary of the industry's efforts to achieve more efficient manufacturing methods. Wire feed equipment so ideal for home use was originally developed to speed up welding production.

Arc welding light fabrications is a growth area for the welding industry. Welding with small wire feed units joins sheet metal faster, at less cost, and more easily than folding and/or riveting. The au-

tomotive industry needs to build lighter more fuel-efficient vehicles so high-strength, low alloy (HSLA) steels are now commonplace. These heat-sensitive steels must be welded only with wire feed units and small-diameter wire electrodes. Other welding methods produce too much heat. The automotive repair industry has responded by purchasing thousands of small wire feed welding outfits. These outfits operate from existing 220-volt (and some even 115-volt) electrical line hookups. Automotive exhaust specialty shops have also been quick to profit from small wire feed units.

Besides equipment improvements, new solid wire electrodes are now available as small as .023 inch. This permits welding material as thin as 20 gauge (.036 inch). Flux-cored electrodes are now as small as .035 and .045 inch. With these cored electrodes you can make high quality welds with thin steel in the 14-to 16-gauge range.

Wire feed welding, like stick welding, is also arc welding. The same safety precautions discussed in Chapter 10 apply. Other similarities with SMAW are found in certain of the operating principles.

WIRE FEED WELDING

"Wire feed welding" is a blanket term that really includes several specific processes. Most home metal projects are light mild steel fabrications enabling the home welder to select from a menu of five distinct wire feed arc welding methods. The process used depends on the plate thickness, workpiece location, and personal preference. Wire feed welding at home is done with:

- Gas Metal Arc Welding (GMAW), sometimes called "MIG," with 3 possible variations: short circuit transfer, globular transfer, and spray transfer
- Flux-cored Arc Welding (FCAW) with 2 variations: self-shielded and gas-shielded

The same basic equipment is used for all five variations. There are few welding jobs that cannot be done with them. As stated earlier, the fastest way to begin home welding is with a cutting torch and a wire feed system.

Some limited commercial use of automatic wire feed welding systems took place in the 1920s, but the practical hand-held (semi-automatic) wire feed gun was not seen until the 1950s. Through the 1950s and 1960s, improving wire feed processes replaced a great deal of Shielded Metal Arc Welding. By the mid-1980s, smaller and more convenient units were ready. Home welders saw in these wire systems the expanded potential to work with lighter, more heat-sensitive, and even unusual metals. Today all major manufacturers produce a variety of wire feed equipment designed with you in mind. This equipment is sturdy and easy to use.

There are similarities between wire welding and SMAW. (If you have not already done so, you might read or reread about arc welding at the beginning of Chapter 10.) A power source that looks similar to a SMAW power source transforms and delivers the particular type of power needed for wire feed welding. Like SMAW, heavy connectors and cables deliver high welding current to the electrode.

A few equipment differences are necessary to weld with the relatively small wire electrodes. Figure 11-1 illustrates the equipment required for welding with the five-wire feed variations used in home shops. Shielding gas apparatus is used for four of these. A good deal of wire feed welding is done without shielding gas, using small diameter self-shielded, flux-cored electrodes. Figures 11-2 and 11-3 show two wire feed welding systems. The *wire feeder* and power source can be combined into one package.

A power source for wire feed welding has a different type of output than one used for SMAW. Constant voltage (CV) output is used for wire and constant current (CC) for the SMAW stick electrodes. (Appendix A says more about these differences.)

The home welder requiring portability away from line power can use wire feed welding with a motor generator unit and a *spool gun* as shown in Fig. 4-20. Engine-driven units are expensive but do the job no matter how remote the location.

HOW WIRE FEED WELDING WORKS

The same operating principles underlie all five home shop wire feed processes. You can use an assortment of wires for a variety of applications with little modification of technique. For example, increased voltage produces a wider and "wetter" bead. This is true for all process variations with solid and flux-cored wire. It even works the same for aluminum and other metals.

Constant Voltage

Power sources for wire feed welding are nearly always constant voltage (CV) units. Sometimes the term constant potential (CP) is used. (See GMAW and FCAW in Appendix A.) CV and CP mean the same thing. After all, voltage or electrical "pressure" is the potential or force behind all electrical processes. Voltage is steady or constant and so is the arc length. This means that weld bead width is uniform even if you move the gun slightly closer to or farther from the plate. Such torch-to-work distance is very significant to the welding operation and more must be said about it.

Besides voltage, the wire speed you select is

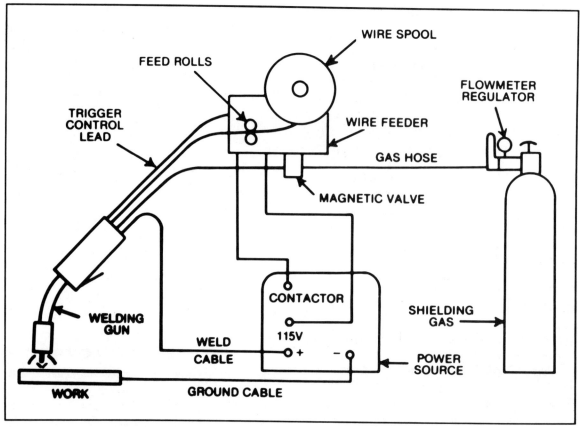

Fig. 11-1. Schematic diagram of wire feed welding equipment. (Courtesy of Alloy Rods.)

also constant, helping to make wire feed welding easy. If voltage and wire feed speed are correct for the job type, if the torch is moved at a fairly even rate, and if the torch is held an even distance from the plate, the resulting weld will be a good one—everytime.

Theory of Operation

The wire feed welding operation starts when you press the trigger switch on the gun (torch). Several things happen almost at once. The wire feeds out and is also energized. As it strikes the plate, an arc forms and welding begins. Shielding gas (if used) also begins flowing through the gun and onto the plate.

Wire electrode, carried on a spool or coil, is unwound and fed by drive rolls (Fig. 11-4). Wire travels towards the gun via a flexible sleeve, or *liner*

into a drilled copper *contact tip* or *tube* in the end of the gun (Fig. 11-5). The tip is energized through a large welding cable connected to one side of the power source. This happens when the *contactor* in the power source is closed by the trigger signal. The contactor is a relay (mechanical or solid state) that connects the power source transformer to the input electrical line to supply output energy for welding.

The wire sliding through the energized tip is also energized—that is, connected to the insulated side of the power source. Like stick electrode welding, the contact tip is a sort of "sliding electrode holder." The welded article is connected to the other (ground) side of the power source. When wire reaches the plate, the high amperage arc welding circuit is completed as described for SMAW in Chapter 10. The small-diameter wire heats quickly,

Fig. 11-2. A power source and detachable wire feeder. This unit can also be used for SMAW with the coated stick electrodes. (Courtesy of Lincoln Electric.)

Fig. 11-3. A small 200-amp power source and light weight wire feeder is easy to move around the shop. (Courtesy of Hobart Brothers.)

Fig. 11-4. A wire drive mechanism showing the drive tensioning adjustment.

affording fast and accurate location of the weld.

The arc, and the molten metal around it, is shielded from atmospheric contamination by an invisible gas cloud and/or a smoke envelope. Shielding is essential to all welding operations to prevent the formation of harmful oxygen and nitrogen compounds.

CONTROLLING THE WIRE FEED ARC

There are three main variables or controls of the ongoing wire feed arc welding operation: voltage, wire feed speed, and (electrical) wire stickout. There is some overlap of functions but each variable mainly controls a specific aspect of the welding condition.

Voltage and Control of the Wire Feed Arc

The arc is easy to control with a wire feeder be-

cause arc length is mainly controlled by the voltage setting on the power source. Once set, voltage remains nearly constant. Think of voltage as electrical "pressure," and it appears that wire feeding into the arc floats on the cushion of the arc. With higher voltage (pressure) the arc length is longer and the wire appears to float higher or farther from the plate.

Changing arc voltage is like moving a torch flame closer to or farther from the plate. Moved back, torch heat fans out and a wider surface area is heat affected. A narrower area is heated with the torch closer to the plate. Stated simply: more voltage, wider bead; less voltage, narrower bead. Fortunately, setting voltage is not difficult. If it is not right, things just don't work. With a little experimenting and practice, the best voltage can be quickly found.

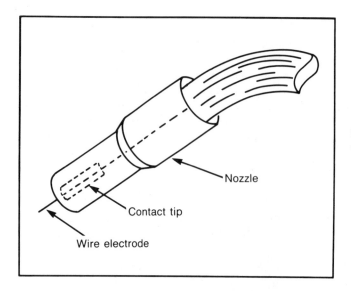

Fig. 11-5. The contact tip, located in the front end of the mig gun, energizes the wire electrode.

Nozzle

Contact tip

Wire electrode

Voltage (and arc length) must be correct to maintain an even melt rate of the electrode. With too little voltage, there is not enough arc energy to melt the feeding wire; with high voltage, the arc is too long and can burn all the way back to (and even into) the contact tip. Excessive voltage burns the tip and stops welding before it starts. When you first set things up, you might not know exactly how fast the wire will be consumed. There is a little set-up trick that avoids burning the tip. It is explained later in the chapter along with how to use each of the wire feed variations. Remember:

- Voltage increase: longer arc length and wider heat fan
- Voltage decrease: shorter arc length and narrower heat fan

Wire Feed Speed and Control of the Welding Amperage

Wire is fed at a steady rate by the drive rolls. This rate is measured in inches per minute (IPM) of *wire feed speed*. A knob, usually on the feeder, controls the feed rate. Increasing wire speed increases amperage flowing through the tip, onto the wire, and across the arc. This increases heat in the arc zone (and also somewhat shortens the arc length). If you lower the wire feed speed, both amperage and heat input are lowered. (Lowering wire speed also causes some increase in arc length.)

Excessive wire feed speed causes *stubbing*— the wire hits the plate at a faster rate than the arc can properly melt. The gun is pushed back from the plate and often stubs of wire fuse off and stick to the plate (Fig. 11-6). Remember:

- Wire feed speed increase: more amps, more heat input, shorter arc length
- Wire feed speed decrease: less amps, less heat input, longer arc length

Voltage-Amps-Voltage-Amps . . .? At this point things sometimes get confused. It was just stated that arc length is controlled mainly by the welder's voltage setting. Then a few lines later it was stated that wire feed speed also causes a change in arc length. So which is it? Actually *both* voltage and wire speed affect arc length, but voltage has the greater effect.

Another related fact is that both voltage and wire speed affect amperage or heat input to the plate. But wire speed has the greater effect.

The controls are best adjusted in a set order: first voltage and then wire feed speed. Voltage might then need more adjusting (trimming) to obtain the desired welding condition. Sometimes several back-and-forth resettings are needed to smooth things out. Remember:

- Voltage is the *main* control over arc length; set it first.
- Wire feed speed is the *main* control over amperage or heat input; it is set after the voltage.

Torch-to-Work Distance (Electrical Stickout)

Considerable control of an ongoing wire feed welding operation is exercised by moving the gun closer to or farther from the plate. This changes the amount of electrode wire sticking out from the contact tip. This distance, often called *electrical stickout, electrode extension,* or simply *(wire) stickout* (Fig. 11-7), affects energy distribution in the weld zone. Welding amperage changes with the stickout.

In the torch, welding current flows from the contact tip to the sliding wire. The short length of wire extending from the tip also carries current into the arc and then to the plate. A lot of amperage is needed for welding. This passes easily through large copper cables and the copper contact tip which offer little resistance to the current flow. However, the steel wire electrode extending from the tip is not as good a conductor as copper and, being small in cross-section, heats up carrying the amps. Stickout wire is a high electrical resistance for welding current. In fact, the stickout wire gets so hot that it is close to melting. This *electrode preheat* is the heating of the wire before it melts in the arc.

Stickout distance is determined in part by how far the contact tip is preset inside (or outside) of the *nozzle*. The nozzle focuses shielding gas. Many torches feature an adjustable nozzle that slides over an insulator, making it easy to adjust how much the tip is extended or recessed (Fig. 11-8). Stickout is also determined by how far away you hold the torch from the plate.

Stickout and Bead Size

It makes sense that if the stickout wire is suffi-

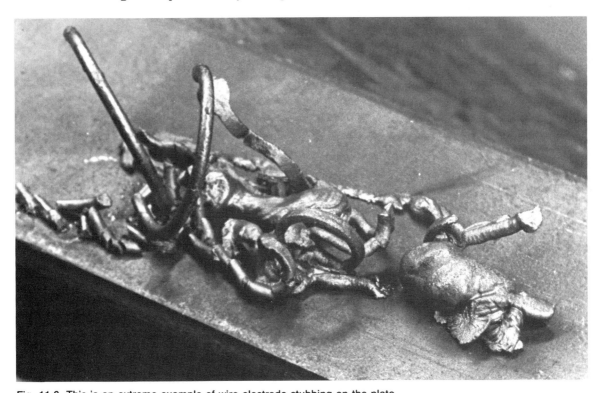

Fig. 11-6. This is an extreme example of wire electrode stubbing on the plate.

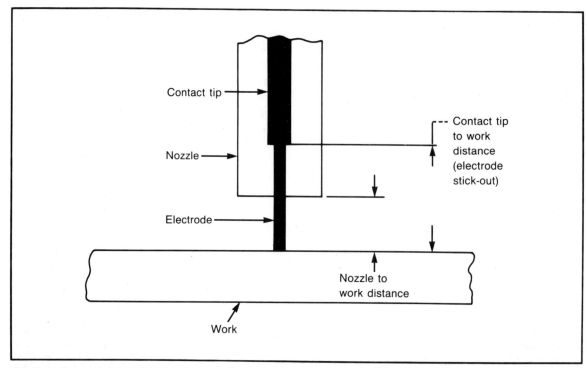

Fig. 11-7. Electrode stickout is a major control over the weld deposit. (Courtesy of Alloy Rods.)

ciently preheated, it will melt easily in the arc. Preheated stickout wire always ends up on the plate—one way or another. With the right conditions this wire becomes a respectable weld bead, but it could also go to waste as spatter or stubs. To deposit a large bead, use a long stickout; smaller beads, shorter stickout.

There is more worth considering about stickout. As it increases, additional energy is used to preheat the extending electrode. This energy is shifted from the arc to the preheated wire. This relationship could be observed by watching an ammeter while changing stickout. Most home welding power sources do not have meters, but you can still see changes in the deposit and heat effect with changes in the stickout. Figure 11-9 compares two beads run with different stickouts.

As stickout is increased, amps drop; so heat input to the plate is lowered. This deposits more with less amperage or, "more wire deposited colder." This fact is useful for welding joints where fitup is wide. A larger gap between plates can be bridged

by lengthening the stickout.

With more stickout the deposit is larger but does not penetrate as far into the plate as with shorter stickout. If, however, more energy is gained by increasing both voltage and wire speed (amps), the larger weld bead will be hot enough to also achieve good penetration.

Electrical stickout is often underrated as an important control for wire feed welding. Skilled welders use stickout, voltage, and wire feed speed to match the weld to the job.

Summary of Basic Wire Feed Controls

Three basic variables (controls) of the wire feed arc have been discussed: voltage, wire feed speed, and electrical stickout. While each does its own thing, they also are interrelated to some extent. You will find it easiest to deal with only one variable at a time by considering its primary function in arc control. These are summed up as follows:

● Voltage—the main control of arc length—affects bead width and wetness

- Wire feed speed—the main control of amperage
- Electrical stickout—affects bead size and penetration

GAS METAL ARC WELDING (GMAW)

Every welding process can be thought of as a multi-purpose tool. GMAW joins steel as thin as 18-gauge exhaust tubing and as heavy as 1/4-inch plate.

Advantages

GMAW advantages over SMAW and oxyacetylene were discussed in Chapter 8. Those of greatest interest to the home craftsman are:

- Learning ease
- Little waste of filler metal
- Very little smoke and fumes
- Versatility in metal types and thicknesses
- Highly controlled heat input

Gas metal arc welding is done with solid electrode wire that is uncoated or nearly so. In the home shop, .035-inch diameter wire is the most common

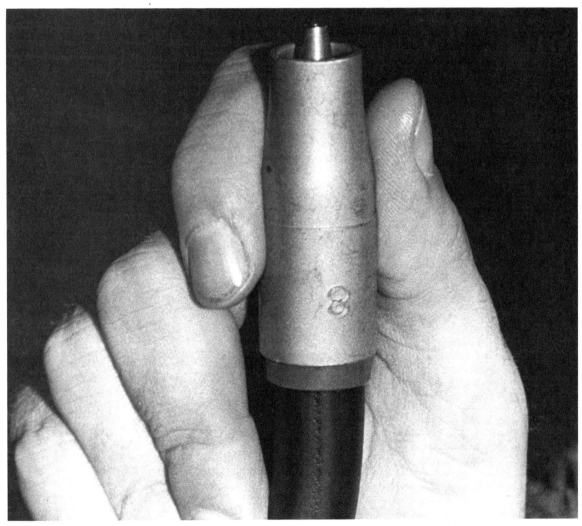

Fig. 11-8. An adjustable nozzle, slid back, allows the contact tip to extend form the end of the gun.

size; but .030 inch and even .023 or .024 inch are sometimes used. GMAW also requires an externally supplied shielding gas. With home-type, single-phase equipment, C-25—a blend of 25 percent CO_2 and 75 percent argon, is usually recommended over straight CO_2. Direct current with the electrode positive (DCEP or "reversed polarity") is supplied from a constant voltage (CV/CP) power source. Ac is too rough for wire, due to its inherent arc outages and part-time straight polarity. Because no flux or chemical cleaner is used, GMAW must be performed only on clean or nearly clean metal. Little, if any, slag removal is needed. In summary, GMAW characteristics:

- Solid electrode wire
- Shielding gas
- DCEP
- Work must be clean
- Little slag

GMAW VARIATIONS

As mentioned earlier, the three GMAW variations are:

- Short circuit transfer, also known as "short arc," "microwire," "wire weld," and "dip transfer"
- Globular transfer, called "CO_2 welding"
- Spray transfer, simply called "spray"

Each transfer variation is associated with a particular combination of voltage, amperage, and shielding gas which causes the electrode metal to *transfer* to the plate in a distinct way.

Short Circuit Transfer

Short circuit transfer is low voltage (17 to 20 volts), low amperage (75 to 120 amps), and therefore, the lowest heat input GMAW process variation. Low heat input solves many melt-thru problems and reduces distortion. Electrode filler metal is transferred to the plate with a rapid series of shorted contacts. With each short circuiting, some wire sticks to the plate. A fuse action follows in which the wire melts and is thus pinched off from the electrode (Fig. 11-10). Sixty to 200 short circuits per second produce a smooth bead of finely overlapped deposits.

Sheet metal fabrications as thin as 18 gauge, and the root (first) passes in prepared (bevelled) joints of heavier plate or pipe are often welded this way. Short circuit transfer is effective and easy to use in any position (Fig. 11-11). Electrode metal does not cross an open arc, so gravity has little effect.

Welds in lighter metal must be small. Recall how weld size is partly controlled by wire stickout. A short stickout is obtained by locating the contact tip flush with the end of the nozzle. To reach an

Fig. 11-9. The wider bead at the left was run with a short stickout length, while the higher narrow bead on the right was run with a longer stickout.

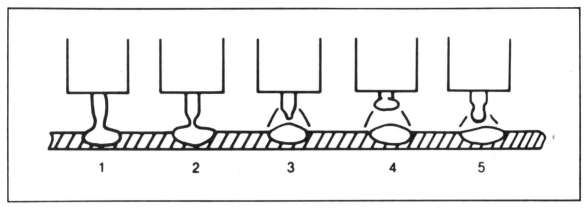

Fig. 11-10. Short circuit transfer ("short arc") is a rapid series of fuse actions each of which leaves a small deposit on the plate. There is no metal passing across the arc. (Courtesy of Alloy Rods.)

Fig. 11-11. Short circuit transfer is often used in manufacturing to produce strong and attractive structures.

inside corner, the tip can be extended up to 3/16 inch. This adjustment is called *tip extension*. The tip is recessed for the GMAW globular and spray transfer modes. Setting tip extension (or recess) is easy with adjustable nozzles as was shown in Fig. 11-8. When the nozzle is not adjustable, the tip is often held in a clamping *collet* which is loosened to change the in-or-out location of the tip as shown in Fig. 11-12. If neither nozzle nor tip is adjustable, the manufacturer makes tips of different lengths. If you are in a bind for the correct length of tip, a little hacksaw action can modify a tip or nozzle.

Besides mild steel, short circuit transfer also joins heat-sensitive materials like high-strength low-alloy steels (HSLA). The use of these metals is in-creasing. HSLA steels are used extensively in auto bodies and frames for lighter, more efficient vehi-cles. As a consequence, most automotive weld repairs are made with low-heat short circuit trans-fer. Home shop wire feed welding units are avail-able with timers (Fig. 11-13) to limit the *arc on* time (and heat input) to produce arc spot welds. Heat input can also be kept low by substituting silicon bronze filler for steel wire. This technique, increas-ingly popular with body shops, is similar in effect to braze welding—with reduced heat input.

Short Circuit Limitations

Short circuit transfer, with low heat and no flux, is subject to producing *cold laps* and lack of fusion.

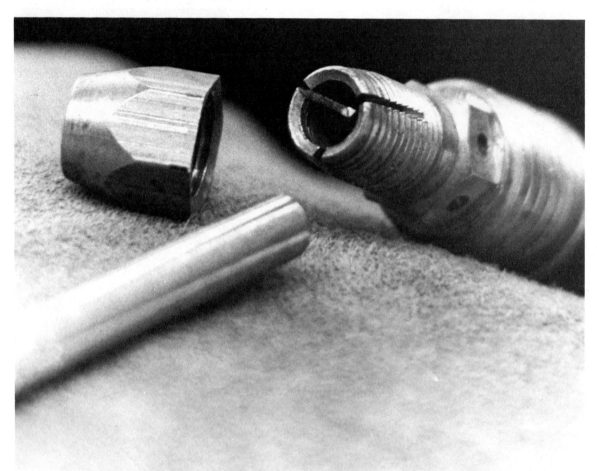

Fig. 11-12. The threaded fitting is a clamping collet which secures the slide-in type of contact tip as the tapered nut is tightened.

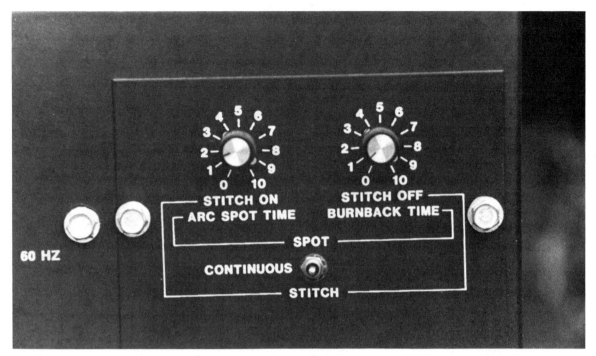

Fig. 11-13. An arc timer, often built into a power source, allows a uniform weld deposit to be made repeatedly.

Also, low short-circuit voltages can cause difficult arc striking and arc outages. To avoid porosity (Fig. 11-14), the base metal must also be very clean—free from rust, paint, and other surface contaminants. These problems can be largely avoided if you assess your weld quality by first running test beads on scrap material. This literally takes only a minute or two, and saves rework time—not to mention frustration from weld failures.

Globular Transfer

Globular transfer is so-called because large "globes" or "globs" of melted electrode are carried across the arc. These roughly spherical droplets are larger in diameter than the wire from which they are melted (Fig. 11-15). This is like metal transfer with SMAW. (Recall that no electrode metal crosses the arc in short circuit transfer.) Globular droplets are susceptible to atmospheric contamination so the contact tip is recessed into the nozzle at least 1/8 inch.

Globular transfer is a higher voltage and higher amperage GMAW variation than short circuit transfer. About 26 volts and 150-200 amps are used with globular transfer. Because heat input at these amperages is high, it is only suited for welding sections 3/16 inch and thicker. In earlier days, the term for globular transfer was CO_2 Welding.

At home, even on heavier plate, GMAW globular transfer is usually not the best process choice. Spatter is great because the large droplets splash into the weld puddle. Gravity also plays havoc with the slow-freezing globular deposits, so it is difficult to run in any position besides flat. Like short-circuit transfer, the plate should be really clean. A better all-around job is usually done with flux-cored wires. Still, globular transfer does produce deep penetration using low-cost CO_2 shielding gas and the same wire used for short-circuit transfer.

Spray Transfer

The spray transfer method of GMAW is, like globular transfer, a higher voltage and amperage process than short-circuit transfer. Spray transfer is suit-

193

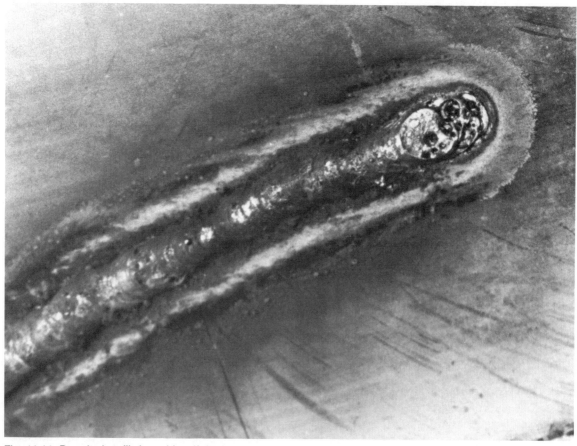

Fig. 11-14. Porosity is a likely problem if the base metal is not clean and free of paint, etc.

able for welding heavier steel thicknesses in the 1/4-to 3/8-inch range. Voltage and amperage are about the same as those used for globular. Globular and spray transfer differ in the size of the melted droplets. Spray droplets are much smaller than the electrode (Fig. 11-16).

The finer spray droplet is achieved with argon-rich shielding. Mild steel spray requires a shield of at least 85 percent argon with the balance CO_2 or pure oxygen (O_2). Even with this mixture there is a transition current below which the droplets are large (globular) and above which spray transfer occurs. With .035 electrode, about 28 volts and 160 amps are required for spray. Argon, unlike straight CO_2, provides little cooling of the gun. Consequently, the 10-minute amperage rating, or *duty cycle*, of the gun must be cut in half. This is not likely

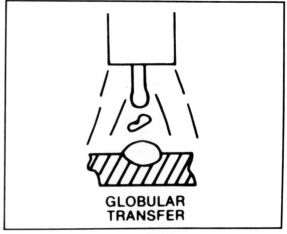

GLOBULAR TRANSFER

Fig. 11-15. Globular transfer involves large drops of weld metal passing across the arc. Considerable splashing or spatter is involved. (Courtesy of Alloy Rods.)

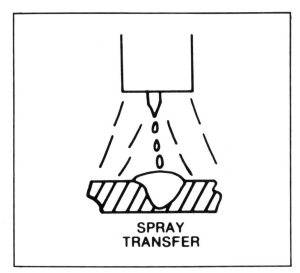

Fig. 11-16. Spray transfer involves fine droplets of weld metal passing across the arc. The deposit is smooth and largely free of spatter. (Courtesy of Alloy Rods.)

to be a major concern for the home craftsman, but knowing this could save burning up a gun.

Spray transfer produces little spatter, and weld appearance and penetration are excellent. The arc zone is hot and the weld puddle is "wet." Spray transfer is limited to primarily the flat position, although a decent downhill weld can be made on a slightly inclined surface. It is effective on carefully positioned outside corner joints.

GMAW Summary

Gas metal arc welding for the home shop is mostly used in the short circuit transfer mode. The use of globular and spray is limited. The most suitable electrode size is .035 inch. The combination of voltage, wire feed speed (amps), and proper gas determines the type of transfer. Tables 11-1 and 11-2 compare these and other facts about the GMAW processes.

FLUX-CORED ARC WELDING (FCAW)

Flux-cored arc welding (FCAW) was first regarded as a form of gas metal arc welding. This is understandable because nearly the same equipment is used for both. FCAW also works like GMAW. Recent improvements in all-position flux-cored electrodes have justified a separate FCAW category. The main advances have been with control of the flux elements and better manufacturing methods for small tubular wires. Only recently, flux-cored wires were limited to welding 1/4-inch and heavier steel—and then only flat or horizontal. Few thought that home craftsman could do precise welding with

Table 11-1. Shielding Gases, Processes, Applications, Flow Rates.

GAS	PROCESS	APPLICATION	FLOW RATE IN CUBIC FEET/HR
CO2	short circuit transfer w/solid electrode	18-gauge to 3/16-inch plate	25 cfh
	globular transfer w/both solid and tubular electrodes	3/16- to 3/8-inch plate	35 cfh
C-25	short circuit transfer w/solid electrode	better quality, smoother welds on 18-ga to 3/16-inch plate	35 cfh
	globular transfer w/both solid and tubular electrodes	better quality, smoother welds on 3/16- to 3/8-inch plate	35 cfh
95-5	spray transfer w/solid electrode	3/16- to 3/8-inch plate	35 cfh

Table 11-2. Wire Feed Systems, Costs, Advantages, and Disadvantages.

WIRE FEED SYSTEM	COST	ADVANTAGE	DISADVANTAGE
1. combined conventional power source and feeder	$1200 to $1800	low cost, compact, proven performers	limited gun reach from power source
2. conventional power source (CV) with separate push feeder	$1500 to $2100	traditional design with improvements made over time	(minor) extra exposed cables and connectors
3. conventional power source (CC and CV) with stick and TIG abilities, separate push feeder	$1800 to $2100	added stick and TIG abilities	(minor) extra circuitry
4. inverter power source with push feeder	$1400 to $1700	easy to store and transport, low line draw, efficient, very smooth arc	(minor) potential electronic vulnerability
5. inverter pkg same as above with stick and TIG abilities	$2100 to $2300	same as above with added abilities	same as above
6. engine driven power source	$1200 to $2300 plus wire feeder	ability to weld almost anywhere	cost, noise, fumes

flux-cored wires. Many do-it-yourselfers now do most of their mild steel work with these electrodes.

The same constant voltage (CV/CP) power source and wire feeder used for solid wire GMAW is used for welding with the tubular FCAW wires. To avoid crushing tubular wire, drive rolls with teeth or *knurls* are often substituted for the smooth rolls used with solid wire (Fig. 11-17).

FCAW Advantages

Home welders are discovering advantages with flux-cored electrodes over both SMAW and GMAW. Several advantages are:

- Easy to use
- Versatile—same size wire electrode used for different metal thicknesses
- Very high quality weld deposits
- Absolutely gorgeous welds
- Ability to bridge gaps in fitup
- Controlled heat input
- Base plate does not have to be absolutely clean

Manufacturing Flux-Cored Electrodes

Tubular flux-filled wire is an engineering marvel.

A bit of magic is involved in making and filling these tiny electrodes. Alloy Rods Corporation, a developmental pioneer, helps explain the magic with the following short description of how flux-cored wires are made.

Manufacturing flux-cored electrodes requires close controls. Because the weld metal is a combination of the metal sheath and the flux ingredients, both must be closely checked for size and chemical composition before fabrication begins.

The space within the wire is limited, so particle size of the ingredients becomes very important, so that the particles will "nest" together. Flux ingredients must be totally mixed or blended and measures taken to prevent segregation of the elements before fabrication.

Most flux-cored electrodes are manufactured from a flat metal strip, which is passed through a mill where forming rolls progressively shape it into a U shaped section. A metered amount of granular flux is fed into the formed strip. It then passes through the closing rolls which form the strip into a tube, and tightly compress the core material (Fig. 11-18). The tube is then pulled through a series of drawing dies that reduce it to its final size, and further compress the flux to lock it in place within the tube.

196

During manufacture, close control is necessary to assure that flux voids do not occur throughout the entire length of the wire. Also the surface of the wire must be smooth and free of contaminants that can be detrimental to feeding and welding current transfer to the wire. The wire must be carefully wound on spools, coils, or into drums, so that kinks or bends do not occur. Spools and coils are usually packaged in plastic with some sort of desiccant material to absorb moisture within the package, and are then placed in a cardboard carton for protection.

FCAW VARIATIONS

Some tubular wires are designed for use with and others without shielding gas. The flux ingredients determine if gas is to be used.

Self-Shielded Flux-Cored Electrodes

The self shielded electrodes have been greatly improved since the late 1970s. Dispensing with a shielding gas saves on the expenses of the gas, a cylinder, the regulator-flowmeter, connecting hose, and fittings. You also avoid lugging the gas supply around. For outside work in a mild cross breeze, the self-shielded types have an advantage in that a curtain or wind screen is not required.

Small self-shielded flux-cored electrodes are made in .045-, .052-, and .068-inch diameters. For home use, the .045 wire is the most useful. It enables you to weld metal as thin as 18-gauge exhaust tubing or as heavy as 1/4-inch plate. These electrodes use DCEN or "straight polarity" (see Fig. 10-3). The electrode runs hot to adequately vaporize the flux and uses a healthy 1/2-inch stickout. Penetration is shallow. This could be a disadvantage on heavier sections, but it is a blessing when working on thin material or to bridge wide gaps.

Self-shielded electrodes perform well with less than ideal fitup because the gun can be moved in or out over a wide range to vary the wire stickout. This is possible with no shielding gas to lose, and affords an unparalleled degree of control over heat input. For example, if the tightness of a joint varies, the gun is held close (about 3/8 inch) for the tight sections and moved back (up to 1 inch or more) with a wide fitup. (Fig. 11-19). The gas nozzle can even be dispensed with for a better view of the arc. It is, however, a good idea to use some sort of insula-

Fig. 11-17. Flux-cored wires are driven with knurled drive rolls. The teeth grip the wire lightly and thereby avoid crushing the tubular wire.

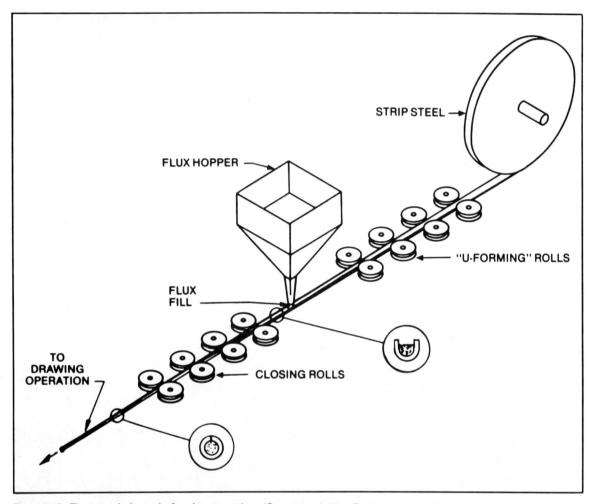

Fig. 11-18. Flux-cored electrode forming operation. (Courtesy of Alloy Rods.)

tor on the end of the gun.

Because it is so versatile, the .045-inch self-shielded flux-cored electrode has been called an "inside-out E-6011" stick electrode. Penetration is less, but like E6011, the .045 wire allows you to weld over light rust, paint, and even zinc or aluminum coated steel. This helps with repairs if sand blast or wire brush cleaning are impractical. Fast repairs are often made by just connecting the ground, turning on the power, and welding.

Self-Shielding Disadvantages

Self-shielding electrodes might sound almost too good to be true, but there are a few flies in this oint-

ment. With some wires, the welds tend to look cloudy. All weld impurities do not always make it entirely into the slag on the bead surface. For this reason, "single pass only" is often recommended. If you need larger multiple pass welds and strength is critical, grind the "skin" from all but the final cover pass.

There is considerable smoke associated with the self-shielded wires. This means planning ahead for proper ventilation and positioning yourself out of the smoke. Spatter level is higher than some other welding methods, but it can be minimized with the right voltage setting. These wires typically leave a powdery residue that can be a minor nui-

198

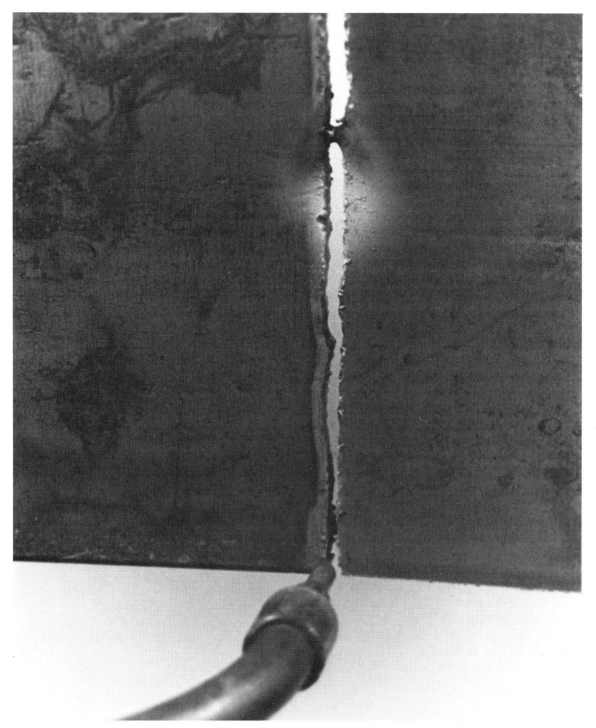

Fig. 11-19. Welding a joint with uneven fitup is easy to do with the self-shielded wires because the stickout can be varied over a wide range without loss of a shielding gas.

sance to remove from complex assemblies.

Some users have felt inspired to make disparaging claims such as, "welding with the self-shielded wires is like squirting oatmeal into a fog bank." In other words; operator appeal is not always high. This was more the case with early wires, but then self-shielded weld metal never does flow out quite like the gas-shielded wires—yet, it is this fast-freeze characteristic that gives it its unique advantages.

Despite such objections, self-shielded wires are convenient, versatile, and dependable. In fact, many home craftsmen do most of their welding with them.

Gas-Shielded Flux-Cored Arc Welding

The gas-shielded flux-cored process (often called "Dual Shield") was, for many years, used only to weld heavy steel fabrications. Since its inception in the 1950s until the late 1970s, it was suitable only for flat and horizontal use. The most popular wire size was 3/32-inch operated at 350 amps and above—strictly a heavy plate proposition. Material thicknesses welded by hand-held operations were often 1/2 inch and greater.

With the development of a faster freezing puddle and smaller electrodes, vertical and overhead welding became practical. The use of gas-shielded flux-cored electrodes expanded to small shops and more field fabrications. By the early 1980s a tiny .035-inch diameter wire was introduced (Fig. 11-20), allowing the home craftsman to finally use this splendid welding process. Gas-shielded flux-cored wires operate with GMAW equipment, except that special (knurled and grooved) drive rolls are recommended to avoid crushing the tubular wire.

Probably the greatest single advantage to using the gas-shielded flux-cored electrodes is the high weld quality. These wires also provide a stable arc, produce little spatter, can be used in single or multiple passes, and are extremely easy to slag. All this probably sounds pretty good; it is.

Gas-Shielded Flux-Cored Wire Disadvantages

There are limitations with gas-shielded flux-cored wires. If you have been welding without shielding gas, you will need to obtain gas equipment. You will also need to transport the gas, and possibly provide shields to keep from losing it to drafts. The minimum amperage for the .035-inch wire is 100. This means 14 gauge is the thinnest *practical* limit to weld with these wires.

WIRE ELECTRODE CLASSIFICATIONS

The home craftsman will use only a few of the many different wires. The American Welding Society publishes filler metal specifications for solid and flux-cored wire electrodes. Most manufacturers adhere to these. There is a range of wires with different operating characteristics within the broad specification. As with stick electrodes, an easy-to-read code identifies each wire.

Solid Wires

An example of a GMAW wire is ER70 S-3. Each part of this description has significance. Figure 11-21 shows the elements of a solid wire electrode description. The "E" means electrode, the "R" (if used) means this wire is alright for use as a filler rod (with the Gas Tungsten Arc Welding process). The "70" means minimum tensile strength of the weld deposit times 1000 psi.

Electrode wires are either Solid or Tubular; S or T. Therefore the "S" in the electrode description means bare solid wire. The final number indicates the wire chemistry. In this case, 3 specifies that ER70 S-3 deposits a moderate amount of manganese and silicon.

For the home shop, ER70 S-3 is a "fits-all" wire suitable for any low-carbon or mild steel. If you get involved with auto body repair, be sure to use ER70 S-6 with more manganese and silicon. This wire works well with the high-strength low-alloy steels.

Tubular Wires

E71 T-1 is a flux-filled tubular wire electrode. Figure 11-22 explains the designation. Here, as with the solid wires, "E" specifies electrode, but the "7" means minimum tensile strength times 10,000 psi. The first "1" indicates use in *all-positions*. The "T"

Fig. 11-20. The .035-inch Dual Shield II gas-shielded flux-cored electrodes permit top quality welding of the lighter structures found in the home shop.

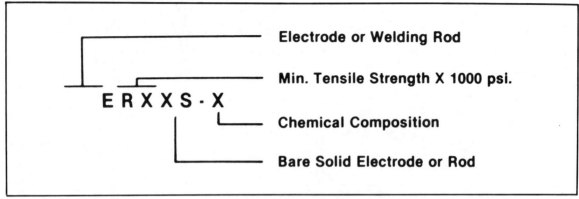

E R X X S - X

- Electrode or Welding Rod
- Min. Tensile Strength X 1000 psi.
- Chemical Composition
- Bare Solid Electrode or Rod

Fig. 11-21. Letter-number designations for solid carbon and low alloy steel wire electrodes. (Courtesy of Alloy Rods.)

indicates tubular wire. The final "1" indicates that shielding gas—either CO_2 or C-25 shielding gas is required. Actually only T-1, 2, or 5 are gas wires, the rest are self-shielded types.

The home craftsman is likely to use only E71 T-1, which is great for most work. E71 T-5 might be advisable if the base metal is not very clean. The self-shielded wires, E71 T-GS or T-11 are suitable for home use (Fig. 11-23). "G" means the electrode is not made to operate in accordance with any particular specification, "S" means single pass.

SHIELDING GASES

There are 3 shielding gases often used to weld mild steel in the home shop: carbon dioxide (CO_2), C-25 (a mix of 25 percent CO_2 and 75 percent argon), and 95-5 (95 percent argon and 5 percent oxygen). Gases affect the deposited weld bead (Fig. 11-24).

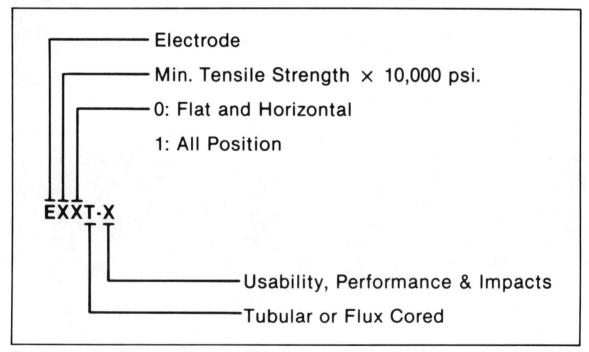

EXXT-X

- Electrode
- Min. Tensile Strength × 10,000 psi.
- 0: Flat and Horizontal
- 1: All Position
- Usability, Performance & Impacts
- Tubular or Flux Cored

Fig. 11-22. Letter-number designations for flux-cored carbon steel wire electrodes. (Courtesy of Alloy Rods.)

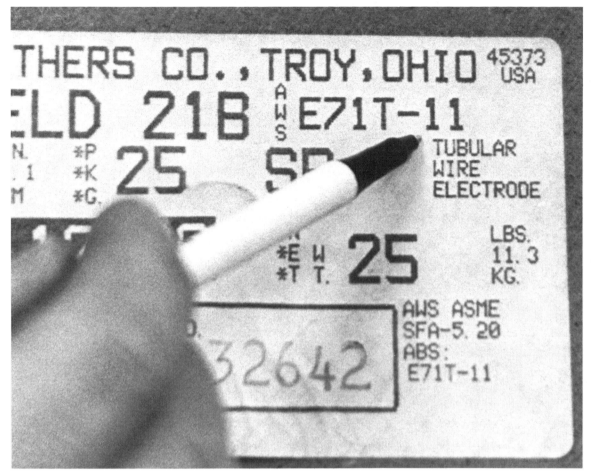

Fig. 11-23. The AWS electrode classification is marked on the spool. This is all-position flux-cored wire. Because the final number is not 1, 2, or 5—this is a self-shielded wire electrode.

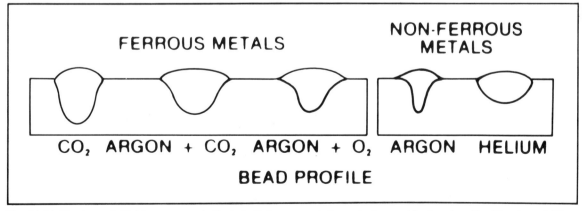

Fig. 11-24. The type of shielding gas used affects both the amount of penetration and the bead profile. (Courtesy of Alloy Rods.)

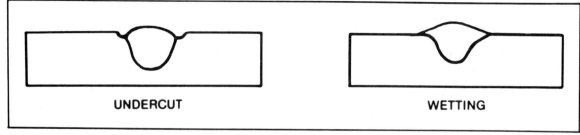

| UNDERCUT | WETTING |

Fig. 11-25. An undercut bead as depicted at the left would result with insufficient oxygen in the shielding gas. The wetter bead at the right would be hotter, with a greater percentage of oxygen in the shielding gas. (Courtesy of Alloy Rods.)

For example, just 1 or 2 percent of pure oxygen added to argon produces a "wetter" bead with less undercut as shown in Fig. 11-25.

Carbon Dioxide

Two GMAW variations—short circuit and globular transfer, and gas-shielded flux-cored welding of mild steel—can be done using inexpensive CO_2. It is purchased in cylinders containing 50 pounds (by weight) of pressurized liquid CO_2 (Fig. 11-26).

Welding grade CO_2 is dryer than that used for carbonated beverages. CO_2 is an abundant by-product of several different manufacturing processes, hence its low price.

As the pressurized cylinder contents are released in use, the liquid turns into a gas. This chills (refrigerates) the cylinder outlet valve and connections. Regulator freeze-up results if the gas withdrawal is too rapid. A *regulator-flowmeter* permits a controlled release of the CO_2 gas from the

Fig. 11-26. This is the label on a cylinder of carbon dioxide. Notice that it contains liquefied gas. Carbon dioxide is the only welding gas usually sold in this form for home use.

upper end of the cylinder. A flow rate of less than 25 cubic feet per hour or 12 liters per minute avoids freeze-up. At higher flow rates, freeze-up prevention is needed. Even if you are happy using CO_2 now, look ahead towards using other gases and mixtures. Regulator-flowmeters designed for other gases can be used for CO_2 with an adapter. A light bulb prevents freezing.

Mixed Gas

The C-25 gas mix is widely used on steel for better weld qualities with both solid and tubular electrodes. CO_2 is less expensive, but it produces a rough weld with considerable spatter. This is especially true with single-phase home shop equipment. C-25 can cost about four times as much as CO_2 but produces smoother welds and less spatter. Figures 11-27 and 11-28 show deposits made with each gas. Freeze-up problems are also eliminated with C-25. For home use with low gas consumption, costs are unlikely to be prohibitive. C-25 is worth the expense.

Whenever a gas other than straight CO_2 is used, the welding gun is subject to overheating. This is not a likely problem with short home type welds. For longer welds, gun damage is avoided by reducing the arc on time.

Other Shielding Gases

To get the solid wire arc into spray transfer, a gas shield rich in argon—in fact 85 percent minimum—is needed. A very popular blend for spray transfer on mild steel is 95-5—95 percent argon and 5 percent pure oxygen. Incidentally, pure argon is not satisfactory as a shielding gas for wire welding the steels. With argon alone, the bead piles up with poor side wetting action. Adding oxygen heats the weld puddle—letting it flow out better.

There are *proprietary gas blends* carrying exclusive marketing rights. These claim advantages over both CO_2 and C-25. Linde Division of Union Carbide produces *Stargon*, a classy substitute for CO_2 or C-25.

Shielding Gas Regulator-Flowmeters

A regulator-flowmeter, as the name implies, is a combination device. The pressure regulator part lowers and maintains pressure. The flowmeter controls gas volume per unit of time flowing into the welding zone. These devices are straightforward and largely trouble-free. Regulators are preset to allow 30 to 50 psi ($35 kg/cm^2$) of reduced gas pressure at the flowmeter inlet. The flowmeter is calibrated in cubic feet per hour (cfh) or liters per minute (liters/min). The flow is adjusted by a knob and read with a float in a calibrated glass tube (Fig. 11-29).

Carbon dioxide cylinders have flat seats at the regulator attaching fitting. These require a sealing washer of Teflon or other such rubber substitutes (Fig. 11-30). Higher pressure cylinders for other gases have a dished sealing seat for a rounded regulator-flowmeter fitting. These brass-to-brass seats require no separate seal (Fig. 11-31).

How to avoid having to buy two regulator-flowmeters? Purchase a higher pressure type equipped with multiple flowmeter scales. It can be attached to the flat-style carbon dioxide seat with an adapter as shown in Fig. 11-32.

With straight CO_2 you must make some provision to avoid regulator freeze-up. For short circuit transfer, you can often avoid difficulty with the gas flow rate under 25 cfh. However, even with short circuit transfer, a higher flow rate might be needed to offset cross breezes. With the hotter globular transfer welding mode with solid wire or a gas shielded flux-core wire, you should use at least 35 cfh—maybe even as much as 50 cfh. Such flow rates cause regulator freeze-up in a matter of minutes. Special non-freezing regulator-flowmeters work well but are solely for CO_2 use. To weld with any other gas or gas blend, you would need to buy another regulator-flowmeter. Allow yourself the option to work with other metals besides mild steel—all of which require high pressure gases and regulator-flowmeters.

Said again, the best route is to use a high-pressure regulator-flow meter with a CO_2 adapter. A lightbulb positioned as shown by Fig. 11-33 prevents freeze-up.

Gas Solenoid

A gas valve, normally held shut by a spring and

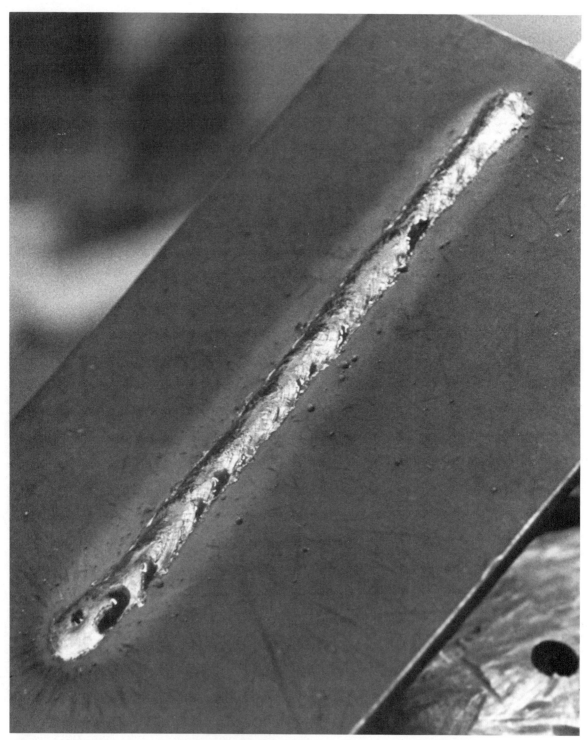

Fig. 11-27. A GMAW bead made with carbon dioxide shielding gas. There is a fair amount of spatter.

Fig. 11-28. A GMAW bead made with C-25 mixed shielding gas. Notice the little amount of spatter.

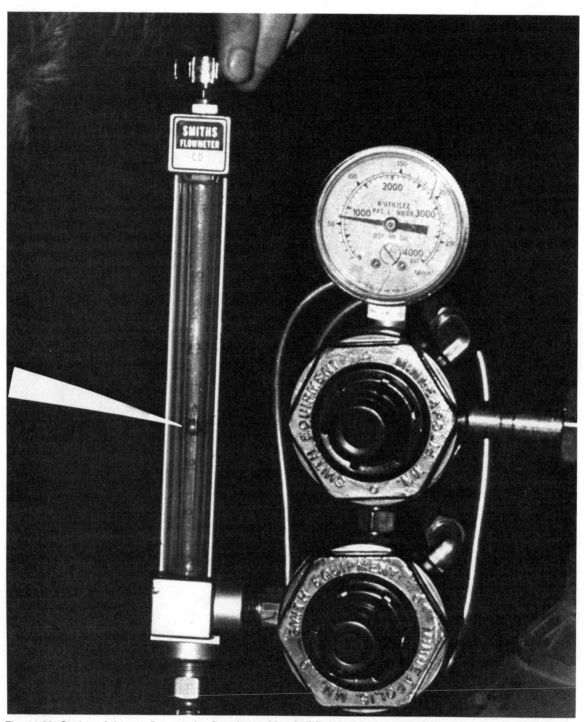

Fig. 11-29. Setting of the gas flow rate is often done with a ball float inside the graduated glass tube of a flowmeter.

Fig. 11-30. Carbon dioxide cylinder fittings are flat and require the use of a sealing washer.

opened by an electromagnet (solenoid), is often located between the gas supply and the hose to the gun. This is the *gas solenoid* (Fig. 11-34) and works with any gas. With self-shielded wire, you can save wear and tear by disconnecting the control coil if it is accessible, otherwise don't worry about it.

Table 11-1 lists each of the shielding gases used in home shops for welding mild steel along with its uses and appropriate flow rate. Remember to increase the flow in a breeze.

WIRE FEED WELDING
SYSTEMS FOR THE HOME SHOP

Most welding equipment is made for 3-phase commercial use, but there is still a wide variety of single-phase wire feed systems suitable for home use. A system refers to a power source and wire feeder with connections, a torch or gun, wire electrode, and provision for shielding gas. As discussed earlier, your final equipment decisions must be based on the capacity of the electrical line to your

Fig. 11-31. Higher pressure gases use a rounded seat at the cylinder and regulator/flowmeter connection. These connections are direct brass-to-brass and require no sealing device.

Fig. 11-32. An adapter allows the use of a high pressure (argon type) regulator on a cylinder of carbon dioxide.

210

Fig. 11-33. The heat from a lightbulb can be used to prevent carbon dioxide regulator freeze-up at higher flow rates.

shop, your budget, and your goals. The following paragraphs attempt to present wire feed options open to the home craftsman. This section concludes with a summary table outlining costs, advantages, and disadvantages of each system.

Wire feed systems for home use, using single-phase line power, include:

1. Combined power source and wire feeder (refer to Figs. 4-17 through 4-19). These units are convenient and, at 200 amps, adequate for any home job. They have been around long enough to have proven themselves and are relatively inexpensive.

2. Conventional power source and separate wire feeder as was shown in Figs. 11-2 and 11-3. This configuration permits using the feeder away from the power source. If the feeder is mounted on top of the power source, the unit is compact and portable.

3. Conventional power source and separate wire feeder with stick and limited TIG capability. The equipment is similar to option 2 except the power source has both CV/CP and CC output characteristics.

4. Inverter-converter power source and separate feeder. These relatively new power sources offer unequaled compactness and portability. They produce a very smooth arc.

5. Inverter-converter power source and separate feeder with stick and limited (not for aluminum) TIG capability. Nearly the same as option 4 but extra circuitry allows for the use of SMAW and scratch-start GTAW.

6. Engine-driven power sources as were shown in Figs. 4-10 and 4-11. These are unavoidably expensive with their gasoline or diesel engine, rotating generating component (generator or alternator), and mounting hardware. Sometimes called "poppin' johnnies," they produce a smooth arc and

Fig. 11-34. The gas solenoid is an electrically controlled valve that opens only when the trigger on the gun is depressed.

212

most come with 110-volt auxiliary outlets. Most engine-driven units are designed for SMAW use and require adaptation to use the wire feed processes.

If the range of options does not seem wide enough yet, you can also choose from 3 basic wire feeder designs. These are:

1. Push units (Figs. 11-2 and 11-3) with the advantage of simple design. Seldom a problem with .035-inch mild steel electrode, push-fed wires smaller than .035 do tend to jam or kink in the liner (wire conduit) feeding into the gun. Therefore, gun leads for push feeders are limited to 10 feet.

2. Push-pull units for use with easily kinked wires. These require 2 motors with the associated expense and complexity, but they do feed positively even around corners. While 25-foot and longer guns are available, 12 to 15 feet is more common. These units are quite popular with aluminum welding shops. They are not always recommended for use with the flux-cored electrodes.

3. Spool guns for use with either CV/CP or CC power sources (Fig. 4-20). These are portable although bulky. They use small 1-or 2-pound spools of solid wire only. Spool guns are handy if you change wire *types* often, but small spools also empty quickly. These guns require special equipment to use SMAW (CC) type power sources. They are most popular for welding aluminum.

Table 11-2 summarizes the main facts about wire feed systems, their costs, advantages, and disadvantages. Actual prices can vary with sales volumes and shipping costs.

Table 11-3 summarizes the types of wire feeders, their costs, advantages, and disadvantages.

While finding your way through this maze of options, remember what you want to do with your equipment. The combined power source and wire feeder using self-shielded electrodes probably packs the most bang for the buck. The listings are inclusive to provide information for those do-it-yourselfers who are starting from scratch and for others who may already have part of a system or have located a deal they can't refuse.

So far this chapter has described the process, consumables, and equipment necessary for wire feed welding. The next step is to put it all together and do some welding.

SETTING UP THE EQUIPMENT

To get things rolling with your wire feed system, there are a few things to assemble and adjust: connect the power source to the feeder, load electrode, and ready the gas supply. These setup operations plus a few adjustments are nearly the same for all brands and designs of wire feed equipment.

One precautionary note: before turning the power source "on," be sure that the work (ground)

Table 11-3. Feeder Types, Costs, Advantages, and Disadvantages.

	FEEDERS	COST	ADVANTAGE	DISADVANTAGE
1.	push design	$450 to $600 (plus $100 to $150 gun)	straight forward tested design	small and soft wires tend to jam
2.	push-pull units	$1200 to $2300	positive wire feeding even around corners	more mechanism, higher cost
3.	spool guns	$700 to $1000	portability, practical to keep variety of smaller wire spools	higher wire costs, frequent need to change empty spools

clamp is on the plate (never touching the power source) and that you know where the gun is. This eliminates arc damage to the equipment, the workpiece, and you.

Power Source and Feeder Connections

The on-off switch on the power source must be in the "off" position before inserting the power source plug into the line receptacle. It should also be "off" while making any other electrical connections. With a direct short circuit a CV/CP machine does not know when to quit. Severe damage could result if the machine were to discharge before you are ready.

Figure 11-1 shows the 3 electrical connections between the power source and the wire feeder/gun. These are the large welding power cable and two smaller wire sets. The smaller wires are the 115-volt feeder supply and the contactor control connection. With a self-contained system these connections are more or less internal and permanent.

Often the 115-volt feeder supply and contactor control wires are bundled together in a *wye* cable and equipped with special Amphenol-style end connectors (Fig. 11-35). To avoid damage, be sure to only turn the knurled outside ring (the largest in diameter) of an Amphenol connector. One hundred fifteen volts supplies the wire drive and control circuits as well as the gas solenoid.

The *contactor* is the device inside a constant voltage power source that connects (and disconnects) the main transformer to the electrical supply line. It is either a mechanical relay (Fig. 11-36) or a solid-state device. Relays "click" as the trigger circuit is closed; a solid state contactor works silently. The contactor control connection ensures that welding energy is ready only when the trigger is depressed.

Protect cables from burning, bumping, cutting, walking upon, etc. Connectors should be covered when not in use to exclude dust and dirt and to protect the connecting surfaces.

All electrical connections must be tight. Remember both ends of the welding work (ground) cable and the 3 wire sets just discussed. Wire feed welding uses lower voltages than SMAW, so good connections are especially important for a smooth arc. After a while, the spring-loaded ground clamp might become pitted and oxidized and require filing.

Connecting Shielding Gas

If you are using shielding gas, secure the cylinder to a solid support with a safety chain—then remove the safety cap. Wearing eye protection, stand to one side and "crack" the cylinder valve to remove any dirt. If (and only if) using CO_2, insert a new sealing washer onto the flat regulator attaching surface. Support the regulator-flowmeter while hand tightening its cylinder connection. Finish tightening with a properly fitting wrench. Use no sealing compounds.

The fullness of a CO_2 cylinder can be estimated by watching for a jet of ice particles when cracking the cylinder valve (Fig. 11-37). If no ice, the cylinder is close to empty and should be replaced.

Connect a clean gas hose to the flowmeter. Make sure that the flowmeter valve is closed, and then open the cylinder (slowly at first) all the way. Next, open the flowmeter valve enough to ventilate the flowmeter and hose with the shielding gas. This removes any dirt or small critters that might have moved in. If you forget and open the cylinder with the flowmeter valve open, you could launch the float through the top of the glass tube. This is considered bad form.

The free end of the gas hose is now connected to the solenoid. Again, observe cleanliness. Like most valves, the gas solenoid can leak with just a speck of dirt on the seat. Open the flowmeter a half turn or so and check for gas leaks. If you suspect a leak, apply liquid soap and watch for bubbles (Fig. 11-38). A small leak causes problems; not only can gas be lost, but air can be drawn in with the shielding gas. Discard worn hoses and replace damaged fittings.

Shielding gas flow is set with the power source "on" and gas flowing all the way through to the gun nozzle. On the smaller home systems, this means that you must pull the gun trigger that also energizes the gun. Use special care to avoid any destruc-

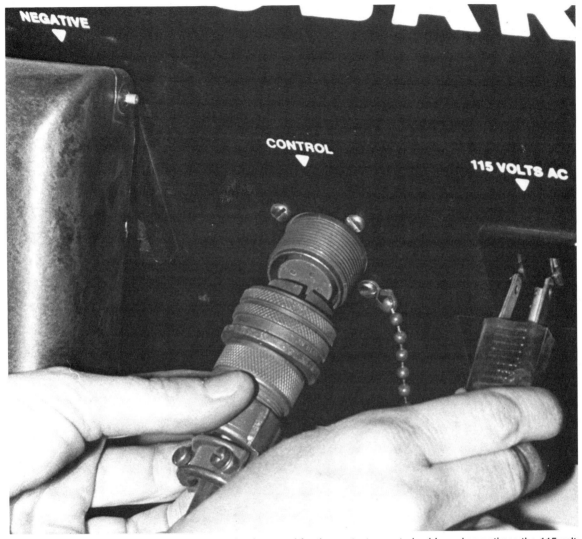

Fig. 11-35. A multiple-pin Amphenol connector is often used for the contactor control cable and sometimes the 115-volt supply as well. In this case the 115V is supplied through a household-type plug.

tive arcing at the gun. If wire is already loaded, turn the wire feed speed all the way down to prevent "hot" wire from snaking about.

Remember a light bulb for straight CO_2 over 25 cfh. If your style gun has O ring gas seals where it attaches to the feeder, check them at this time (Fig. 11-39).

Loading Wire Electrode

Before loading wire, use an air hose to blow out the gun wire liner (Fig. 11-40). Every fourth wire change, remove the liner and inspect it for damage (Figs. 11-41 to 11-46). The liner slides in or out easily only if the gun cable is on a flat surface. Blow out the gun conduit and the liner separately. This simple operation almost guarantees freedom from wire jam-ups. To slide the liner back into the gun conduit, it might be necessary to slightly "whip" the cable (Fig. 11-47). Check your feeder for the correct size drive rolls and wire guides.

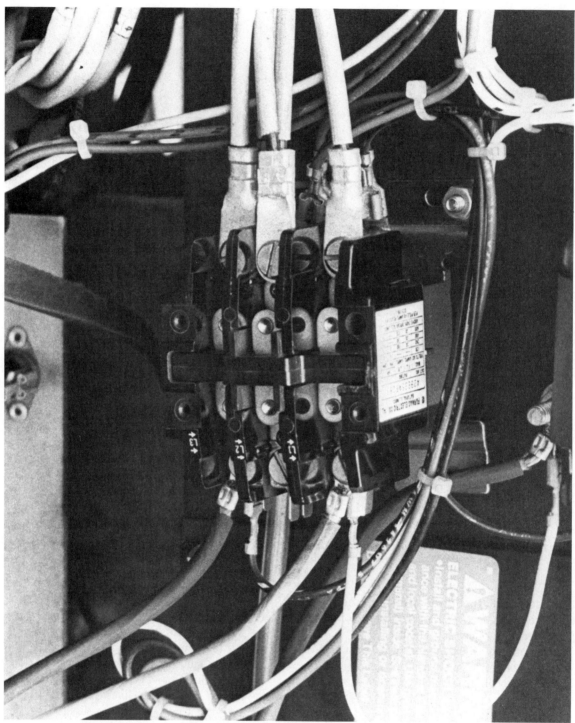

Fig. 11-36. The contactor is a type of heavy-duty relay between the large line input wires at the bottom and the transformer above. It operates (closes) when the gun trigger is depressed to begin welding.

Fig. 11-37. This carbon dioxide cylinder is nearly full as shown by the ice particles.

Fig. 11-38. A leak like this is hard to miss!

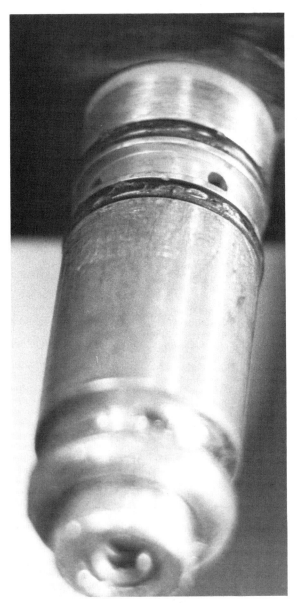

Fig. 11-39. These **O** ring gas seals need to be replaced.

Leave the power source off until power is needed to feed the wire. Also turn off the gas to prevent waste while loading the wire.

Wire electrode, solid and tubular, is most convenient on a spool. These spools are carried on an adjustable spring-loaded hub. The hub needs enough drag to prevent coasting or wire overspooling as you release the gun trigger to stop welding.

When loading the spool, be sure that the pin on the hub enters the matching hole on the spool (Fig. 11-48). Unless this is done, the spool has rotational "slop" that allows the wire to surge as it feeds. Replace the spool retaining nut or clip.

The hub drag adjustment is correct if the spool stops after only an inch or two travel. This measurement is taken at the outside of the spool.

The wire is next threaded through the drive rolls and into the gun liner. First, back off or release entirely the drive roll tension. Then remove the contact tube (this avoids a possible wire blockage). Replace the nozzle to protect the front end of the gun from accidental arc damage.

Unhook the wire from the side of the spool. Hang onto it! Otherwise it will spring out and unwind. Use sharp cutting pliers to cut off any kinks or bends from the end of the wire (Fig. 11-49). Insert the wire through the drive rolls and a few inches into the liner.

The drive rolls are now tensioned. Check that the ground cable is connected to the workpiece and that the gun is safely located. Turn the power on and set the wire feed speed to about 1/2 of the maximum setting. Slacken off the drive tension until the wire can be held back while the drive rolls slip. Note from Fig. 11-50 that it is possible to hold the wire back, adjust tension, and press the trigger all at once. Increase tension gradually until the wire cannot be held back. An additional 1/8-turn provides the required wire drive tension. The tension can be tested further by allowing the feeding wire to slide onto a gloved hand (Fig. 11-51).

Devise some way, perhaps with a clothes pin and small cloth, to lightly wipe the wire as it feeds. This keeps the liner clean.

Preparing the Front End of the Gun

The front end of the welding gun consists of the gun tube, diffuser, contact tip, insulator, and nozzle. These part names vary somewhat with different manufacturers.

Nozzles and Insulators. The gas shielded wire modes require an insulated gun nozzle. During welding, you should be able to drag the nozzle directly on the plate without shorting out. Some

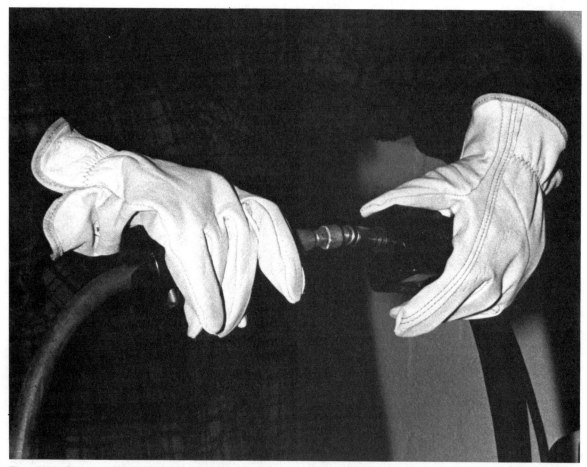

Fig. 11-40. Blow out the gun liner at each wire change.

nozzles have a built-in insulator. However, the most convenient arrangement is a separate insulator and an adjustable "Tweco-style" slide-on nozzle as seen in Fig. 11-8. Tweco Products pioneered this design, but now similar parts are made for most gun brands. The insulator should be snug on the threaded gun tube. Avoid rotating the nozzle on the insulator. This causes the tension springs (Fig. 11-52) to cut into the insulator grooves, leaving the nozzle to fit loosely on the end of the gun.

Contact Tips. The copper contact tip or tube is a critical, yet often forgotten, component of the welding circuit. This small item delivers high amperage to rapidly moving wire. There are a couple of designs for securing contact tips. Figure 11-53 shows both threaded and unthreaded varieties. As

you might well imagine, tips wear out. A modified tip design with a notched end (Fig. 11-54) is claimed to last a long time. The slightly angled platform at the end makes more positive contact with the wire.

Do not scrimp on tips; keep a few spares. The hole in the tip wears egg-shaped after an hour or so of use (Fig. 11-55). Excessive drive tension makes the wire harder and rougher, thus accelerating tip wear.

An interruption to the smooth feeding of the wire electrode causes *burnback* of the wire from the arc into the tip (Fig. 11-56). This stops welding because you will need to repair or replace the tip. A worn tip seizes wire in the small side of the egg-shaped hole. Worn tips also allow the wire to bounce—causing arcing and pitting to the wire, fol-

220

lowed by snagging and burnback.

The literature for your welding gun specifies which tip goes with which size of wire. The size is usually stamped on the tip. Generally, for mild steel, a tip orifice .008 to .013 inch larger than the wire is required. For .045-inch flux-cored wire, use a .060 tip. This allows smooth passage even if the wire is slightly flattened. With an undersized tip the wire seizes as it heats up and expands. If the tip is oversize, the wire bounces as if the tip were worn out. An incorrect tip leads to burnback and operator frustration.

A tip with spatter buildup (Fig. 11-57) attracts yet more spatter, which eventually snags the wire. Spatter accumulated inside the nozzle allows the nozzle to be shorted against the plate, causing dam-

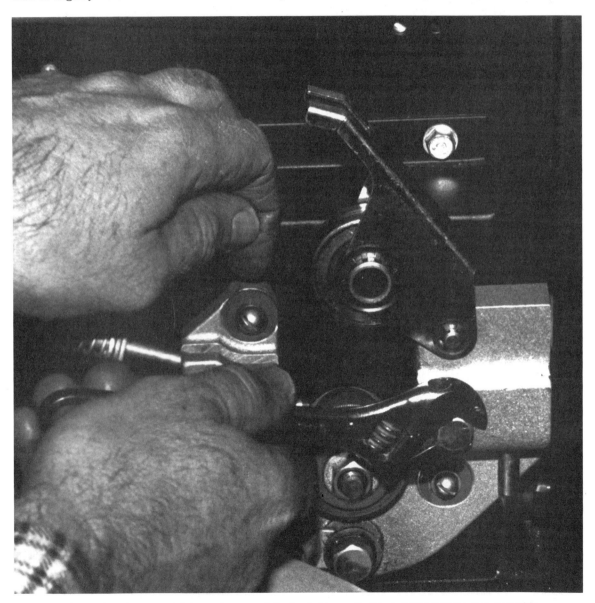

Fig. 11-41. Loosening the gun retaining set screw at the feed drive unit allows removal of the gun assembly.

Fig. 11-42. Loosening the set screw that secures the liner at the gas diffuser in the gun.

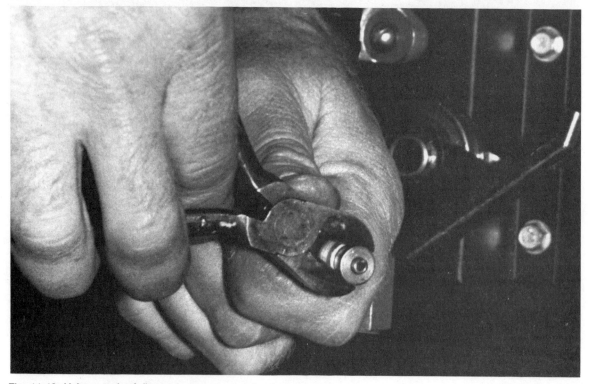

Fig. 11-43. Using a pair of diagonal pliers to start the liner out of the gun conduit.

222

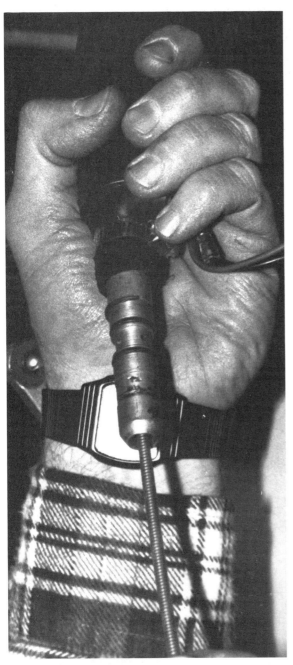

Fig. 11-44. Remove the liner from the straightened-out gun cable. Do not force it or it will be damaged.

age to the nozzle and extinguishing the arc (Fig. 11-58). Spatter removal is easiest if the nozzle and tip are coated with an anti-spatter compound. If the

tip is slightly damaged from spatter and/or burnback, remove it from the gun. File away all traces of spatter, clear the orifice with tip cleaners, and place it back in service.

Necessary Spare Gun Parts. Keep these spare parts for the front end of the gun:

- Tips—they always wear out every couple of hours
  - Insulator—replace if loose or cracked
  - Diffuser and nut (if used)
  - Nozzle

USING WIRE FEED WELDING

This section explains some wire feed welding techniques. It is divided into five areas for the five different wire modes suitable for home use. Right now, you may be interested in only one of these areas and be eager to weld; if so, study that area and read about the rest later. The five areas include:

- Welding with Self-Shielded Flux-Cored Electrode
- Welding with Solid Wire and Short Circuit Transfer
- Welding with Solid Wire and Globular Transfer
- Welding with Solid Wire and Spray Transfer
- Welding with Gas-Shielded Flux-Cored Wire

Before you can weld with any of these methods, you must first adjust the equipment to obtain the correct heat input for the job you wish to do. Basically, this means setting voltage and wire feed speed/amps. Be sure that you have read the opening material in this chapter describing the functions of voltage, wire feed speed/amps, and stickout. In addition, you should be familiar with the information in Chapter 10 concerning proper use of the welding hood.

The easiest way to adjust the heat input is by following a system. The following has proven to be a good one:

- Make approximate voltage and wire feed speed settings.

Fig. 11-45. Coil up the liner as you remove it to prevent it from hitting the shop floor.

● Try welding on scrap material similar to your actual project.

● Fine-tune the voltage and wire feed speed as required.

● Make a practice weld as nearly identical as possible to the actual weld for the job; if possible, test this weld.

Once you are familiar with your equipment, you will be able to quickly adjust it for a variety of jobs.

How to Calculate Wire Feed Speed

Small power sources do not have ammeters—yet it would be nice to know the amperage because it is a major contributor to heat input. However, wire feed speed is also an accurate gauge of heat input and is easy to measure.

To determine wire feed speed, run wire out for six seconds (1/10 of a minute). Measure the wire length and multiply by 10. This is the wire feed speed in inches per minute (IPM). Find this setting again by noting the position of the wire speed control knob. Save time in the future with a record of your settings.

Repeat Performances

How can you later duplicate an ideal welding condition? Or, put another way, how can you duplicate heat input? Answer: Use the same *variables*. These include:

● Wire type and size
● Voltage
● Wire feed speed/amps
● Stickout distance
● Travel speed
● Shielding gas

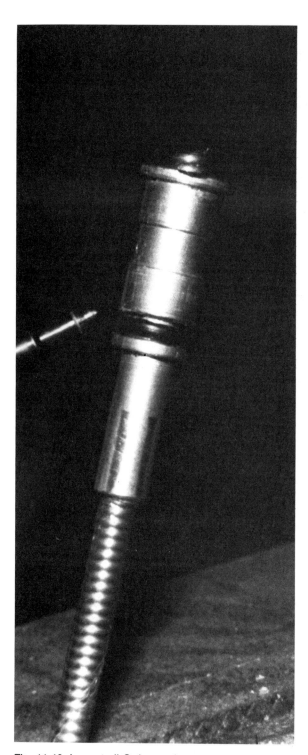

Fig. 11-46. Inspect all O ring seals.

To practice wire feed welding, use uncoated and unpainted mild steel plates about 6 or 8 inches square. Anything smaller heats up too quickly. Use 16-gauge (1/16 inch) to 3/16-inch plate for the self-shielded and short circuit methods; 3/16-to 3/8-inch plate for globular, spray, and gas-shielded flux core.

Voltage and other settings are summarized for each of the 5 wire feed modes later. Table 11-4 helps to compare these settings.

WELDING WITH SELF-SHIELDED FLUX-CORED ELECTRODE

Load the wire feeder and set tensions as explained earlier. Check all electrical connections. Use DCEN (straight polarity) with the ground cable connected to the positive (+) terminal of the power source (Fig. 11-59). E71 T-GS .045 wire is used here. Make your settings and proceed with test welds as follows:

- Set the voltage to the middle of the maximum range, or to about 21 volts. Self-shielded wires are quite voltage-sensitive, so you might need to readjust the fine voltage control.
- Set the wire feed speed to approximately 60 IPM as just explained.
- Test your settings by attempting to run a 2-inch long bead on the plate; start with a stickout of 1 inch or more and *gradually* move the gun to the normal 1/2-inch stickout—this gives you time to react, should the arc burn back up the wire. Avoid sharp turns in the gun cable which bind the wire. If the arc burns back, lower the voltage slightly and try it again. If the wire *stubs* or pushes the gun away from the plate, increase the voltage slightly and try it again.
- Evaluate the bead and the heat effect to the plate. The operation should be smooth. Readjust the voltage and wire feed speed if required.
- Vary the drag angle to see what effect this has on the bead. Avoid a push angle with the self-shielded wires because it causes porosity.

Fillet Welds

Once you can run a respectable-looking bead, tack

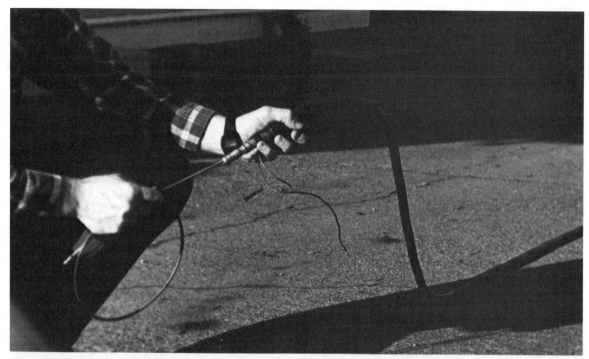

Fig. 11-47. It is usually necessary to whip the gun cable as the liner is reinserted.

Fig. 11-48. The pin on the spool hub must line up with a hole in the wire spool. This ensures that the spool will not slip on the hub.

226

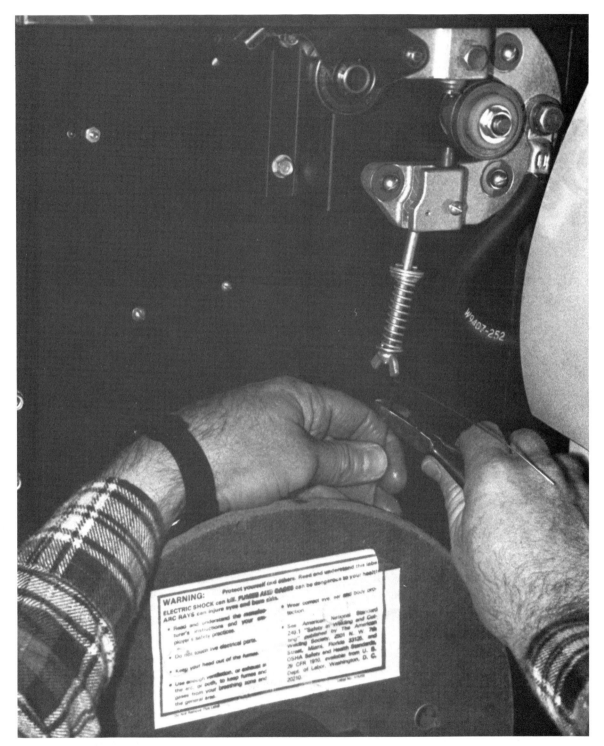

Fig. 11-49. Discard the kinked or bent section of wire.

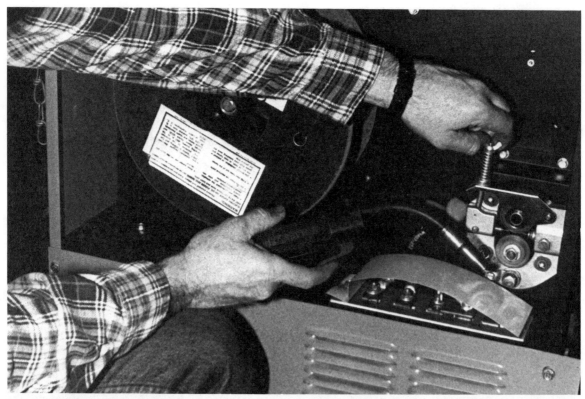

Fig. 11-50. Place your left forearm as a light drag against the wire spool and your left hand on the drive tension screw. Pull the trigger to start the drive rolls. Increase the tension until the spool moves positively. Do not overtension the drive rolls!

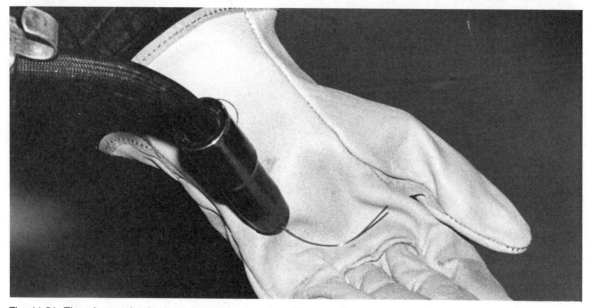

Fig. 11-51. The wire tension is about right when the wire slides steadily onto a gloved hand.

Fig. 11-52. The tension springs on an insulator float in grooves.

some strips together to make full open-corner joints similar to those shown in Fig. 10-48. Make the tack welds small to avoid interfering with the final weld. To make the best quality welds with the self-shielding process, it is essential to remove all slag from tacks and prior passes because the *scavenging* or washing-out ability of these wires is low.

Place the practice corner joints so that you can weld the entire length by travelling vertically downward. As with stick welding, body positioning is important to make this a continuous operation. The completed fillet weld should fill the joint.

Experiment by changing the stickout for different portions of a joint to see how this affects the amount of fill. Experiment with another joint to see how changing travel speed affects the fill. It might be necessary to make two passes if the position causes a large deposit to sag.

Also practice lap and T joints with both continuous and intermittent or "skip" welds. Try welding with the work in different positions. Remember: only change one variable at a time to see what effect it has upon the finished bead.

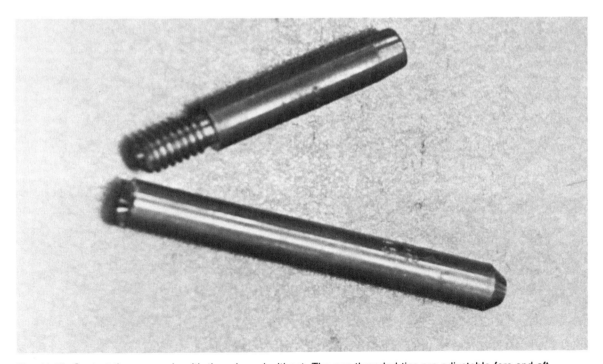

Fig. 11-53. Contact tips are made with threads and without. The non-threaded tips are adjustable fore and aft.

229

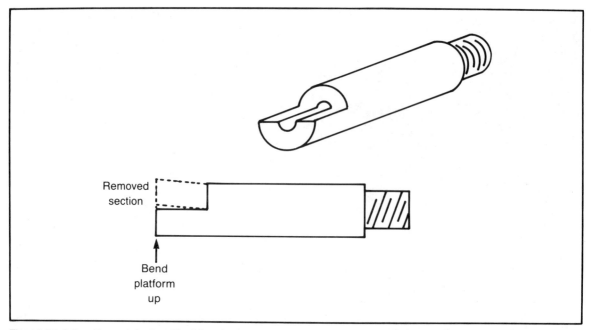

Fig. 11-54. A tip with a notched end is claimed to have an extended lifespan. Although not always commercially available, these are easy to make using a saw and file.

Butt Joints/Poor Fitup

Self-shielding wires have the unique advantage of bridging wide gaps. Tack up a butt joint as was shown in Fig. 8-4. Where fitup is tight, use a shorter, 1/2-inch stickout; where wider, use more— up to 1 inch or more. This amazing ability is possible, of course, only because there is no danger of losing shielding gas with a gasless process.

Welding Heavier
Sections with Self-Shielded Flux Core

To weld plate heavier than 1/4 inch, first make a short practice fillet weld on a test joint and break it apart. This simple test shows if the heat input was adequate for proper fusion. The break also tests if the metals are compatible. A weld should not just snap off the base metal. Rather, it should tear with some difficulty before completely failing. Basically, weld fractures are either brittle and sharp or ductile and ragged; ductile is better.

Preheating

Heavier plate (over 1/4 inch) should not be welded

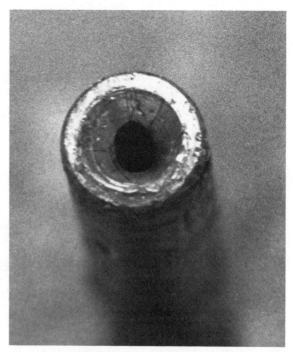

Fig. 11-55. This tip shows considerable wear and should be near its retirement.

Fig. 11-56. Wire burnback has blocked the end of this tip. This is caused by an interruption to the steady feeding of electrode or by excessive arc voltage.

metal to about 150 degrees Fahrenheit—easily accomplished with an oxyacetylene torch.

There is another benefit to preheating: it removes hydrogen-bearing moisture from the plate surface. As you move a heating torch across a cool plate, notice how water appears and then flashes away. If welding on cold plate, some of that moisture could be trapped within the weld. This leads to a sort of delayed-action cracking called *underbead cracking* in which trapped hydrogen works outward from the inside of the weld.

WELDING WITH SOLID WIRE AND SHORT CIRCUIT TRANSFER

Load the feeder with E(R)70 S-3 electrode,

Fig. 11-57. The buildup of spatter on the front end of the gun must be removed frequently.

without considering the cooling rate of the weld; welds generally should cool slowly. If the surrounding plate is ice cold, the weld zone loses heat rapidly. A sudden quench can form hard and brittle zones, and often causes cracking. To slow down the cooling of the weld, preheat the surrounding

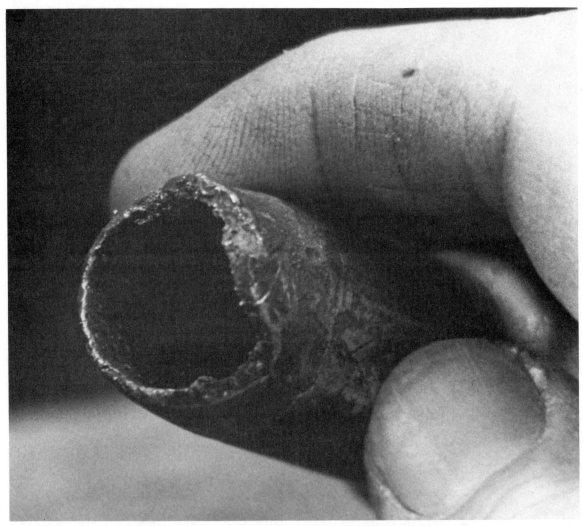

Fig. 11-58. This nozzle damage was caused by a buildup of spatter which bridged the nozzle insulator. As the nozzle touched the plate a short circuit was formed. Prevent this by removing spatter often.

.035-inch diameter, and set tensions as explained earlier. Check all electrical connections. Use DCEP (reverse polarity) with the ground cable connected to the negative (–) terminal on the power source (Fig. 11-60).

As also explained earlier: connect and adjust a supply of shielding gas and, if using carbon dioxide, avoid regulator freezeup. Tighten all connections and see that the gas flow to the nozzle is uninterrupted.

For your initial short circuit transfer practice,

adjust the nozzle and contact tip so the tip protrudes about 1/8 inch beyond the nozzle. This provides good visibility of the welding action and helps to keep the stickout length short.

Keep a pair of wire-cutting pliers handy. The arc is easier to restrike if the melted end of the electrode is cut off. (Be careful because the cut-off piece of wire is sharp and can fly some distance.) Long-nose cutting pliers offer the advantage of acting as a tool to remove accumulated spatter from the inside of the nozzle.

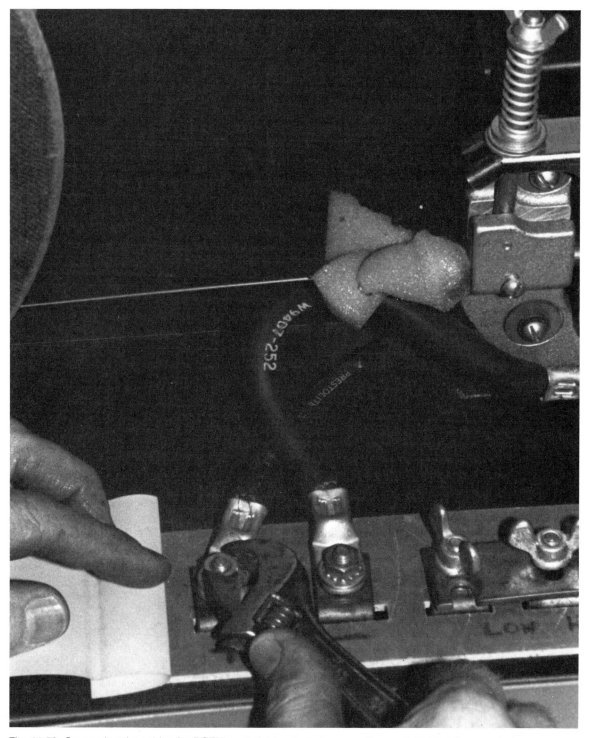

Fig. 11-59. Connecting the cables for DCEN or straight polarity for use with a self-shielded flux-cored wire.

Table 11-4. Wire Processes, Voltages, Amps, Wire Speed, Stickout.

WIRE PROCESS	ARC VOLTAGE	AMPS/WIRE FEED SPEED IN INCHES PER MINUTE	(ELECTRICAL) STICKOUT
short circuit transfer*	17-21	55 to 100 amps/ 130 inches per min	3/16 to 5/16 inch
globular transfer*	23-28	135 to 150 amps/ 290 inches per min	1/4 to 1/2 inch
spray transfer*	26-30	180 to 195 amps/ 320 inches per min	1/4 to 1/2 inch
self shielded flux core**	18-21	50 to 150 amps/ 60 inches per min	1/4 to 1-1/2 inch
gas shielded flux core***	23-26	90 to 180 amps/ 120 inches per min	3/8 to 5/8 inch

     * ER70 S-3, .035-inch diameter electrode
   ** E71 T-11, .045-inch diameter electrode
 *** E71 T-1, .035-inch diameter electrode

Make your settings and proceed with test welds as follows:

• Set the voltage to the middle of the maximum range—about 20 volts.

• Set the wire feed speed to approximately 130 inches per minute as explained earlier in this chapter.

• Without pulling the trigger, practice moving the gun smoothly across a 16-gauge scrap plate. Use a *lag* or drag gun angle. You need to move the gun quite rapidly, so give yourself every advantage and drag the nozzle on the plate (Fig. 11-61). Avoid sharp turns in the gun cable. Use "tip dip" or anti-spatter compound and frequently remove spatter from the tip and nozzle.

• Test your settings by attempting to run a bead 2 inches long on the 16-gauge scrap plate. Start with a long wire stickout of 1 inch or more and *gradually* move the gun inward to the normal 1/2-inch stickout—this gives you time to react should the arc back up the wire. If the arc burns back, lower the voltage slightly and try it again. If the wire *stubs* or pushes the gun away from the plate, increase the voltage slightly and try it again.

• Evaluate the bead. It should look like the example in Fig. 11-27. Also examine the heat effect on the backside of the plate (Fig. 11-62). Readjust wire speed and voltage if needed. The pale green, glasslike silicate slag must be chipped off before more passes are made. Welding over slag forms inclusions.

Once you can run a decent bead, experiment with the shielding gas turned off. This shows you what it is like to run out of gas. Also try to make some small deposits on the plate about the same size as tack welds. Do this by increasing the voltage or your short circuit tacks will probably look more like wads than welds (Fig. 11-63). Doing all your tacking at once makes the trip to the voltage control less of a chore. Just turn the voltage back down when you want to resume continuous welding.

Short circuit transfer is the low heat input GMAW variation. The arc voltage must be set low or the weld will transform into the globular transfer mode. Short circuit transfer makes an unmistakable rapid snapping. This comes from the series of short circuits, each accompanied by a pinching off of the electrode like miniaturized lightning and thunder about 100 times per second. If you have doubts about what short circuit transfer is like, adjust the voltage and wire feed speed alternately to the lowest settings that still maintain a continuous arc—and you have it.

Fillet Welds

Tack several strips approximately 2- × -6-inch ×

14-gauge together into open corner joints. With the tacks on the inside, the outside of the joint will be unimpaired for smooth welding. Such joints are ideal for welding. If you tack weld the outside of the joint, remember to remove the silicate slag.

Locate these corner joints to permit welding in a vertical downhill position. Adjust the nozzle flush with the tip and you can slide the nozzle along the V groove of the joint (Fig. 11-64). This avoids short-

Fig. 11-60. GMAW, including short circuit transfer, uses DCEP or reverse polarity.

Fig. 11-61. Dragging the nozzle lightly on the plate helps make a uniform weld deposit.

ing out the tip on the work. Try to maintain at least 1/8-inch stickout—less than this allows spatter to bridge between the tip and the plate, causing the wire to stick to the tip and burnback. Varying the gun travel angle varies the stickout. Remove spatter from the gun after every weld and cut off the melted wire ends.

Alter voltage both slightly higher and lower to see what effect this has. You should find that higher voltage produces a "wetter" and smoother bead. However, at a certain increased voltage the arc length becomes too long to permit short circuit transfer. Also, burn-thru is likely. Make a note of

the optimum voltage and wire feed speed settings. With these experiments, remember to change only one variable at a time. Compare your results to Fig. 11-65.

Practice welding lap joints in much the same manner as the outside corner joint. With T joints, however, you slide the nozzle back to reach into the root of the joint without excessive stickout (Fig. 11-66). If you don't reduce stickout, the bead will be too convex. Also see what happens when you *push* the gun along the T joint. Pushing the gun produces a flatter bead than you will obtain with a drag travel angle.

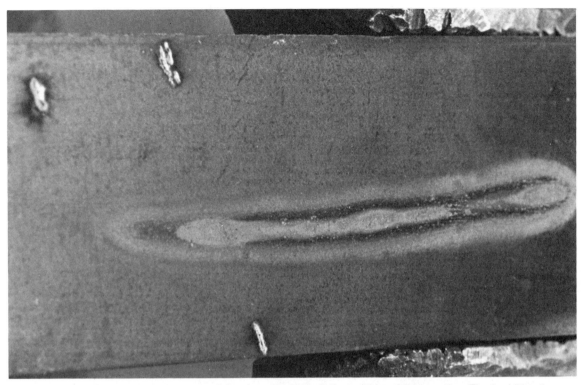

Fig. 11-62. The heat effect on the backside of the plate should be distinct and free from burn-thru. The 3 additional scars were caused from arcing between the plate and a rough table top.

Fig. 11-63. The properly made tack at the left is small and shows good fusion. The tack at the right not only looks like a wad of Juicy Fruit but also has about the same holding strength.

Fig. 11-64. Welding an outside corner joint is simplified by dragging the nozzle in the groove formed by the joint.

Butt Joints/Poor Fitup

Short circuit transfer can bridge moderately wide gaps. This is done by combining two techniques:

- Increase the stickout a small amount.
- Weave the gun slightly with a brief pause at the sides of the joint.

Try tacking up a butt joint as was shown in Fig. 8-4. Before welding this joint, you just might want to practice welding with a slight weave (no more than 1/4-inch wide) on a flat plate.

Remember that short circuit transfer is a "cold" process and is therefore subject to producing cold laps (Fig. 11-67). Always test your welds

with practice plates. Also, with no flux, the base plate must be clean. This means not only free of dirt and oil, but also paint, plating, and even mill scale. If you must weld with short circuit transfer on slightly contaminated surfaces do the following:

- Remove loose surface contaminants.
- Use E(R)70 S-2 wire (rather than S-3).
- Increase gas flow rate to 35 cfh.
- Increase voltage and wire feed speed.
- Take extra care to protect yourself from fumes.

WELDING WITH SOLID WIRE AND GLOBULAR TRANSFER

Load the wire feeder and set tensions as explained

Fig. 11-65. The corner fillet weld should just fill the joint.

Fig. 11-66. This tip is extended to reach the root of the **T** joint to avoid using excessive wire (electrical) stickout.

Fig. 11-67. Under impact, a cold-lapped deposit will simply lift away from the surface of the plate.

earlier. Check all electrical connections. Use DCEP (reverse polarity) with the ground cable connected to the negative (–) terminal of the power source.

Connect and adjust a supply of shielding gas as also explained earlier. Use a flow rate of at least 35 cubic feet per hour and, if using carbon dioxide, make some arrangement to prevent regulator freezeup—a mixed gas avoids the problem. Tighten all connections and see that the gas flow to the nozzle is uninterrupted.

Globular transfer requires a wire stickout of 3/8 of an inch or more. (The nozzle is held about 1/4 inch off the plate.) This means adjusting the nozzle to *recess* the tip.

Keep a pair of wire cutting pliers handy. The arc is easier to strike (and restrike) with the melted electrode end cut off. (Be careful, the cut-off piece of wire is sharp and can fly some distance.) Long-nose cutting pliers can also serve as a tool to remove accumulated spatter from the inside of the nozzle.

Living with Spatter

The hot and violent globular transfer arc produces a lot of spatter (Fig. 11-68). Use anti-spatter compound on the tip and nozzle. There are several other ways to protect the work from spatter. Aerosol anti-spatter compounds are suited for both gun and work, but with a large area this is expensive. Paint-on protection compounds are far more economical. These are available as a plaster-like wash and as a clear liquid.

Part of larger surfaces can also be protected from spatter with plywood, sheet metal, or even fiberglass blankets. If you are making a quick weld near a thread, it might be possible to use ordinary paper masking tape. By the time it burns up, the welding is finished. Tubing slipped over an outside thread also makes an effective spatter shield.

Making Test Welds

Be sure to use at least 1/4-inch-thick plate—heavy enough to handle the high globular heat input. The flat position is best for welding. Globular welding in the horizontal position is possible, but the flat is so much easier. Make your settings and proceed with test welds as follows:

● Set the voltage to the high range setting of about 26 volts.
● Set the wire feed speed to about 280 inches per minute as explained earlier.

● Test your settings with a short bead on a test plate. Using a drag angle and a long stickout of 1 inch or more, *gradually* move the gun inward to the normal nozzle-to-work distance of 1/4 inch. This provides time to react if the arc burns back up the wire. Avoid sharp turns to the gun cable. If the arc does burn back, lower the voltage slightly and try it again.

● If the wire *stubs* or pushes the gun away from the plate, increase the voltage slightly and try it again.

● Evaluate the bead and the heat effect upon the plate. Readjust the wire feed speed and voltage if required until the bead goes down uniformly. Do not weld over the glasslike slag. This is silicon and iron oxide and does not float out of later beads. It is removed by light chipping; wear eye protection.

● Run several more beads with the nozzle slid both in and out to vary stickout. Use the same electrode angle, travel angle, travel speed, and nozzle-to-work distance for each bead.

Once you can run a decent bead on flat plate, try welding with the gas off in order to see the symptoms of an inadequate gas shield. Restore the gas, and using 1/4-inch plate or heavier, tack a T joint together. Place the joint in the flat position and weld a short distance on one side only. Break the joint apart. As it fractures, the weld should tear rather than break cleanly. It should penetrate well into the base metal. If necessary, increase the heat input with more voltage and wire feed speed.

Globular transfer tends to produce undercutting at the toes of the weld. Correct this by slightly oscillating the gun, allowing some filler to spill back into the undercut areas. Avoid wide weaves and travel quickly enough to avoid overlapping.

Welding a T joint horizontally with solid wire globular transfer is a minor challenge. It is difficult

Fig. 11-68. A globular deposit is seldom pretty but usually effective.

to deposit a bead of any size without both under-cutting and overlapping. Slightly oscillating the gun helps, along with:

- Lower voltage and wire feed speed
- Using a push rather than a drag travel angle
- Using multiple passes
- Faster travel
- Allowing the plate to partially cool between passes

The commercial use of solid wire globular transfer has declined. Yet, it is there in an instant if you are already using solid wire for short circuit transfer. Many former applications are now done with small flux-cored wires.

WELDING WITH SOLID WIRE AND SPRAY TRANSFER

Load the wire feeder and set tensions as explained earlier in this chapter. Check all electrical connections. Use DCEP (reverse polarity) with the ground connected to the negative (–) post of the power source.

Connect and adjust a supply of shielding gas as also explained earlier. You have a choice of shielding gas mixtures with spray transfer on mild steel, but the argon content must be at least 85 percent. This means that the popular C-25 gas mixture (25 percent CO_2 + 75 percent argon) will not allow spray transfer. The 98-2 blend (98 percent argon + 2 percent oxygen) or 95-5 blend (95 percent argon + 5 percent either oxygen or CO_2 are popular. Tighten all connections and see that the gas flow to the nozzle is uninterrupted. Use at least 35 cfh.

Spray transfer, like globular transfer, requires a wire stickout of 3/8 inch or more. The nozzle is held about 1/4 inch from the plate. You can slide the nozzle to recess the tip.

Keep wire cutters handy. The arc striking is easier if the melted end of the electrode is cut off. Be careful, the cut-off piece of wire is sharp and can fly some distance. Long-nose cutting pliers have the advantage of doubling as a tool to remove accumulated spatter from the inside of the nozzle.

Brighter and Hotter

Spray transfer naturally forms a longer arc. This, along with the high amperage (current density) needed for transforming globular into spray transfer, produces a lot of arc flash. You might want to use a filter lens as dark as shade 12. Ask your welding equipment supplier about different lenses. Additional amperage and arc length also radiate more heat energy than the other GMAW processes; your hands warm up fast. Aluminized gloves and glove shields reflect this energy away from your hands.

An argon-bearing gas also makes the welding gun run hotter. Guns have a continuous amperage rating based on carbon dioxide shielding. Carbon dioxide cools the gun but argon and argon mixtures do not. A safe rule of thumb is to halve the amperage rating when using argon. This does not mean, however, that you must buy a larger, more expensive gun. If you limit the arc on time, you can overload a gun. The rating is also based on a 10-minute interval. For example: your 200-amp gun, with argon now derated to 100 amps for non-stop welding, would still be safe blasting away with 200 amps for five solid minutes. Five continuous minutes of wire welding in a home shop? Not likely unless you rig up some special automated operation.

Making Test Welds

Be sure to use at least 1/4-inch-thick plate—heavy enough to handle the high heat of spray transfer. Also position work for flat position welding. Spray transfer welding of steel in the horizontal position is possible but difficult.

Make your settings and proceed with test welds as follows:

- Set the voltage to the high range—about 26 volts.
- Set the wire feed speed to about 320 inches per minute as explained earlier.
- Check your settings with a bead on a test plate. A drag angle is used only for this initial test; spray transfer is normally done with a leading or push angle. Start with a long wire stickout of 1 inch or more and *gradually* move the gun inward to the normal nozzle-to-work distance of 1/4 inch. This

gives you time to react, should the arc burn back up the wire. Avoid sharp turns to the gun cable. If the arc burns back, lower the voltage slightly and try it again. If the wire *stubs* or pushes the gun away from the plate, increase the voltage slightly and try it again.

● Evaluate the bead and the heat effect upon the plate. Readjust the wire speed and voltage as needed. Do not weld over the glasslike slag. This slag does not float out of any later beads. It is removed by light chipping; wear eye protection.

● Run several more beads with the nozzle slid both in and out to vary the stickout. Use the same electrode angle, travel angle, travel speed, and nozzle-to-plate distance for each bead.

Once you can run a decent bead on the flat plate, try welding with the gas off to see the symptoms of an inadequate gas shield. Restore the gas and, using steel at least 1/4-inch-thick, tack up a T joint. Position the joint flat and weld a short distance on one side only. Break the joint apart. The weld should tear rather than break cleanly. Figure 11-69 shows how the weld should remain attached to the base metal. If necessary, increase the heat input with more voltage and wire feed speed.

Even more than globular, spray transfer produces undercut at the weld toes with the extreme arc length and high heat. If you have this problem, oscillate the gun slightly—pushing more filler into the undercut areas. Avoid wide weaves and travel fast enough to avoid overlapping.

As mentioned, spray arc welding a T joint in the horizontal position is difficult without both undercutting and overlapping. A slight, almost "nervous," jiggle of the gun helps to fill the undercut. The key is a super-fast travel speed that is often impractical for hand-held use. These other techniques might prove helpful with spray transfer in the horizontal position:

● Lower the voltage and wire feed speed.
● Use multiple passes.
● Change stickout distance.
● Allow the plate to partially cool between passes.

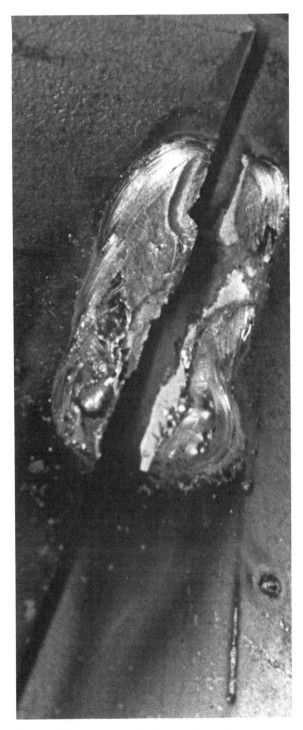

Fig. 11-69. This test break shows sufficient fusion to the member plates of the T joint.

Even though the spray process has substantial limitations, it does produce an attractive bead. Figure 11-70 shows a fillet. Like solid wire globular transfer, spray use has been displaced by small flux-cored wires. This trend will likely continue.

WELDING WITH
GAS-SHIELDED FLUX-CORED WIRE

Load the wire feeder with E71T-1 .035-inch electrode, and set tensions as explained earlier. Check all electrical connections. Use DCEP (reverse polarity) with the ground cable connected to the negative ($-$) post of the power source.

Connect and adjust a supply of shielding gas as also explained earlier. Use a flow rate of at least 35 cfh. The best welds are made with C-25 mixed gas which makes all-position welding easier with a faster freeze. If you use straight CO_2, make some provision to prevent regulator freeze-up. A mixed gas avoids the problem. Tighten all gas connections and see that the gas flow to the nozzle is uninterrupted.

The nozzle is adjusted to recess the tip 1/8 to 3/16 inch and is held about 1/4 inch away from the plate. This gives the proper stickout. Spatter is not much of a problem with gas-shielded flux-cored

wire, but anti-spatter compound on the nozzle and contact tip is advisable.

Keep wire cutters handy. The arc striking is easier if the melted end of the electrode is cut off. Be careful, the cut-off piece of wire is sharp and can fly some distance. Long-nose cutting pliers have the advantage of doubling as a tool to remove accumulated spatter from the inside of the nozzle.

Making Test Welds

Use plate at least 1/4-inch thick for weld testing your equipment settings. Once you have developed a feel for the process, run your practice beads on material the same thickness as the actual workpiece.

Make your settings and proceed with test welds as follows:

● Set the voltage to about 24 volts. These wires are voltage-sensitive, so you might have to readjust the voltage.

● Set the wire feed speed to about 150 inches per minute as explained earlier.

● Test your settings with a 2-inch-long bead on the test plate. Use a drag angle. Start with a long stickout of 1 inch or more and *gradually* move the

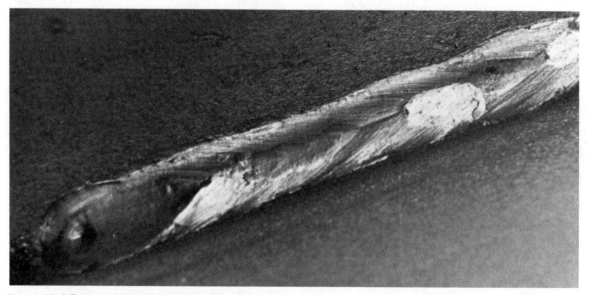

Fig. 11-70. A **T** joint welded with spray transfer. The line running along the face of the bead is formed as the weld deposit contracts, similar to the formation of a crater.

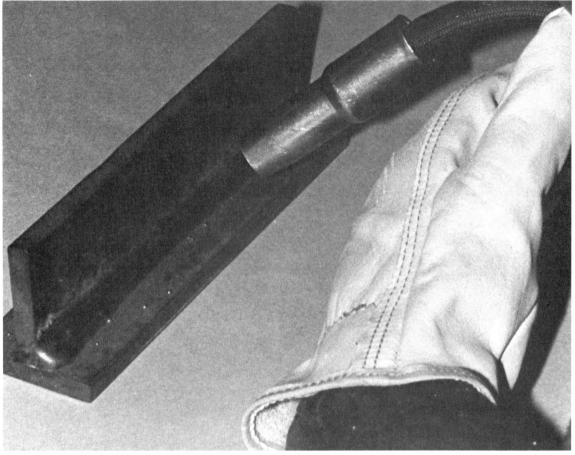

Fig. 11-71. A drag travel angle is normally used with the gas-shielded flux-cored wires, but it is alright to push it to obtain a flatter bead profile.

gun to the normal nozzle-to-work distance of approximately 1/4 inch. This artificially long stickout provides time to react, should the arc burn back up the wire. Avoid sharp turns to the gun cable. If the arc burns back, lower the voltage slightly and try it again. If the wire *stubs* or pushes the gun away from the plate, increase the voltage slightly and try it again.

● Remove the slag and wire brush the bead. Evaluate the bead and its heat effect upon the plate. Readjust the wire feed speed and voltage as required.

● Run several more beads with both pushing and dragging travel angles to learn the effects of changing:

—Stickout distance
—Electrode angle
—Travel angle
—Travel speed
—Nozzle-to-work distance

Once you can run reasonably decent beads on the flat plate, weld with the gas off so that you will recognize the rather subtle symptoms of a gas shortage. Restore the gas and, using 3/16-plate or heavier, tack up a T joint.

Place the T joint in the horizontal position and weld a short distance on one side only. Break the joint apart. As it fractures, the weld should tear away rather than break cleanly as was shown in

245

Table 11-5. Troubleshooting of Wire Feed Welding Systems.

PROBLEM	PROBABLE CAUSE	REMEDY
1. no arc, irregular arc, or arc is difficult to start	a. poor ground or other connection b. end of electrode burned c. defective trigger switch	a. clean ground clamp and other connections feel cables and connections for hot spots b. cut off end of wire for each restart c. replace trigger switch
2. wire burns back or surges	a. improper wire tension(s) b. liner dirty or damaged c. tip contaminated or worn out d. excessive arc voltage e. wire feed speed too low	a. readjust tension(s) b. clean or replace liner c. clean or replace tip d. lower voltage setting e. increase wire feed speed
3. wire stubs	a. wire feed speed too high b. arc voltage too low	a. lower wire feed speed b. increase voltage setting
4. lack of fusion	a. insufficient heat input b. contaminated plate c. plate too cold	a. increase wire feed speed and voltage setting, use slower travel speed b. clean plate before welding c. use preheat
5. excessive spatter	a. arc voltage too high b. incorrect polarity c. inadequate shielding gas d. excessive wire speed	a. lower voltage setting b. check polarity, switch cables if wrong c. restore flow of shielding gas d. reduce wire speed until arc is smoother
6. bead too high	a. excessive stickout b. arc voltage too low c. travel speed too slow d. excessive travel angle	a. reduce stickout b. increase voltage setting c. increase travel speed d. reduce travel angle
7. bead too wide	a. arc voltage too high b. travel speed too slow	a. lower voltage setting b. increase travel speed

PROBLEM	PROBABLE CAUSE	REMEDY
8. undercut	a. arc voltage too high	a. lower voltage setting
	b. travel speed too fast	b. reduce travel speed, oscillate
	c. gun angle too high	c. lower gun angle
9. overlap	a. travel speed too slow	a. increase travel speed
	b. gun angle too low	b. raise gun angle
10. porosity	a. contaminated plate	a. use only clean plate
	b. insufficient gas shield	b. check gas system from cylinder to nozzle
	c. arc voltage too high	c. lower voltage setting
	d. defective wire electrode	d. try another spool, discard or exchange defective wire
11. cracking	a. lack of shielding gas	a. check gas system, increase flow rate, use a curtain
	b. wire and base plate incompatibility	b. test electrode compatibility on a known steel
	c. tacks too small	c. increase size of tack welds
	d. gap too wide	d. run bead(s) along edges of joint to close up gap
	e. craters not filled	e. fill craters by moving gun in tight circular path

Fig. 11-69. The weld should penetrate well into the base metal and remain attached. If necessary, increase the heat input with more voltage and wire feed speed.

Weld another horizontal T joint, using a drag travel angle on one side of the joint and a push angle on the other. Allow the plate to cool between these welds. There should be a distinct difference in the bead profiles. With a drag angle there is also slightly better penetration than with a push angle.

Other Positions

It is easy to weld a T joint in the vertical or overhead positions with this small diameter flux-cored wire. If at all possible, use C-25 shielding gas. Again, the smoothness of your welding largely depends on how you position yourself before welding. Use one hand as a guide by resting it on the plate or a surface in common with the plate. The electrode angle is also kept approximately square to the plate. For overhead work, always use a drag travel angle great enough to permit spatter to fly over the nozzle instead of filling it up. Turn the wire feed speed up slightly to compensate for the additional mechanical drag imposed on the wire. Be sure to wear ear plugs and safety glasses when welding overhead.

Light Metal

The .035-inch diameter gas-shielded flux-cored wire is claimed to be suitable for material as thin as 18-gauge exhaust tubing. This is reasonable only if a high enough travel speed can be maintained. It is, however, quite possible to weld 14-gauge lap, corner, and T joints with this wire. If possible, position thin plates to permit downhill travel. For metal lighter than 14 or 16 gauge, you are nearly always better off using either the self-shielded flux-cored wires or short circuit transfer.

The quality of gas-shielded flux-cored wire welds is hard to beat; they also look great. Multiple passes are no problem and slagging is easy. Gas-shielded flux-cored arc welding is an excellent process for the home welder.

Refer to Table 11-4 for basic information about wire feed processes and the machinery settings used.

WIRE FEED WELDING TROUBLESHOOTING

No weld is ever perfect but the craftsman must always be sure that each weld is of a quality sufficient for its intended purpose. It has been suggested that you test your weld settings on practice joints before welding on an actual project. Continue this practice; it eliminates problems you might otherwise experience with a new batch of steel or a new spool of wire.

There are 11 main problem areas with home type wire feed welding. These are seldom distinct. Multiple problems often result from a single cause. Table 11-5 lists these problems, probable causes, and remedies.

Chapter 12

A Plan of Action

T HE MAIN STEPS TO CONSTRUCT A WELDED
project follow a sequence or formula similar
to that of woodworking:

- Get the idea.
- Formulate a step-by-step plan.
- Make a materials list.
- Gather the materials.
- Make your layout.
- Cut and form parts.
- Fit and assemble.
- Weld it.
- Finish as required.

Take a few minutes to investigate these familiar
steps. In most cases you follow them anyway,
whether or not you are aware of it. You can sometimes abbreviate a step or two, but you cannot avoid
them entirely. With the best plan, you will have little trouble generating material for your scrap bin.
With no plan at all, scrap may be what your shop
turns out most.

GET THE IDEA

Most project ideas seem the result of partly dream-
ing (*remembering*, some argue) of a way to solve a
problem. Other ideas are "discovered" solutions already working that we see and record well enough
to duplicate. You really should not hesitate to bor-
row someone else's good idea; after all, is not imitation the highest form of flattery? Things could get
a bit sticky only if you try to competitively market
your close imitation of a manufactured item. Chapter 13 says more about project ideas.

FORMULATE A STEP-BY-STEP PLAN

Where does idea development stop and plan for-
mulation begin? While looking over welded objects,
you will also be figuring out how to build them. The
plan, (blueprint, shop drawing, hard copy) is a standardized picture of the object (Fig. 12-1). It is handy
for simple projects and essential for complex ones.
You will likely draw many of your own plans. Engineering (graph) paper makes it easy to keep lines
straight and corners square.

Blueprint reading can be a little tricky if you are unaware of the scheme behind the arrangement of views. It helps to visualize the outline and main features of an object from several viewpoints. The "glass-box" analogy is a good explanation. Figure 12-2 shows an object placed inside a glass box with hinged corners. Outlines of the object from different viewpoints are projected onto the transparent planes of the box. As the box is unfolded (Fig. 12-3), the views are arranged in the relative positions as they appear on a blueprint.

Early in your planning, figure out what operations will be needed to build a project. Your project can consist of one or many parts, but it is best to try to group operations. This saves setup time and ensures uniformity. For example, do all possible layout, drill all the holes, or flame cut all you can at one time. Sometimes, as with woodworking, several parts can be made together. By jotting down these operations, you will create your step-by-step plan. Once the steps are laid out in sequence, you might choose a better order or even discover some missing steps.

A workable plan saves repeat operations and means more enjoyable working sessions. You might need to suspend work on a project for a few days or so. With a clear plan, there is little lost motion when resuming work. As you put in more shop time, it gets easier to make plans and estimate time and materials. Have backup copies of plans in case your working copy catches fire. Plastic covers help them survive longer in a shop environment.

MAKE A MATERIALS LIST

To plan a project you also need a good idea of what materials are needed. Give each part a name and list metal shapes, sizes, and how many of each are required. This list of materials is often called the *bill of materials*.

Armed with the bill of materials, a steel supplier's databook or products and services list, and a calculator, you are ready to estimate what must be purchased or gleaned from your scrap supply. The databook has valuable information about metal shapes, weights, and actual dimensions.

GATHER THE MATERIALS

Gathering materials can involve a short walk to your storage rack or scrap bin. It could mean a drive to a local metal vendor. There are do's and don'ts for obtaining metal: Do have it delivered if you don't have a truck and are dealing with large or long items. Wear gloves when handling larger pieces of metal. Figure 12-4 shows the use of vise-grip locking pliers to avoid handling sharp edges.

To carry longer lengths of steel with your vehicle, use strapping or duct tape to prevent them from sliding apart. Secure loads against any fore-and-aft movement. You are responsible for preventing potentially lethal objects from falling onto the road. Never allow any part of a load to protrude from the side of the vehicle and use a red flag to mark any overhang.

Once the metal is at your shop, keep it dry. Leave protective coatings or primers intact until removal is required. The oil on mechanical tubing keeps it rust-free. If you will not finish the project within a very short time, leave the oil. It is more difficult to remove rust than oil. Use a sacrificial pair of gloves to handle oiled metal.

Steel stored directly on concrete collects moisture from the concrete and eventually rusts. Moisture is even worse over grass and bare earth. To store metal on one of these surfaces, first lay down some plastic sheeting as a moisture barrier. Also, place metal on some sort of spacer—wood blocks, for example—to make it easier to pick up. Avoid storing metal overhead in precarious locations. If you intend to keep a sizeable amount of metal, construct suitable racks.

MAKE YOUR LAYOUT

Layout refers to marking cut lines and other symbols on the metal. The shape and accuracy of a project is only as good as the layout. Making layouts can take longer than any other step in the construction of a welded project.

You need to identify areas of the material for parts and for leftovers (remnants). Mark the expendable side of a cut line with "XS" meaning "excess." If you store leftover metal, mark it

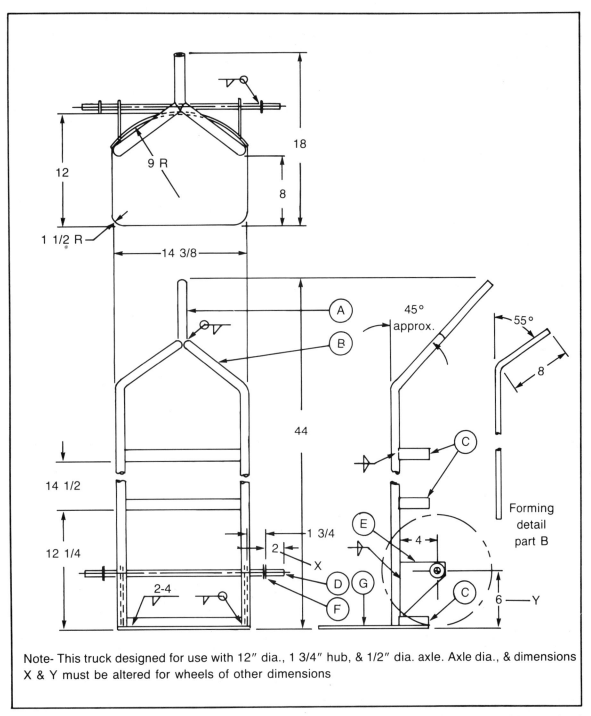

Note- This truck designed for use with 12″ dia., 1 3/4″ hub, & 1/2″ dia. axle. Axle dia., & dimensions X & Y must be altered for wheels of other dimensions

Fig. 12-1. A blueprint shows a project in standard two-dimensional views. Sometimes a perspective view of the completed item is included. (Courtesy of Hobart Brothers.)

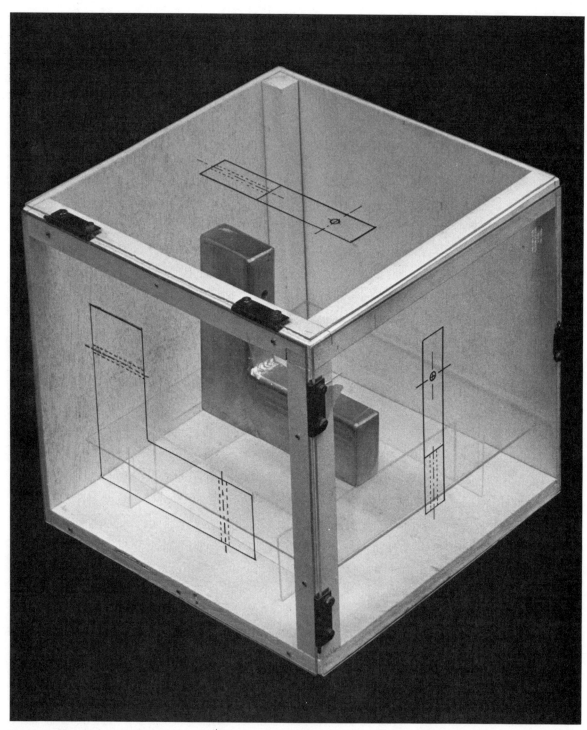

Fig. 12-2. The glass box with front, top, and side views projected onto three of the surfaces. (Courtesy of Bart Attebery Photographic Illustrator, Seattle.)

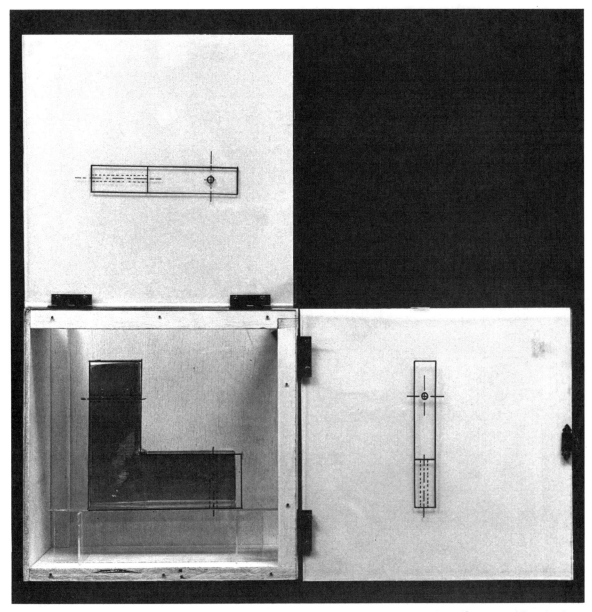

Fig. 12-3. When the glass box is opened, the views are arranged as they are on a blueprint. (Courtesy of Bart Attebery Photographic Illustrator, Seattle.)

"REM" (for remnant). Write dimensions on REMs so they are easy to spot in storage. Sometimes it helps to write the name of the project part ("front," "top," etc.) on the metal.

A center punch mark has uses besides locating starting points for drills. Use punch marks for at least the endpoints of lines and as often along a line as needed for clarity. Chapter 15 covers some specific layout techniques.

CUT AND FORM PARTS

With layout lines as guides, metal is cut and formed

Fig. 12-4. Avoid handling sharp edges by using two pairs of vise grips.

into shape(s). Outline shapes and large openings are mostly flame cut. Small holes are drilled and taps and dies make screw threads.

Forming is shaping metal by bending, rolling, heating, or some other operation. It is often easier, or necessary, to do most of the forming before assembling the parts of a project.

Self-sufficiency only goes so far. The home craftsman is sometimes better off "contracting out" certain cutting and forming services. Chapter 20 covers a few of these.

FIT AND ASSEMBLE

Various project parts, properly sized and shaped, are fit together into the final assembly. Fitup often requires clamps, spacers, bolts, extra hands and/or temporary braces. Hammers, wrenches, pry bars, and special-purpose tools are also used. Chapter 17 describes fitup and assembly. This is sometimes the best part of project construction because the item starts to resemble what it was intended to be.

Small tack welds (Fig. 12-5) temporarily hold

Fig. 12-5. This frame for a race car is properly assembled with small but hot tack welds.

parts of a project prior to the final welding. Tacks are also used to hold certain fitup devices. The placement and quality of tack welds can have a great effect on the appearance and quality of the final project. Tack welds are often best if made with hotter power source settings than those used for final welding.

WELD IT

The tack welded project is permanently joined with one or more of the processes described in Chapters 9 through 11. These final welds are often called *production* welds. With less exacting projects, welding can often be done with disregard for the effects of heat. For other projects, even slight welding distortion can hamper the utility of the project. It is important to see what can happen to shapes with the heat of welding.

Large welds cause the most distortion. Metal expands with heat and contracts as it cools. Allowances for heat should be planned before welding on your project. A classic example of expansion and contraction is seen in plate movement with a familiar welding operation. A fillet weld on just one side of an unrestrained T joint causes angular distortion. Hot weld metal is at its maximum expansion while being deposited. With cooling, it contracts and pulls the plates together (Fig. 12-6). Such distortion can easily ruin a project.

Welds on a similar T joint can balance the contraction one way with contraction in the other. The desired effect is to pull the plates into alignment with the later and greater weld.

With a *weld sequence* you do your welding by

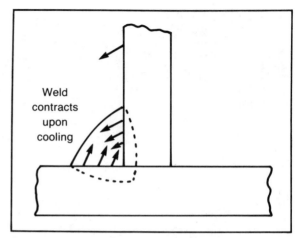

Fig. 12-6. Angular distortion after welding.

zones to avoid an accumulation of heat in any one area. Welding with a sequence can make the difference between a useful or useless project. Sequence and other distortion-control factors are discussed in Chapter 17.

FINISH AS REQUIRED

Wooden articles require sanding and painting to be useable. So too, welded products need some final touches. Even humble metal objects should not be left rough and unprotected. Besides looking bad, metal projects with sharp corners and edges, or even rough spatter, are hazardous. Unpainted or otherwise unprotected projects can rust and eventually fall apart.

Cleaning a welded project means removing those rough corners and edges, and slag and spatter. Chapter 19 covers finishing methods and tools.

Chapter 13
Not Very Creative?

C REATIVITY. MANY SAY THEY DON'T HAVE much of it. Would you? You might if you only compare yourself to famous artists. Yet, the fact that you are reading this suggests that you are creative. What does this mean? A rather cosmic definition says that creativity gives structure to disorder. Well then, what else (besides having fun) are you doing with that torch? By building and fixing you bring structure to disordered metal. That is creativity. You do not have to be a card-carrying artistic genius to be creative.

Creativity means recognizing good ideas from what others have already done. Look for imaginative solutions and save "rediscovering the wheel." Figures 13-1 through 13-5 show a few creative home welded solutions to various problems, and many more are found in Appendix A.

Because we tend towards strict practicality, special efforts to gain artistic talents don't seem to work. This does not mean that you are not unique as a craftsperson. Most enjoyed objects of art are not originals. They are copies made by adept craftspeople, each of whom puts something of him/herself into the copy. Mere mimicry? Not at all! Rather the exercise of creative skills.

In manufacturing, the term *one off* describes the fabrication of *one* item *off* of the blueprint. One-off production is rare, it jeopardizes profits. Such items tend to have a distinction seen as detrimental to the doctrine of industrial standardization—yet distinctiveness is prized by discerning consumers. Home shop projects are usually one-off or custom items.

Even familiar do-it-yourself projects like hand trucks (Fig. 13-6) have a unique character reflecting who made them and how. The hobbyist works in a relaxed atmosphere away from production pressures, where it is a virtue to take time and care.

This is not a futile attempt to explain the essence of art, nor is it a proposal to counterfeit metal sculptures. However, techniques used by others are often attainable, straightforward, and worthy of consideration—if not outright imitation. So, go for it! Find those creative details and adapt or adopt them (Figs. 13-7 and 13-8).

Fig. 13-1. This hand truck can be used with 2 wheels for maneuvering, and with the extra two wheels lowered, moving even very heavy loads is easy on the back.

RECORDING IDEAS

A good idea seen somewhere or other is too often forgotten. A picture *is* worth a thousand words, so always try to keep at least a scrap of paper and a pencil with you. An actual sketch book is probably better. (It helps to know how to sketch.) Cameras are great if you happen to have one at the right time. Often a project can be made from a few photographs.

More than once, this writer has photographically "borrowed" ideas from the sales floor of a store (Fig. 13-9). You might want to try this. First,

Fig. 13-2. A wheelbarrow is handy but requires constant lifting. Adding wheels from a motorcycle salvage makes the job easier.

Fig.13-3. A boat on a bug. Probably not too stable in a strong wind.

Fig. 13-4. This vise swings out for use, and a simple fabricated rack allows it to be secured when not in use.

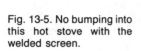

Fig. 13-5. No bumping into this hot stove with the welded screen.

Fig. 13-6. Just because you have seen one hand truck doesn't mean you've seen them all.

put the store personnel at ease about your motives. Tell the manager how you want a snapshot for an absent friend (probably true), or that you are gathering material for your neighborhood consumer research group. (Also true, neighbors do keep track of your projects.) Either story usually works. If a camera is not allowed, make sketches and get key measurements.

Good project ideas are found about anywhere and at anytime. So be ready with your sketchpad and/or camera and start your own idea collection or scrapbook.

Sketches

A sketch is a line drawing that is less formal than

Fig. 13-7. This fixture, made from surplus swivels, permits clamping small parts in just about any position.

a blueprint. Sketches give enough information to build many wood and metal projects. Most good drafting or technical drawing books explain techniques for freehand sketching. These show you how to make straight lines, estimate proportions, and draw circles and shapes. The basic approach is to first draw the outlines of the largest parts of the object represented. Use correct proportions, such as a rectangle twice as long as it is wide. Then fill in the larger areas with the smaller features; again use relative proportions. Sighting an object with a graduated object held at arm's length is very helpful for estimating proportions.

Photos

While a photograph gives useful information, it also can include distracting detail. Shadows provide depth but can also block out important facts. When you take a picture for a project, try to take them "straight on" so that (like sighting for a sketch) relative proportions can be determined.

Considerable accuracy is possible using dividers and a scale to estimate proportions off a photo (Fig. 13-10). A vernier caliper can be used for more precise measurements. Of course, besides proportions, you also must know the dimensions of some part(s) of the object. A sketch with a few key dimensions can be invaluable. Try to place a scale against a part of the photographed object (Fig. 13-11).

MAKING PLANS

We rarely have the luxury of having another project on hand to duplicate. Sets of worthwhile plans are sold through some magazines, but you will often simply have to rely on whatever records you can make.

Fig. 13-8. A most civilizing idea for a cylinder cart.

Fig. 13-9. An idea for a log rack borrowed from a store.

Fig. 13-10. Using dividers to get the relative sizes of parts off a photo. Note the working drawing with dimensions being developed on tracing paper to the right.

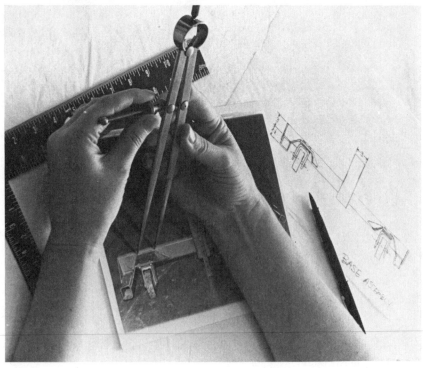

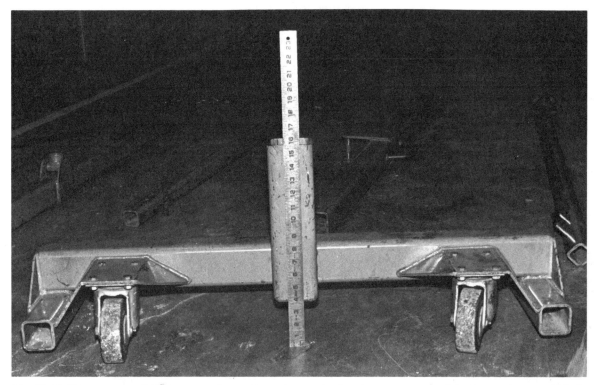

Fig. 13-11. Whenever possible place a scale or other reference in the photo as you take it for later reference. This is part of the base for a knock-down engine hoist.

Project ideas and help abound where craftsmen gather. Don't overlook the advantages of collective action. It might be difficult to find home welders, however. A way to locate these elusive people is to post a note at welding supply houses saying something like: "Home welder wishes to share ideas about project construction, etc." You will get a few calls. Also check your local welding school, regional fairs, street fairs, and craft shows.

For some readers these suggestions could well be unwanted advice. After all, many do-it-your-selfers seek recreational solitude and desire to avoid collective activities.

Magazines catering to the home craftsman often have articles involving welding. If you are a vehicle buff, check magazines specializing in your type of vehicle. Wood and solar heating as well as other alternative energy sources are areas with many home shop projects. Periodicals usually keep up with what is new. The material listed in Appendix C has additional inspiration for your creative talents.

Chapter 14

Tools

T HE HOME CRAFTSMAN IS USUALLY NO stranger to the advantages of having the proper tool for the job. Many times, lacking the best equipment, he or she has agonized through the job the hard way. Suffering might help build character, but it does little to build home welding projects. Fortunately, the tools needed for a reasonably efficient home weld shop are neither elaborate nor costly. Many of these are the same as those used for woodworking.

The tools (besides welding and flame cutting equipment) that hold the starring roles in your home shop productions fall into four somewhat overlapping categories. These are: hand tools, power tools, measuring tools, and marking tools.

HAND TOOLS

Basically, a hand tool is one that does its thing without having to be plugged into the wall or connected to compressed air. This includes a very large number of items. Essential hand tools for the home weld shop are now briefly discussed.

Hammers

A ball peen and at least one heavier type of hammer are basic items to "persuade" metal into desired shapes. Hammers are grouped by the design and weight of the head in ounces. A 12- or 16-ounce ball peen hammer is the handiest all-around size. For heavier work a 4-pound blacksmith's or engineer's hammer is effective. A 15-inch or shorter handle is normally used for these hammers. To make an even heavier impact, use a long-handled 10-pound sledge. A tinner's setting hammer is useful for working light-gauge metal. Special non-marking soft hammers are made with leather, plastic, lead, brass, and even hard rubber heads. A piece of wood placed between a hard hammer and the work can often achieve the same effect.

Replace broken or split hammer handles with high-quality hickory. Be sure that the handle is secure in the head. Also, try to avoid swinging a hammer in line with any other person.

A slide hammer (Fig. 14-1) permits impact pulling. Fitted with different types of ends, these are

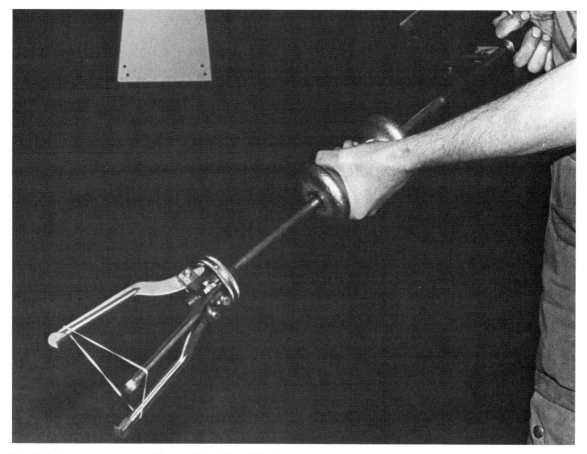

Fig. 14-1. Hammering outward is possible with a slide hammer.

used to remove dents in car bodies or to pull gears and pulleys from shafts. A slide hammer is a good welding project.

Chisels

Metal-working chisels are made from tool steel and must be hard enough to hold a point, and tough enough to absorb impacts. They are heat treated to obtain these qualities. The most familiar is the *cold chisel*, so called because it is intended for use only on cold material. (Yes, there are *hot chisels* to cut hot material.) *Diamond point* chisels and *gouge (round nose) chisels* are handy for removing smaller amounts of metal—like tack welds.

When you sharpen a chisel on a grinding wheel, dip it in water often enough to avoid overheating the point. If the point gets too hot, the hardness is drawn from it. If you can touch it, it is not too hot. Keep "mushrooms" (Fig. 14-2) ground from the driven end of chisels and punches. These can fly off at high speed and cause severe injury.

Wrenches

Wrenches are used with welding activities. Adjustable open end (crescent) wrenches can be used to twist metal and to turn most cable and hose connections. Union Carbide acetylene cylinders take a special cylinder valve wrench. A universal cylinder wrench fits all the common gas fittings.

A pipe wrench is handy for twisting pipes or other round shapes. Strap and chain wrenches wrap around pipe like an automotive oil filter wrench (Fig. 14-3). The strap wrench leaves no scratches or marks on the work.

Fig. 14-2. A mushroomed head is dangerous.

Box, (open) end, and combination wrenches are used to turn nuts and bolts. Socket wrenches speed up the assembly and disassembly of bolted structures. Internal hexagonal or "Allen" screws require hex key wrenches. These are usually sold in sets.

Pliers

Several types of pliers are very useful in the home weld shop. The familiar combination and the mul-tiple slip-joint pliers (Fig.14-4) permit safe handling of hot work. Locking pliers, or vise grips, are available with straight or curved jaws. The curve fits round sections, but reduces the clamping effectiveness for most other shapes. Figure 14-5 shows locking pliers.

There are many types of cutting pliers. The most common include diagonal pliers, side-cutting or lineman pliers, and long-nose pliers with cutters.

Fig. 14-3. A strap wrench for round shapes leaves no mark.

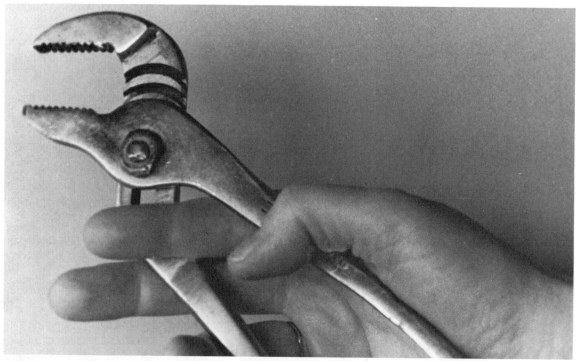

Fig. 14-4. Slip-joint or "channel locks" are also called water pump pliers.

Rather heavy cutting can be done with diagonal and side-cutting pliers. Long-nose cutters are handy for use with a wire feed welding system.

Clamps

Different styles of clamps are used in welding fabrication. The familiar C clamp (Fig. 14-6) has a swivel foot. Many welders attach a piece of angle bar to the foot for an improved grip on rounded shapes. Spring clamps (Fig. 14-7) and adjustable pipe clamps are used to align parts. Variations of the vise grip design (Fig. 14-8) permit welding with the clamp in place. Protect clamps with anti-spatter compound. A universal positioner consists of two locking pliers on a frame with ball joints to aid fitting together delicate or awkward smaller assemblies.

Figures 14-9 to 14-11 show several different types of welding clamps. These are high-quality, useful time savers.

Vises

A vise to hold metal is almost indispensable in the welding shop. Besides clamping, vises are also used for bending, forming, and pressing. Vises are identified by the width of the jaws, the maximum opening, and the depth of the throat. A 4- or 5-inch-wide jaw with an equal opening is about as small as you should go. As a rule, the bigger the better. However, as you might have discovered, a large, good-quality vise is expensive. When buying a vise, you do tend to get what you pay for. Some of the cheaper imported vises will just end up in your scrap bin.

A swivel base raises the cost and can be a detriment. The swiveling feature can work at the wrong times. A vise mounted on a fabricated stand instead of a bench allows work to be placed in it from many different angles (Figs. 14-12 and 14-13).

Hand Drill

In this time of power tools, the hand-powered or

270

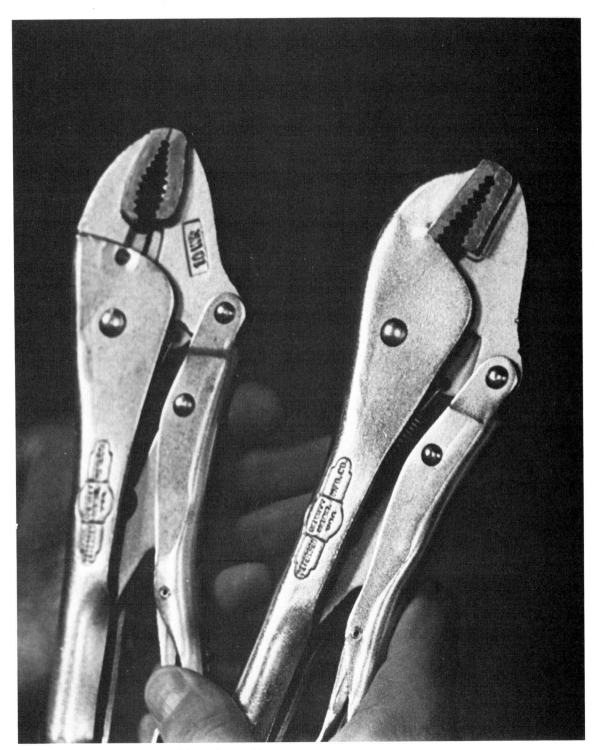
Fig. 14-5. Locking pliers with straight and curved jaws.

Fig. 14-6. **C** clamps are used just about everywhere.

"eggbeater" drill (Fig. 14-14) is still very useful for small jobs. Prices and quality of hand drills vary widely. The drill-holding capacity of the chuck can be as great as 3/8 inch, although the less expensive units are usually only 1/4 inch.

Prying Tools

Welding project fabrication often involves lifting or prying. Levers for this include the familiar gooseneck ripping bar (crow bar), smaller pry bars, and even screwdrivers (Fig. 14-15). By the way, screwdrivers should be twisted and not used for heavy-duty levering. Prying tools, made from alloy steel, are forged and heat treated for strength and durability. High temperatures destroy the usefulness of such tools.

Metal wedges can be used to pry sections of metal projects apart—or to push them together. Sometimes a pulling tool can also be used for prying.

Pulling Tools

A pulling tool is used to draw parts together. Examples of these include come-alongs and chain hoists for heavy jobs.

Files

As with woodworking, a great deal of metal finishing and shaping can be done using files. Files are hand-powered cutting tools. Chapter 16 describes files and their uses. Sometimes the finer-toothed cutters used in drill motors, drill presses, or die grinders are called rotary files. If made of high-speed steel, these can cut softer wood but should not be used for cutting metal. Carbide-tipped cutters can handle the high loads and temperatures of metal cutting.

POWER TOOLS

It is nearly impossible to work very long with welding fabrications without using some power tool. Most are electric, but there seems to be a promotion by tool producers for the home use of air tools. There is some merit in this. If pushed too hard and forced to turn slowly or stall, electric tools end up

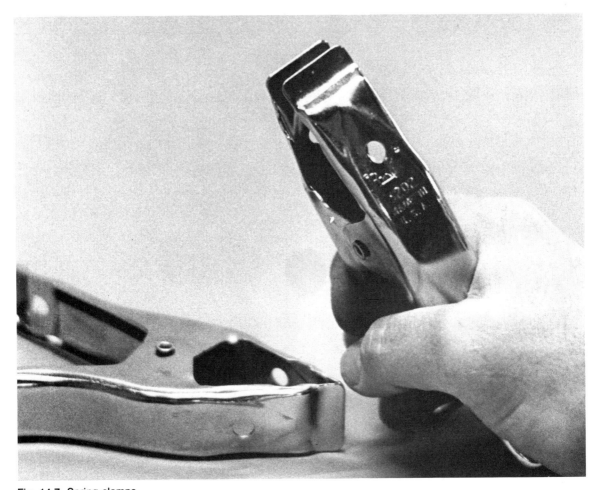

Fig. 14-7. Spring clamps.

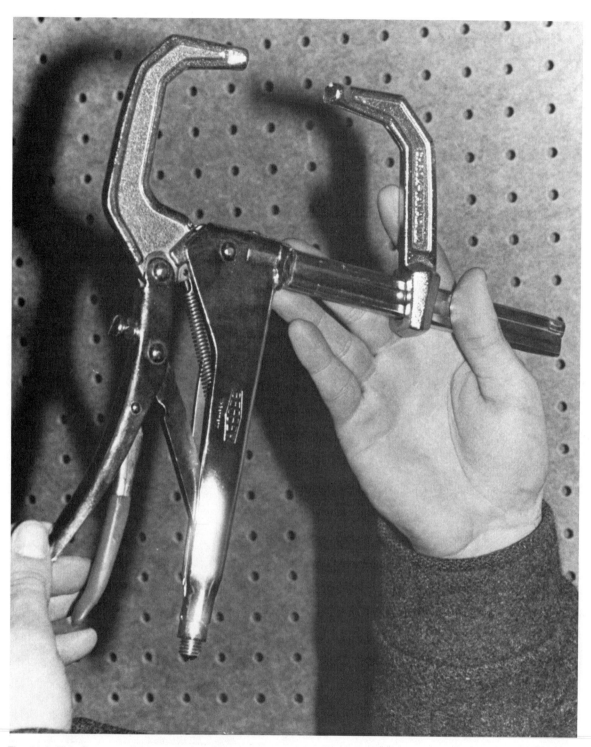

Fig. 14-8. This Bessey welding clamp permits rapid adjustments with the sliding arm.

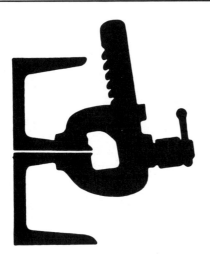

No obstruction from long screw.
Ideal for structural steel fabrication.

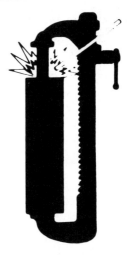

Unaffected by weld spatter.
Screw is shielded and out of the work area.

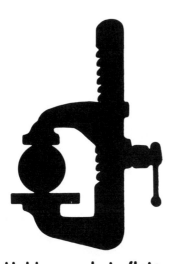

Holds rounds to flats.
Limited movement of moveable jaw pad and grooved face ensure positive grip on round objects.

Will operate in reverse.
Jaw may be reversed to give spreading action.

Fig. 14-9. Welding clamps can be used in many different configurations.

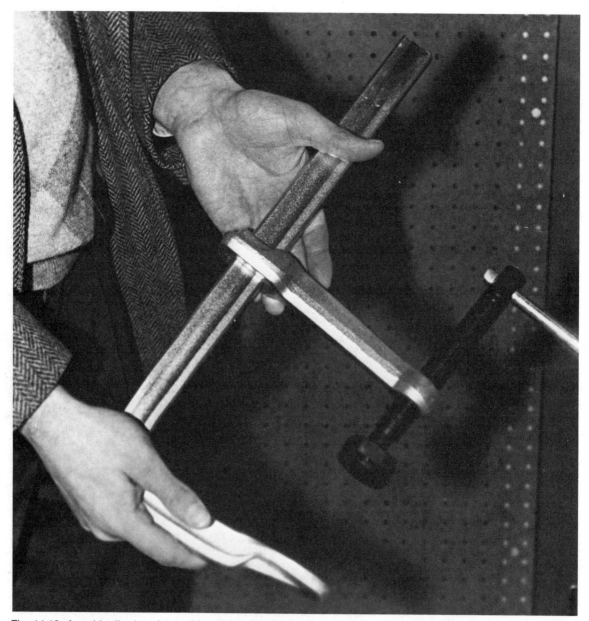

Fig. 14-10. A rapid adjusting clamp with a tightening screw.

with burned motor coils. Air tools, if kept lubricated, are difficult to damage. However, compressed air is not always available.

Disc Sander/Grinder

A disc grinder or sander/grinder, as was shown in Fig. 4-24, is very useful in constructing welded projects. These units are used for both weld preparation and finishing operations. With a standard 5/8-11 threaded shaft, the versatile disc grinder can be fitted with several different types of wheels for grinding, sanding, wire brushing, or even cutting. Most disc grinders used in home shops are electrically powered, although air-driven units are also available. Many users favor the electrical units, claiming they are faster and more powerful than air grinders.

A disc grinder is one tool on which you should splurge and go for the best quality. The amperage draw of a grinder is a fair estimate of its power.

A unit drawing 15 amps will enable you to do some serious grinding without worrying about stalling. The main bodies of grinders and all heavy-duty power tools should be constructed of metal rather than plastic.

A 7- or 9-inch disc grinder is the best all-around size. Be sure that the rated safe RPM of the disc

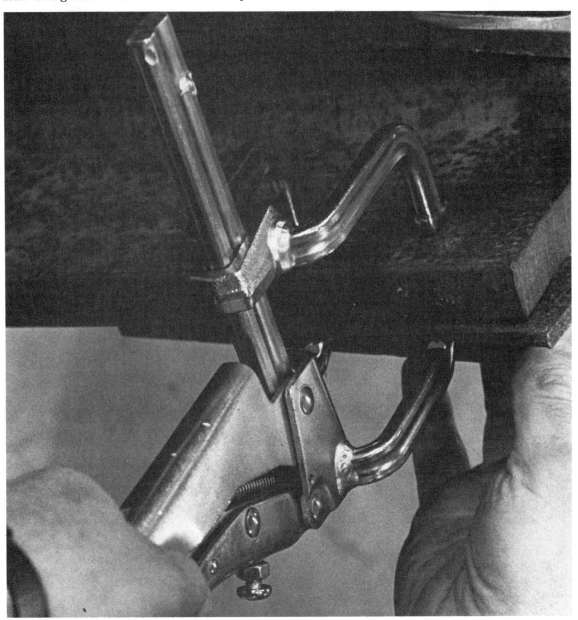

Fig. 14-11. Another hybrid design welding clamp.

Fig. 14-12. A vise mounted on a pedestal is easy to move around. This one is carried on a pipe screwed into pipe couplings at the top and bottom.

Fig. 14-13. This vise stand has levelling bolts.

Fig. 14-14. A small hand drill ("eggbeater") is very handy for small jobs.

(Fig. 14-16) is compatible with your grinder. For fine work in compact spaces the smaller 5-inch grinders pack an amazing amount of power. They also are less tiring to use.

When using a grinder, you must always wear eye protection. A face mask is also recommended. Ear plugs and gloves provide other important protection. When finished grinding, stop the coasting wheel on the work before lifting it; this prevents accidentally grinding something or someone. A grinder should be set down so the switch will not be accidentally pressed (Fig. 14-17).

Provide a "backstop" shield to keep chips and grit from going very far. This makes cleanup easier and prevents fires. You might amaze yourself at just how far grinding material and sparks travel by observing grinding in a dark area. The disc grinder as a finishing tool is covered in Chapter 19.

A metalworking grinder should not be fitted with a polishing pad in an attempt to shine the old family bus. A typical grinder travels at 5000 RPM or more while polishers travel at only 1750 RPM. The higher speed of a grinder produces enough frictional heat to easily burn the paint right off your car.

Die Grinder

Die grinders are high-speed tools that are fitted with special grinding wheels or carbide tipped burring (cutting) tools. The latter resemble woodworking router bits. Die grinders are useful for metal finishing and preparing weld joints. They are indispensable for certain operations.

Die grinders are small to reach tight areas. Air-powered units have a size advantage over more expensive electric units (Fig. 14-18). If kept lubricated with turbine oil, air units are almost indestructible. If stalled, and this can easily happen, an electric die grinder overheats and is permanently damaged.

Because die grinders rotate so fast—over 20,000 RPM—it is most important to know the rated speed of tools mounted on them. If you don't know, don't take a chance and use it anyway. Needless to say, it is dangerous to improvise or adapt grinding tools originally intended for some other use.

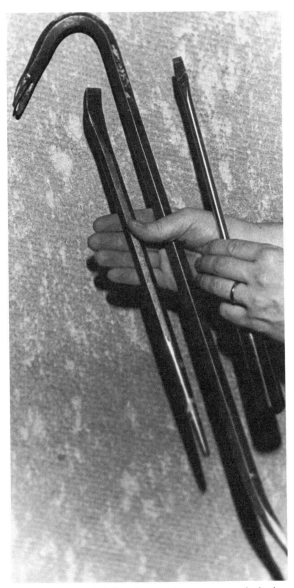

Fig. 14-15. (l.to r.) A locating pry bar, goose-neck ripping crow-bar, and a screwdriver all have their places in the home weld shop.

Bench or Pedestal Grinder

A bench or pedestal grinder (Fig. 14-19) is very useful for light- or medium-duty jobs. The difference between bench and pedestal grinders is what the unit is mounted on. If you make a stand, you turn a bench grinder into a pedestal grinder. Grinder size is determined by the size of wheel used. A 6- to 8-inch diameter wheel, 3/4 or 1 inch wide, is the most common for home use. For heavy grinding, an expensive industrial 10- or 12-inch unit is needed. Do your heavy grinding with a disc grinder.

Grinding wheels are pieces of sharp grit, often aluminum oxide, glued together. There is a large variety of wheels to fit almost any bench grinder. You are, to a certain extent, at the mercy of your tool dealer in selecting wheels. Most have a 5/8-inch hole used with a plastic bushing for the 1/2-inch grinder shaft, but that is where similarities might end. Wheels come in different *grains* like the roughness or smoothness of sandpaper. Like sandpaper, wheel grain is indicated by numbers according to the size of the bonded grit. A 36-grit wheel is rough grain for coarse metal removal, 60-grit leaves a fine finish.

Wheel *grade* indicates the hardness or softness of the wheel—how easily the grains are dislodged during grinding. Grades are differentiated by letter: M means medium, L is softer, N is harder, and so on. For most steel work, a letter grade N or O is used. For sharpening tools made of harder steel like drill bits, use a softer grade of wheel, such as K. Softer wheels shed their dull grains easily and therefore resist glazing when used on harder materials. Dip the work frequently in cool water when grinding heat-treated objects such as drills, chisels, and punches. This prevents "drawing" or tempering the hardness.

Use blotter paper between the mounting flanges and a new wheel to keep the flanges from cutting into the wheel. Avoid overtightening the mounting nut. If a new wheel were cracked it might fly apart, so stand to one side the first time it is started. A wheel cracked in service can also come apart. Replace and reset the tool rest.

If a new or used grinding wheel is out-of-round, grooved, or glazed, use a wheel dresser to restore a flat and sharp surface. With full face protection, use the dresser only as needed.

Do not attempt to grind any soft materials like aluminum or brass—these only clog the wheel. Such metals are either power sanded or ground with special wheels. Wear eye and face protection while grinding. Gloves are a good idea. The tool rest

Fig. 14-16. This 7-inch grinding disc is safe rated to 8500 RPM.

should be set on center and to within 1/16 inch from the wheel (Fig. 14-20). This avoids the danger of the ground object becoming jammed between the wheel and the tool rest. A jammed grinding wheel could shatter. Metal thinner than 11 gauge (1/8 inch) snags and becomes jammed even with the tool rest properly adjusted; so avoid grinding sheet metal. Hold small items with locking pliers. Avoid grinding on the side surfaces of the wheel, because such forces tend to break the wheel.

Bench grinders can be fitted with wire wheel brushes and buffing wheels. The tool rest and the safety shield should be removed to prevent the work from possibly getting caught. A range of coarseness is available with larger or smaller diameter wires. The finest brushes have wires as small as .006 inch and give a burnishing or polishing effect. Coarse brushes use thicker wires measuring .014 inch. Twisted wire brushes permit a very heavy brushing action for removing rust.

Buffing wheels are stitched together fabric discs. Buffing coarseness depends on the stitch pattern and the type of *buffing compound* applied to the wheel. This gives a range of buffing finishes from coarse scratching to very fine polishing.

When using wire wheels and buffing wheels, special care should be taken to use the lower part of the wheel. This prevents the work from being forced into the operator if it should get snagged by the wheel. Always point the work downward for the same reason. Wear leather gloves and a face mask. Wire wheels tend to shed their bristles from centrifugal force. This can easily cause puncture injuries. If you install a wire wheel on your grinder,

remove it when you are finished.

Portable Electric Drill Motors

Hand-held electric drill motors are used for most home shop drilling. A drill motor is called a "drill," so is the drilling tool (bit). A reasonably accurate job can be done with these units. The chuck capacity indicates the largest size of twist drill that will fit into it (Fig. 14-21). All drill motors are not adequate for turning the largest drills that fit into them. Flimsy "throw-away" drill motors have been promoted at incredibly low prices. Some even have plastic housings. If you wish to use these units, buy several because they simply do not hold up.

You should have two drill motors: a smaller unit for light jobs in confined spaces, and a heavy-duty unit. The smaller unit should have a 3/8-inch ca-

Fig. 14-17. An accident waiting to happen. Never set down a grinder with the switch against the bench.

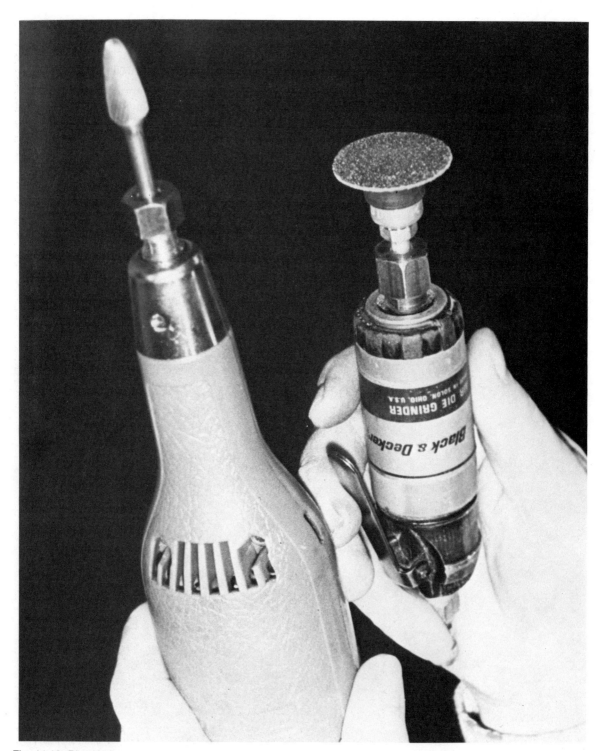

Fig. 14-18. Die grinders.

Fig. 14-19. A bench grinder. Notice the wise removal of the guard and tool rest from the wire brush side.

Fig. 14-20. Set the tool rest to within 1/16 inch from the wheel.

pacity chuck. Avoid drill motors with only 1/4-inch capacities. For drilling steel, a 3/8-inch capacity motor drawing 3 to 5 amps should be satisfactory. A heavy-duty unit with a 1/2- or 5/8-inch capacity that draws 5 to 8 amps enables you to tackle heavier drilling jobs.

Fig. 14-21. A drill chuck is measured according to the capacity of the largest drill it will accept. A 1/2-inch chuck is shown here.

Generally, slower turning drill motors are more powerful. Higher amperage units also tend to have more power. In use, avoid bearing down hard enough to cause much slowing of the motor. Listen to your drill—if it groans, lighten up the feed pressure and let it run free. On the other hand, if you are not making a chip, the drill bit just scuffs and overheats. This is why using an underpowered drill motor is a no-win situation.

More information on the use of drills can be found in Chapter 16. Appendix B contains a tap and drill chart to select the correct drill for a particular size of thread.

Drill Press

Some hobbyists use a drill press with their woodworking. These units work equally well for drilling metal. A drill press solves the problem of maintaining proper feed pressure and the problem of maintaining squareness to the plate. Besides simple drilling, other operations like angular drilling, polishing, and threading are also possible. A drill press vise is handy, but you can safely drill holes in metal by clamping the plate to the drill press table (Fig. 14-22). Use a block of wood under the plate to prevent damaging the table as the drill breaks through.

Follow all the safety instructions of your drill press manufacturer. Avoid loose clothing or unrestrained long hair, and wear eye protection. Always remove the chuck key from the chuck, never touch moving material, and be ready to hit the stop button if anything should go wrong.

MEASURING TOOLS

Most measuring tools used for woodworking work fine for metal fabrication. To measure medium distances, use a tape measure 10 or 12 feet long and 1/2 inch wide. Protect the tape from the heat and spatter of welding and cutting.

Right angles on small layouts are often located with a combination square. It is so called because the scale can be fitted with a choice of three heads for squaring, center finding, and making angular layouts (Fig. 14-23). An ordinary steel square has

Fig. 14-22. A drill press greatly eases the task of drilling holes.

Fig. 14-23. A combination square with (l. to r.) a center finding head, a protractor head, and the squaring head.

legs called the *body* and *tongue* measuring 24 and 16 inches respectively. These are handy for laying out right angles for slightly larger applications.

A pair of dividers is used to scribe circles or arcs, mark off equal dimensions, or to transfer dimensions. Many craftsmen, lacking a special large compass, devise a substitute by taping a pencil to one leg of a pair of dividers (Fig. 14-24). To obtain a span greater than is possible with dividers, pairs of *trammel points* are sometimes used. Many welders have devised their own trammels that clamp onto flat shapes (Fig. 14-25).

TOOLS FOR MARKING

The center punch, struck with a hammer, is essential for marking point locations on metal. A scratch awl, with a handle, is used to scratch lines and mark points. A scribe is just a pointed piece of hardened steel. If scratches are undesirable, dark felt-tip markers provide a background against which lighter scratch marks show up well. Dark blue lay-out die is also available. Silver pencils leave a highly visible line on darker surfaces. These are also available in flat sticks.

Soapstone is used to mark lines to be flame cut. It is available in flat sticks and round rods for use in a metal holder. Center punch marks spaced along a soapstone line provide extra help to follow the cut. A chalk line is made with a string coated with chalk dust. The string is pulled tight from the *chalk line reel* and snapped against the workpiece (Fig. 14-26).

A spirit level is sometimes useful for lining up project parts. The edge of a level is also a straight edge. A *plumb bob* is a weight on a string to help with vertical alignments. The Stanley "Chalk-O-matic" chalk line reel is also a plumb bob.

Angular measurements can be made with an *inclinometer* type of protractor (Fig. 14-27). These handy devices are easier to use and often more accurate than a standard flat protractor.

With time, you will discover better marking devices and improve the ones you have. A popular

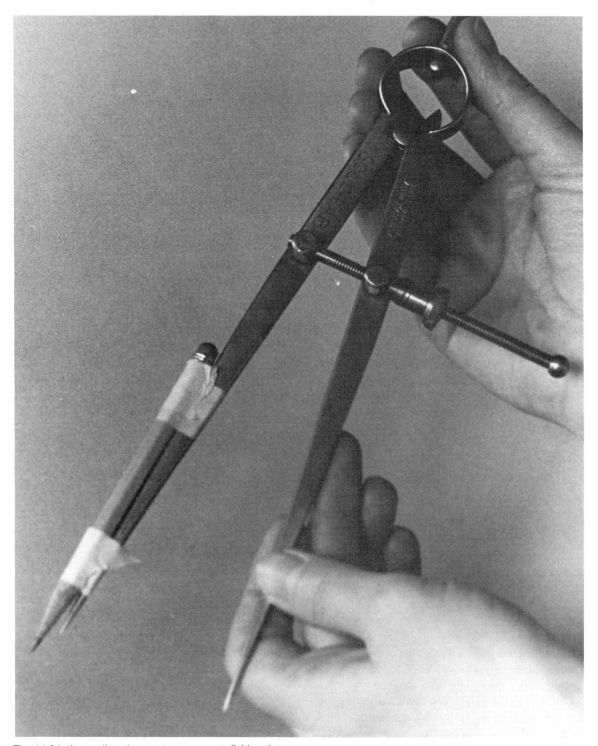

Fig. 14-24. A pencil and some tape convert dividers into a compass.

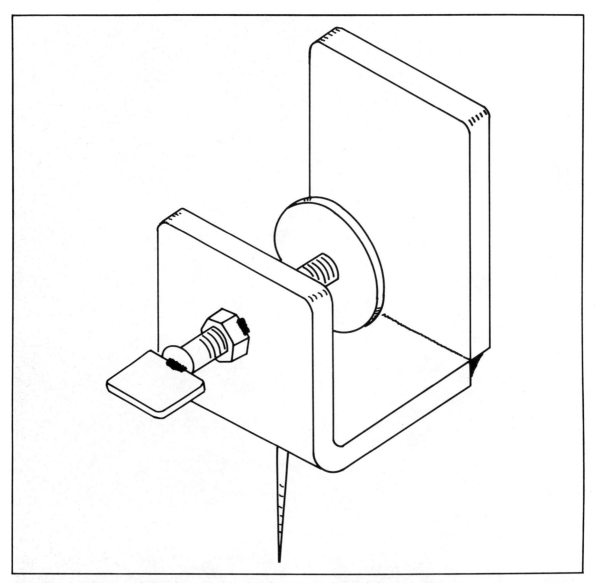

Fig. 14-25. One of a pair of clamping trammel points for swinging large circles and arcs.

homemade marking tool is an ever-sharp heavy duty scribe. It is made by silver brazing a sharp piece of tungsten carbide to the end of an old screwdriver. Tungsten carbide is used for the edges of cutting tools. (Most machine shops have a few scraps lying about.) If you make one of these scribes, use as little brazing heat as possible. Also, do not attempt to resharpen tungsten carbide with an ordinary grinding wheel.

WHERE TO GET TOOLS

The purpose of buying tools is to accomplish a given task with a certain ease and efficiency. The frustration of having an "imitation" tool (usually imported from afar) fall apart just when you need it is one of life's little trials. A little salt is sprinkled in the wound as you realize that, to save a buck, you bought a flimsy tool; and now you need to buy the quality item anyway.

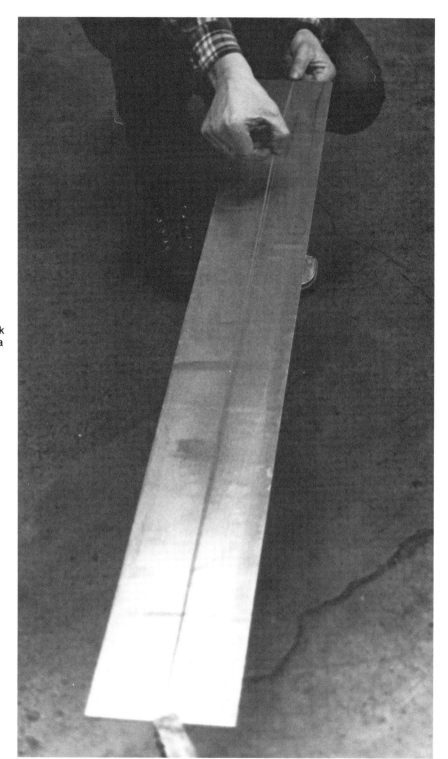

Fig. 14-26. Snapping a chalk line is a way of marking a long straight line.

Fig. 14-27. An inclinometer type of protractor will help ensure that the frame of this race car will be true. A magnetic base makes it useful for both levelling and placing parts at an angle.

On the other hand, if you put off building home welding projects until you can afford top-quality tools and equipment, a lot of time could pass. There could be a couple of solutions to this dilemma.

First, get some tool catalogs and price sheets so that you know the going prices for new tools. As a rule of thumb, you might need to pay only 50 percent (or even less) of the new price for a second-hand equivalent. Also check out pawn shops and second-hand stores for used tools, and do not forget the popular garage and estate sales. Eventually, if you keep your eyes open, you can acquire a collection of name-brand quality tools—some new and some used.

It is safe to say that the do-it-yourselfer is always on the lookout for something to build or improve. He or she is also always searching for the best tool.

Chapter 15

Some Useful Layout Techniques

THIS CHAPTER ILLUSTRATES HOW TO LAY-out, or mark, metal. Layout lines and other marks show where to cut and bend the metal and also where to locate the different parts of a project. Many of these layout techniques are the same as those used to make wooden constructions. There is more than one way to do most layouts. With experience, you can develop a few shortcuts of your own. None of the layouts depicted are difficult, but accuracy is important. It is easier to make many layouts full-size on paper first, and then transfer the important points to the plate.

A WORKING SURFACE

It helps to have a sturdy flat steel working surface like a table top. Enough thickness is needed to prevent warpage from repeated applications of heat. Industrial welding shops sometimes use special heavy surfaces on which to build projects. These are called *platens* or *slabs* (Fig. 15-1). Smaller projects and parts of larger projects are often assembled on a platen.

If you have the space for it, a 1/2-to 1-inch-thick plate measuring perhaps 2 feet on a side, can serve as an ideal platen in the home shop. Figure 15-2 shows such a plate as a table top with 1 1/2 inch pipe for the legs. Layout lines are sometimes drawn directly onto a plate to make patterns. Center punch marks are used to represent key points of the layout.

A blue layout dye can be used to make lines stand out on more exacting layouts. Using a sharp tool like a scribe or scratch awl, lines scratched through the dark dye show up quite well (Fig. 15-3). Layout dye can be removed with a special remover, although lacquer thinner also works well. Be careful, layout dye and removers are highly flammable items to have in a weld shop. Use them only with adequate ventilation—outside whenever possible. Be sure to store them and any rags that might have been used with them in airtight metal containers (a good welding project). A wide felt-tip pen can sometimes be used as a more convenient substitute for layout dye.

Temporary welds can be made directly on your

Fig. 15-1. Heavy iron platens ("acorn plates") are used by commercial fabricators for assembly of welding projects.

heavy plate surface to assemble and hold parts in alignment for final welding. Figure 15-4 shows angle bar clips welded directly to a bench top. The project parts can be clamped to the clips. This is illustrated in greater length in Chapter 17 with the construction of the slotted welding table. Later on, after the assembly and final welding is completed, the project and clips are removed, and any remaining welds can simply be ground off. If you spray some anti-spatter solution on the plate before you do much welding, the cleanup will be that much easier. If too many center punch points have accumulated, they, along with deeper weld scars, can be filled with quick welds and then ground.

LINES AND POINTS

The following rather pragmatic explanation is made with apologies to mathematicians everywhere. When considering the nature of lines, you also must consider points. However, a point is an imaginary intersection of lines so thin they have no measurable width. A discussion about lines and points is something like trying to sort out which came first, the chicken or the egg. So often in sketches, a point is represented by a dot which is a small blob of scribble. This blob covers some considerable area. Remember that a point actually has no dimensions, and neither do lines. For accuracy in your layouts, use only small points and thin lines. A "crow's foot" is often used to mark the location of a point.

A straight line can be drawn through any two points. In fact it would be a good trick to draw one any other way. A *straightedge* serves to guide the marking instrument (pencil etc.) for drawing a

Fig. 15-2. A small table suitable for use in the home shop.

Fig. 15-3. Layout dye is used for intricate layouts on shiny metal; lines scratched through it show up well.

straight line. A scale or any suitable flat, smooth object can be used for a straightedge. Longer straight lines can be made by snapping a chalk line as was shown in Fig. 14-26.

When using a straightedge, be sure to allow for the slight set-over error as the marking instrument glides along the edge. For accuracy, also be sure to look directly at the edge of a scale if measuring a distance along a line.

An arc (part of a circle) can be drawn through any three points—this is explained a bit later in the chapter. To draw a smooth curved line through a number of points that are not part of a circle, a *French curve* is used. The French curve is suitable for smaller constructions. In a pinch, the home craftsman can even use soft solder as a curved drawing guide (Fig. 15-5). Use a batten or flexible flat stick to connect points with a curve on larger constructions.

SQUARE LAYOUTS

Most of your fabricated projects will involve squares and rectangles, simply called *square shapes*.

Fig. 15-4. Angle bar clips welded to a plate can be used to hold parts in alignment.

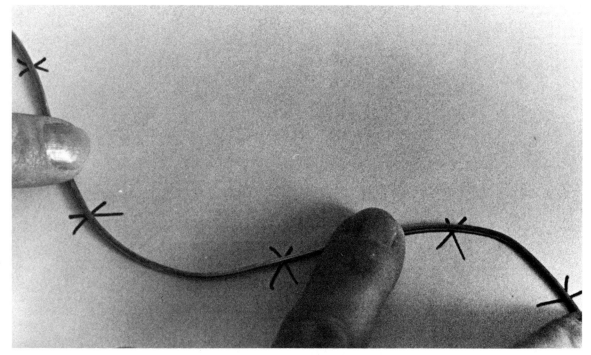

Fig. 15-5. Soft solder wire used to smoothly connect points on an irregular layout line.

It is, therefore, important to know how to quickly and accurately lay out a square shape.

Besides having four 90-degree (perpendicular) corners, squares and rectangles also have equal diagonal or "criss-cross" dimensions. This fact is useful for making a quick check for squareness (Fig. 15-6) and for laying out square shapes. You can lay out a square or rectangle by several different methods.

Laying Out a Square Shape with a Framing Square

Using the inside edges of a framing square, mark out two perpendicular straight lines of the required dimensions. This forms the first corner and two sides (Fig. 15-7). To form the opposite corner, slide the square until the required dimensions on both of the scales are even with the first two lines. Now mark out the remaining two sides using the inside edges of the framing square to complete the layout of the square shape. Always check the accuracy of your square layouts by comparing the diagonal dimensions.

Laying Out a Square Shape by Construction

To lay out a square shape too large for the use of a framing or other type square, a *construction* method is frequently used. Construction layout methods use lines, arcs, and points and very little, if any, direct measuring. For construction you need a means of swinging an arc and a means of projecting a straight line. A compass and ruler or scale can be used for practice on paper.

To make the first square corner, draw a base line OA (Fig. 15-8). Use a pencil compass, a pair of dividers, or trammel points centered at the *origin* point O. Swing an arc I from point A on the base line as shown in Fig. 15-9. The same convenient length of arc is used throughout the construction of this first corner. Use point A as a center and swing a second arc II to intersect the first arc I at point B (Fig. 15-10). Draw a straight line through the two points A and B. Project this line at least twice as far as the distance from point A to point B (Fig. 15-11). Swing one more arc (III) from point B as shown in (Fig. 15-12). The point C formed by the intersection of this last arc III and the projected

299

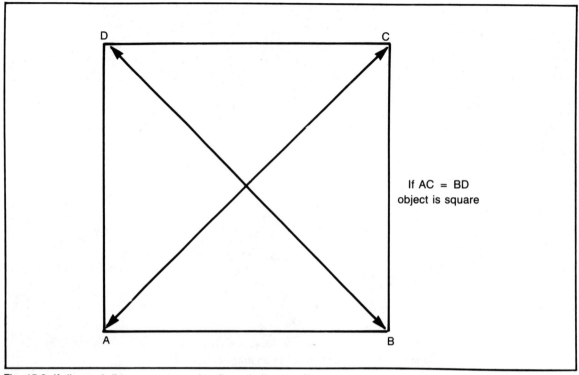

Fig. 15-6. If diagonal distances are equal, a figure is "square."

straight line is on a perpendicular line above the origin point. Draw a straight line through the origin and the final point to complete the first square corner (Fig. 15-13).

Mark off the lengths of the sides along the lines forming the first square corner. This locates the second and third corners as shown in Fig. 15-14.

The final corner can be located at the intersection of two arcs. One arc, swung from point O is equal to the diagonal distance from the second and third corners. The other arc swung is equal to the distance from point O to the third corner (Fig. 15-15).

Laying Out a Square Shape by Measuring

Using a practical application of the Pythagorean Theorem about triangles, a first square corner can be laid out. This is sometimes called the "3-4-5" method because ratios of 3:4:5 are used for the lengths of the sides. If the legs of a right angle measure 3 and 4 units, the diagonal distance be-

tween their endpoints measures 5 units (Fig. 15-16). This is also true for other sets of sides as long as the 3-4-5 ratio is maintained. For example, 15-20-25 will also work because the ratio is increased five times. Another example using a factor of 10 would use lengths of 30, 40, and 50 units. Always remember to use the same factor multiplied by 3, 4, and 5.

To lay out a square corner all you need is an original point A on a base line. Mark point B at a distance of either 3 or 4 units along the base line from the end point A (Fig. 15-17). Using point B as center, swing an arc with your compass, dividers or trammels opened to 5 units (Fig. 15-18). With point A as center, swing a second arc with the compass, dividers or trammels set to 3 or 4 units—whichever was not used in the first step. See Fig. 15-19. Connect the original point to the intersection of the two arcs to complete the first square corner as shown in Fig. 15-20.

When the diagonal method might not be practical, this "3-4-5" method can also be used to check

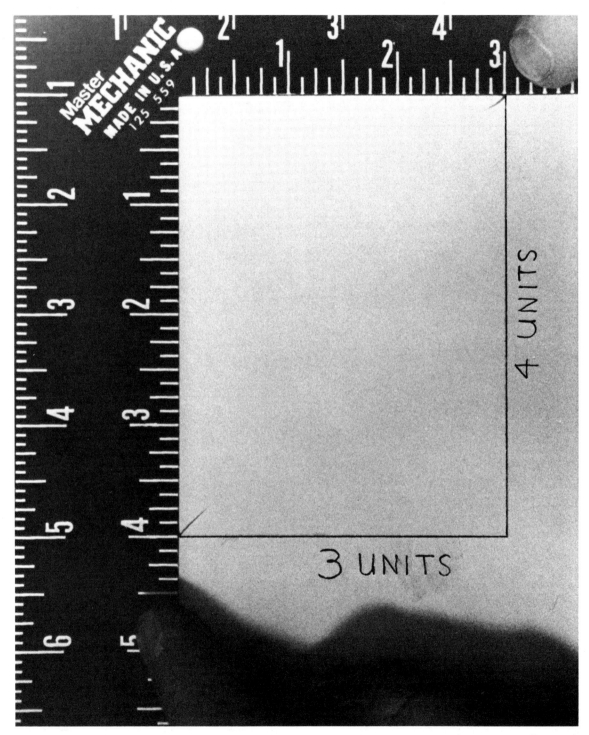

Fig. 15-7. Laying out a square shape with a framing square. Here the second pair of sides is found using the scales on the square.

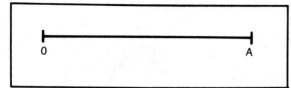

Fig. 15-8. Base line OA.

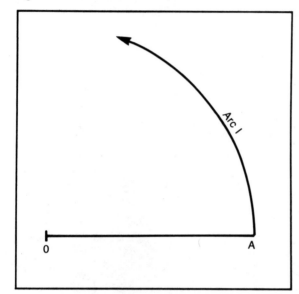

Fig. 15-9. Arc I, centered at O, is swung upward from point A.

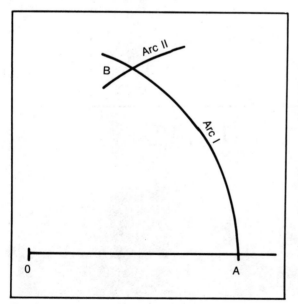

Fig. 15-10. Arc II, centered at A, is swung to intersect Arc I at B.

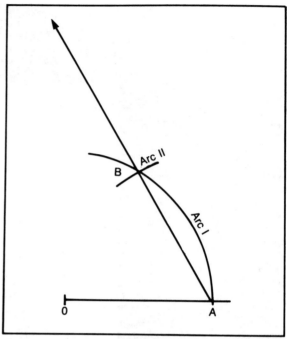

Fig. 15-11. Project a straight line through A and B, extend as shown.

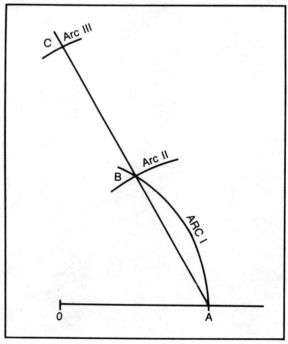

Fig. 15-12. Arc III, centered at B, intersects the projected line at C.

302

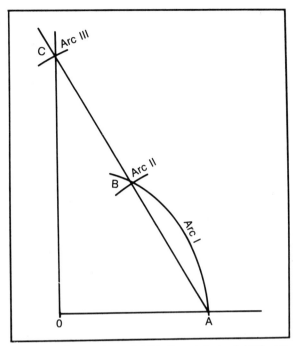

Fig. 15-13. Connecting point C to the original point O forms the perpendicular for the first square corner.

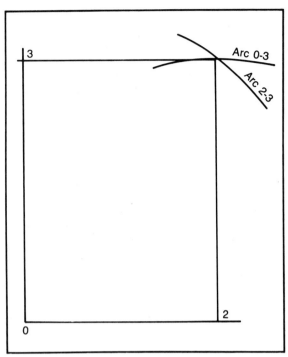

Fig. 15-15. Final corner of a squared shape is located at intersection of an arc equal to the diagonal distance and an arc equal to either of the first two sides.

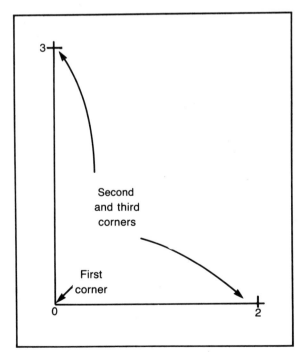

Fig. 15-14. Mark off the required lengths for the second and third corners.

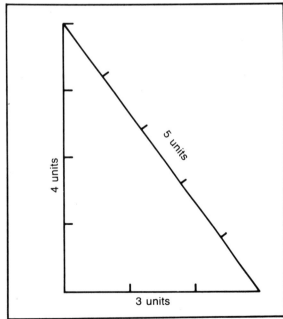

Fig. 15-16. The ratios of the sides of a right triangle.

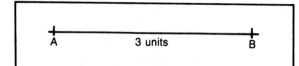

Fig. 15-17. Marking out the first side of a square corner to either 3 or 4 units long.

a corner for squareness. With the use of only a tape measure, considerable accuracy is possible.

Laying out a Perpendicular at a Point on a Line

It is often necessary to make a perpendicular line from a point on a line, and sometimes the use of a square is not practical. If the point is at or near the end of a line, either of the two preceeding methods can be used.

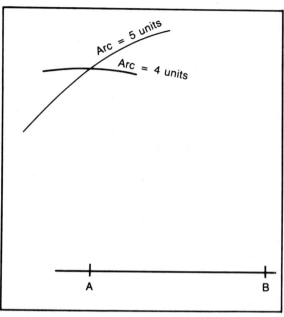

Fig. 15-19. Swing a second arc equal to either 3 or 4 units (whichever has not already been used) to intersect the first arc.

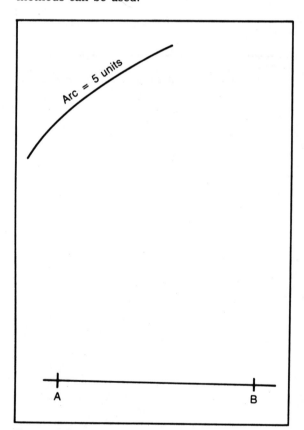

Fig. 15-18. Swing an arc equal to 5 units from an end point, here B is the center for the arc.

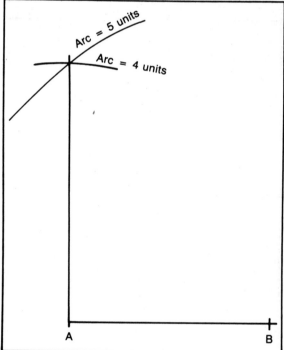

Fig. 15-20. Completing the square corner.

304

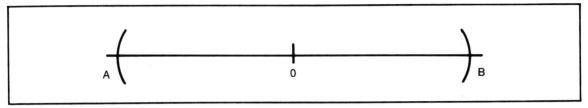

Fig. 15-21. Points A and B at equal distances from O.

To layout a perpendicular from a point O on a line that is not near or at the end of the line, use the following method. Mark off points A and B on the line at convenient equal distances on opposite sides of the original point O (Fig. 15-21). Open your compass, dividers, or trammels to somewhat more than half the distance between A and B. Using A and B as centers, swing two arcs which intersect

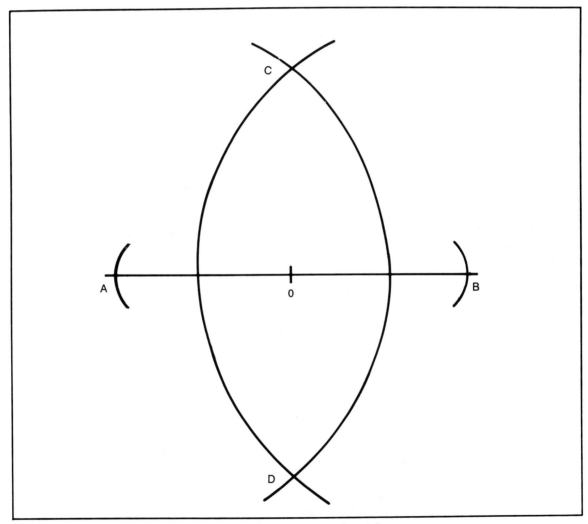

Fig. 15-22. Swing two arcs, one each from A and B, to intersect at C and D.

305

at C and at D (Fig. 15-22). A line drawn from C to D through 0 is perpendicular to the original line as shown in Fig. 15-23.

DIVIDING LINES, ARCS, AND ANGLES

Straight line segments and arcs can be easily divided into any number of equal parts using construction methods. This is far more accurate than attempting to mark off numerical divisions, especially if the lengths involve oddball fractions not on a scale. For example, imagine trying to divide a line measuring 17 7/16 inches into 12 equal segments with a scale.

Bisecting a Line or Arc

To bisect or divide a line segment AB into two equal parts, open your compass, dividers or trammels to somewhat more than one-half the approximate distance between the two endpoints A and B (Fig. 15-24). Using A and B as centers, swing two arcs that intersect at C and D. A line drawn from C to D will bisect the original line segment. There is even a bonus: this bisecting line is also perpendicular to the original line.

An arc is also easy to bisect. Use basically the same procedure as just outlined for bisecting a line segment (Fig. 15-25).

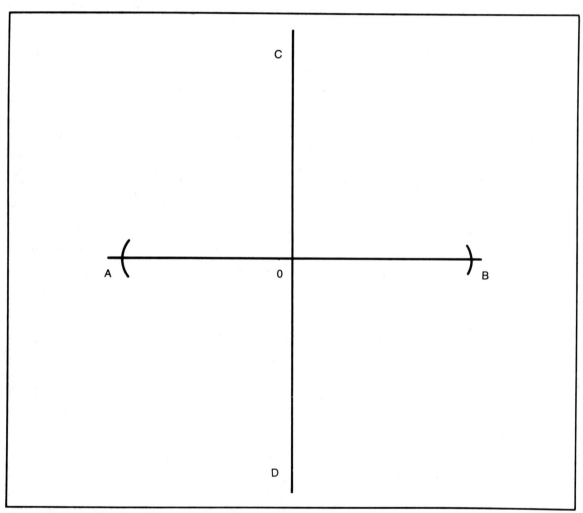

Fig. 15-23. Connecting C and D forms a perpendicular to the original line (and also bisects the distance from A to B.)

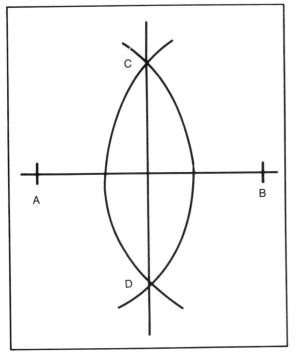

Fig. 15-24. Two arcs greater than half the estimated distance from A to B and swung from A and B, will intersect at C and D. Connecting C and D forms a perpendicular bisector.

Dividing a Line into any Equal Number of Parts

Before looking at how to divide a line segment into an equal number of parts, answer these basic questions:

- How many cuts are needed to divide something into two parts?
- How many cuts are needed to divide something into five parts?

Do you see that you always make one less cut than the number of parts? Realizing this will help in keeping track of the divisions. The line segment in these explanations can very well represent a piece of steel.

To divide a line segment AB into a certain number of equal parts, draw a perpendicular BC at one end of the segment (Fig. 15-26). Position the end of a scale at A, the end of the line to be divided.

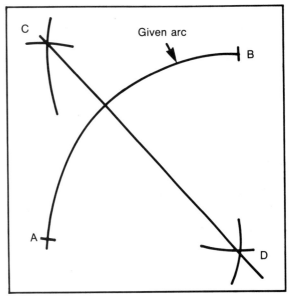

Fig. 15-25. Using the end points A and B of a given arc as centers, two more arcs greater than half the distance will intersect at C and D. A line from C to D will bisect the given arc.

Move the other end of the scale until the desired graduated line on the scale is even with the perpendicular line BC as seen in Fig. 15-27. The graduated line used on the scale is determined by the desired number of divisions. For example, if you wish four divisions, use a value on the scale that is easily divided by 4. With the scale in place, a

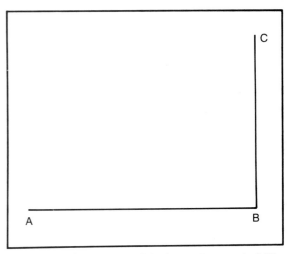

Fig. 15-26. Perpendicular BC is drawn at one end of AB.

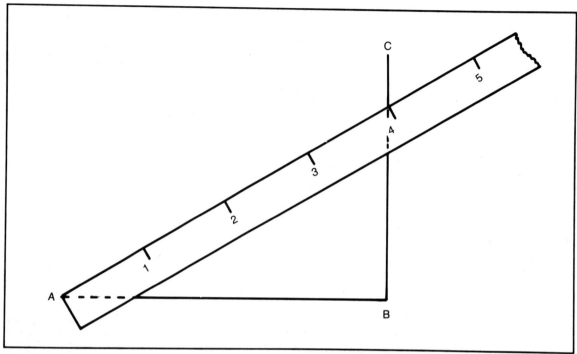

Fig. 15-27. A slanted scale lined up for 4 divisions.

slanted line is drawn through B to the perpendicular AC. Figure 15-28 shows how additional perpendicular lines drawn from the original line segment through divisions on the slanted line will provide the desired number of equal parts.

There is a distinct advantage in using the construction method of line division just outlined. It never was necessary to know the length of the original line.

Bisecting Any Angle

Any angle can be easily bisected. The point V where two straight lines come together to form an angle is the *vertex* of the angle (Fig. 15-29). Using the vertex as a center, swing an arc of any convenient radius to intersect the sides of the angle at points A and B. Using the points A and B as centers, swing two more arcs to intersect at point C. A line drawn from the vertex through point C will bisect the angle.

If, in turn, the bisected half of the angle were also bisected, the original angle would now be quar-

tered. Using this general procedure, accurate parts of angles can be obtained.

Trisecting a 90-Degree Angle

A special case exists with a 90-degree angle which

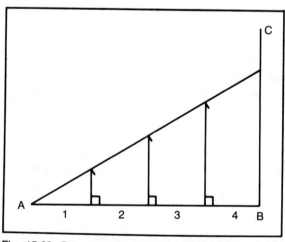

Fig. 15-28. Perpendiculars from the original line to the slanted line divisions also divide the original line into 4 equal parts.

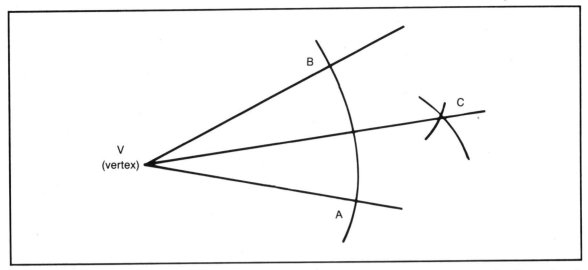

Fig. 15-29. An angle can be bisected by first swinging an arc with the vertex V as center to intersect the sides at A and B. Secondly, using A and B as centers swing two more arcs to intersect at C. A line from the vertex V to C will bisect the angle.

permits easy trisection or dividing into three equal parts. To trisect a 90-degree angle, use the vertex V as center and swing an arc to intersect the sides of the angle at A and B as before. Using the same radius, swing an arc with A as center to intersect the original arc at C (Fig. 15-30). With B as center and using the same radius, swing another arc to intersect the original arc at D (Fig. 15-31). Draw lines from the vertex through points C and D to trisect the 90-degree angle.

Divisions of a 90-degree angle are useful. Trisecting forms both 30-and 60-degree angles.

Other Angular Divisions

Sometimes the home craftsman needs to divide angles of larger fabrications into other than the standard divisions already discussed. Five, seven, or nine divisions can be approximated with a protractor—but accuracy suffers.

One method for dividing an angle involves a

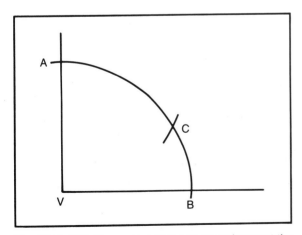

Fig. 15-30. With V as center, swing an arc to intersect the sides of the right angle at A and B. With A as center swing another arc at same radius to intersect first arc at C.

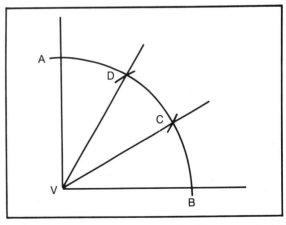

Fig. 15-31. Intersect original arc at D with an arc using B as center. Lines from the vertex to C and D trisect the 90-degree angle.

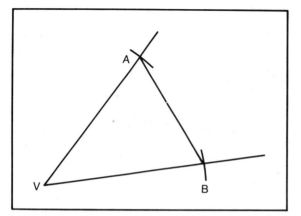

Fig. 15-32. Mark off A and B at equal distances from V.

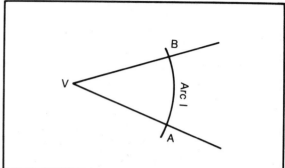

Fig. 15-34. Using Vertex V of the given angle as center, swing Arc I to intersect the sides at A and B.

technique very similar to that used for dividing a line into equal parts. With the vertex V as center, set your compass, dividers, or trammel points to a convenient distance and mark both sides of the angle as was shown before. Draw a straight line AB between these points (Fig. 15-32). Divide this straight line into as many divisions as required— five are shown in this example (Fig. 15-33). Lines drawn from the vertex through these divisions will divide the angle into equal parts.

COPYING AN ANGLE

To copy an angle, only three points are required.

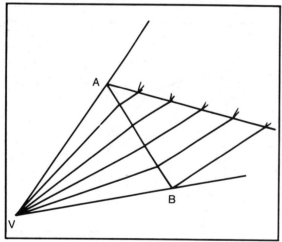

Fig. 15-33. Divide line AB equally as shown before. Lines drawn from the vertex V to the line divisions will equally divide the angle.

Using the vertex V of the original angle as a center, swing an arc I of any convenient radius to intersect the sides of the original angle at points A and B (Fig. 15-34). Draw a straight line for one side of the new angle, establish a new vertex point V', and mark off distance V'A' equal to the radius VA (Fig. 15-35). Using V' as center, swing an arc II through point A as shown in Fig. 15-36. Set your dividers equal to the distance from A to B, and using A' as a center swing an arc III to intersect arc II at B'. A line drawn through the new vertex point V' and point B' will complete the copy of the original angle.

ROUND LAYOUTS

Round layouts involving circles or parts of circles are often needed. The home craftsman should, therefore, be familiar with a few of the more basic round layout techniques.

Parts of a Circle

Figure 15-37 shows the major parts of a circle. Most, if not all, of these might already be familiar. Chord and tangents, usually less well-known, are useful in making constructions.

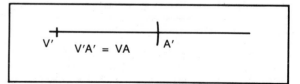

Fig. 15-35. Mark out V'A' for one side of the copy angle.

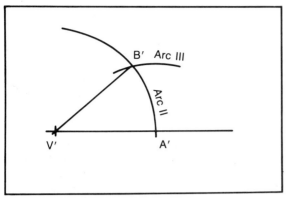

Fig. 15-36. With V' as center, swing Arc II from A'. With A' as center swing Arc III to locate B'. Connect V' to B' to complete the copy of the angle.

The *tangent* line touches the outside or *circumference* of the circle at only one point. The idea of a tangent is important for blending a straight line into an arc. For example, in order to properly radius a corner with a cutting torch (using a *circle cutting attachment*), it is necessary to know exactly where the radius cut should start and stop (Fig. 15-38). The *tangent points* locate exactly where the straight and radiused lines blend. Knowing the tangent points also permit an accurate location of a center from which to swing the torch. If you were to cut with an incorrect center, one of the two problems shown in Fig. 15-39 would be the result.

Locating the Center of a Corner
Radius/Locating Tangent Points at Corners

There are four types of corners—square, acute, and obtuse (Fig. 15-40). In order to know where a corner radius starts and stops, it is necessary to locate the tangent (T) points. For each type of corner, parallel construction lines are first drawn at a distance

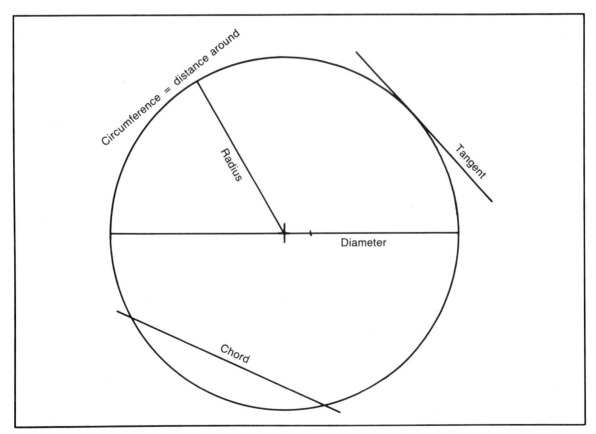

Fig. 15-37. Parts of a circle.

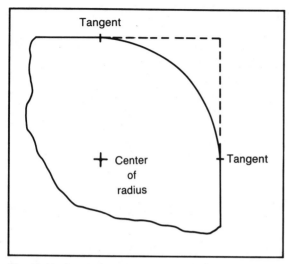

Fig. 15-38. A radiused corner begins and ends at the tangent points.

from the edges equal to the desired corner radius (Fig. 15-41). Where the parallels intersect is the center of the radius as shown at C.

The tangent points of a corner are perpendicular to the parallel construction lines. Figure 15-42 shows how to locate the tangent points (T) with a square slid along the straight portions of a corner. Use this method to locate the center and tangent points of any corner.

Finding the Center of a Circle

Sometimes it is necessary to find the center of a circle. If the circle is not too big, the center finder head of a combination square will do the job. However, you might need to find the center of a circle too large for a center finder to be accurate. It is an easy construction to locate the center of a circle us-

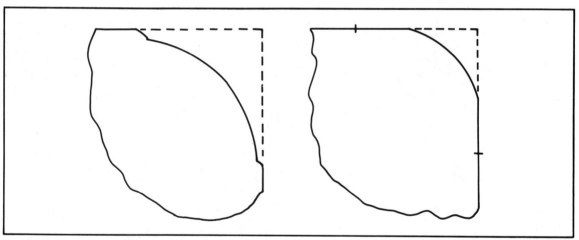

Fig. 15-39. Examples of mislocated centers.

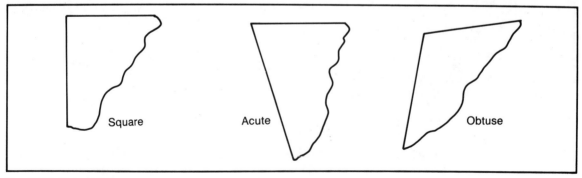

Fig. 15-40. Types of corners.

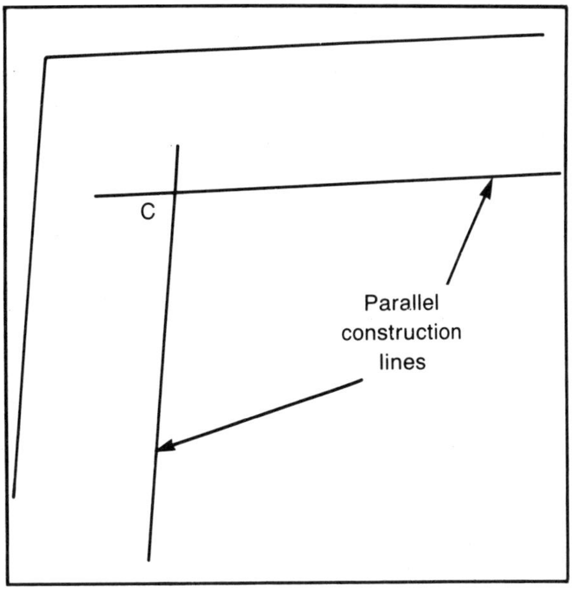

Fig. 15-41. The center C of a corner radius is located at the intersection of parallel construction lines.

ing only a straight edge and a compass, dividers, or trammels.

To find the center of the circle, draw two chords at different locations as shown in Fig. 15-43. Find the perpendicular bisectors of both chords (Fig. 15-44). (This was explained earlier.) Project the chord bisectors inward. Where they intersect is the center C of the circle as shown in Fig. 15-45.

As a check on your accuracy, you can use a third chord and perpendicular bisector. This bisec-

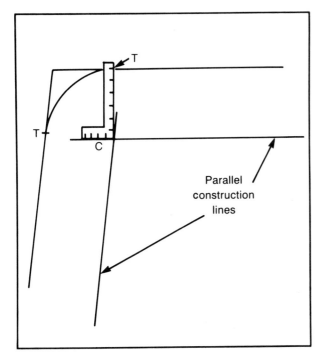

Fig. 15-42. Locating the tangent points with a square.

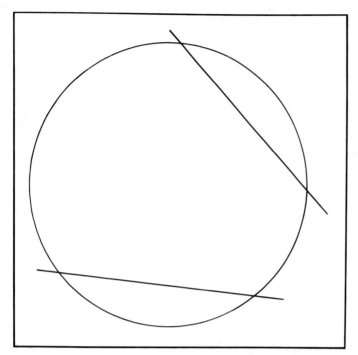

Fig. 15-43. Two chords are drawn as a first step in locating the center of a circle.

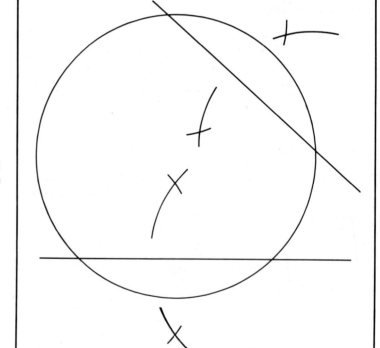

Fig. 15-44. Swing arcs to find perpendicular bisectors of both chords.

tor should also pass through the same center point.

Finding the Center of an Arc

Find the center of an arc with the same technique. Use the perpendicular chord bisectors as you would use to find the center of a circle as shown in Fig. 15-46.

If you need to burn a curved slot with a circle cutting attachment (see Chapter 16) at a radiused corner, you might need to relocate the center of the arc or radius. Figure 15-47 shows a curved slot cut in an adjusting bracket.

Dividing a Circle into Equal Parts

A circle can be divided quite easily into 3, 4, or 6 equal parts. Multiples of these (8, 12, and 16 parts) are also easy to find. In order to divide a circle, it is first necessary to know the location of the center.

While a circle can be divided using degrees and a protractor, the construction methods are usually more accurate. All that is required is a straightedge and a compass or dividers.

Dividing a Circle into Four Equal Parts

To divide a circle into four parts, first draw a di-

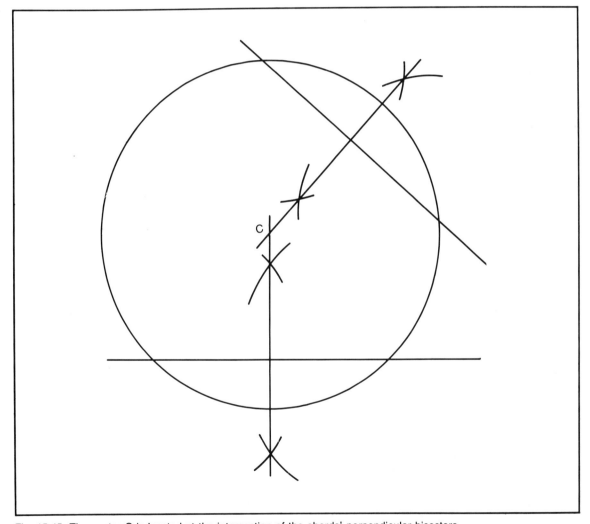

Fig. 15-45. The center C is located at the intersection of the chords' perpendicular bisectors.

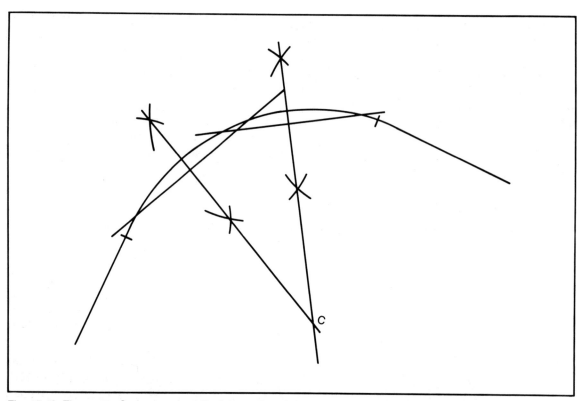

Fig. 15-46. The center C of an arc is at the intersection of the perpendicular bisectors of two chords.

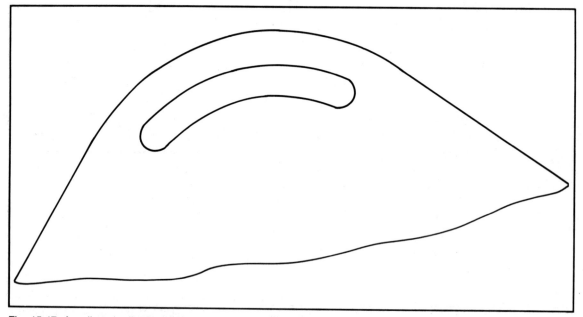

Fig. 15-47. A radiused adjusting slot.

ameter (Fig. 15-48). Swing arcs from the end points of the diameter to intersect at A and B above and below the center. A line drawn through the two arc intersections which should also pass through the center will divide the circle into four equal parts (Fig. 15-49). The accuracy of your layout can be checked by comparing the chord distances between each of the four quarter points.

Dividing a Circle into Eight or 16 Equal Parts

To divide a circle into eight or 16 parts, first divide it into four equal parts. Bisect one of the quarter circles by swinging arcs from the quarter points to intersect as shown at A in Fig. 15-50. A line drawn through the intersection and the center will bisect the quarter circle forming two eight-part divisions. Use your dividers to "walk around" the rest of the circle to obtain the balance of the eight divisions.

Sixteen equal parts can be obtained in a similar manner by further bisecting one of the eight-part divisions. Again check the accuracy of your layout by walking your dividers around the circle.

Dividing a Circle into Six, Three, or 12 Equal Parts

It is very easy to divide a circle into six equal parts. All that is needed is to set your dividing instrument

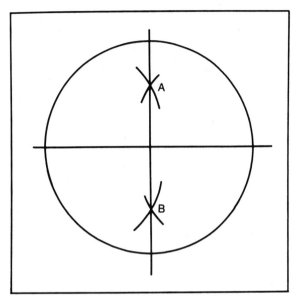

Fig. 15-49. Arcs swung from the endpoints of the diameter intersect at A and B. A line through A and B divides the circle into four equal parts.

equal to the radius. Simply walk the dividers around the circle marking off the six divisions (Fig. 15-51).

To divide a circle into three equal parts, follow the same procedure as for six divisions, but only

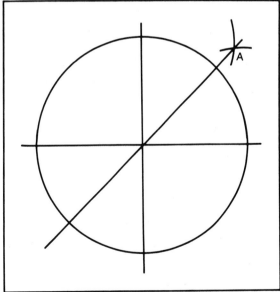

Fig. 15-50. Additional bisections render 8 or 16 equal parts of the circle.

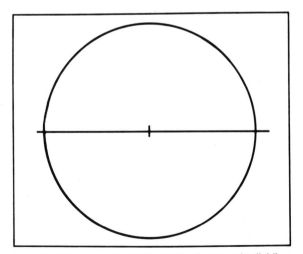

Fig. 15-48. Drawing a diameter is the first step in dividing a circle into 4 equal parts.

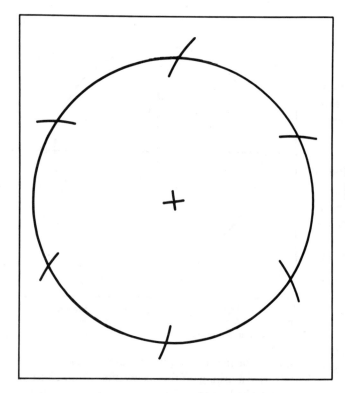

Fig. 15-51. A circle is divided into six equal parts by starting anywhere and stepping off the radius around the circumference.

use every other division. A circle can be divided into 12 equal parts by simply bisecting one of the six-part divisions.

Other Divisions of a Circle

While there are constructions for other divisions, they are a bit complex and outside the scope of this book. If you are faced with one of these use a protractor. Just be careful to use a sharp pencil or marking tool and double check the accuracy of your divisions by walking your dividers around the circle.

A Pattern for Dividing Circles

If two or more circles share the same center, they are *concentric*. Once you have divided a circle into a set number of equal parts, the dividing or "spoke" lines can be projected outward (Fig. 15-52). These lines can be used to divide any size of concentric circle into equal parts.

Use a piece of light (22-or 24-gauge) sheet metal to make a pattern for dividing circles. Lay out a "target" pattern of concentric circles with

spoke lines for various divisions. It is probably easiest to first do the layout on paper and then transfer it to the metal with punch marks through the paper. This pattern can save having to make the same layout repeatedly.

If you drill tiny holes through the target pattern, it can also be used on top of a plate for transferring points. The center point and the desired division points of a circle can be marked on the plate using a scratch awl.

Drawing an Arc Through
Any Three Points Not in a Straight Line

The trick to drawing an arc through any three points X, Y, and Z that are not in a straight line is to first create two "chords" by connecting the points (Figs. 15-53 and 15-54). If the chords are now bisected with perpendiculars as described earlier, the bisectors will intersect at the center of the arc as shown in Fig. 15-55.

Using the newly-found center C, swing an arc through the original three points (Fig. 15-56). This

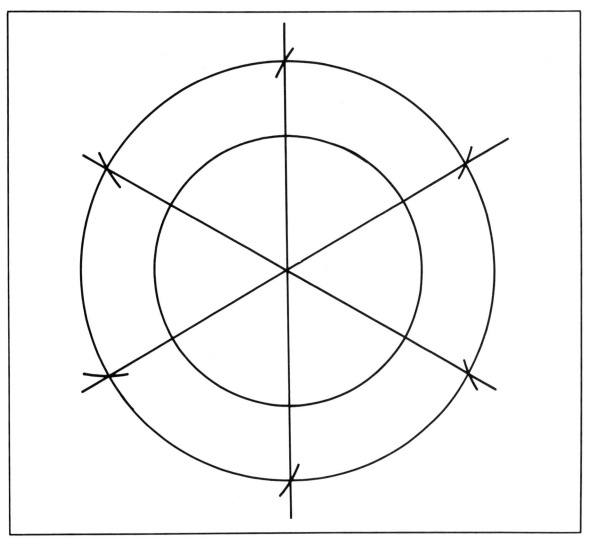

Fig. 15-52. Concentric circles divided by (shared) ''spoke'' lines.

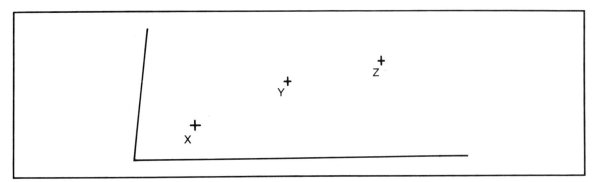

Fig. 15-53. Three points not in a straight line.

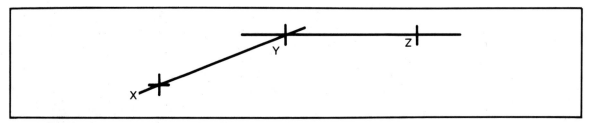

Fig. 15-54. Connecting the points forms chords of the arc-to-be.

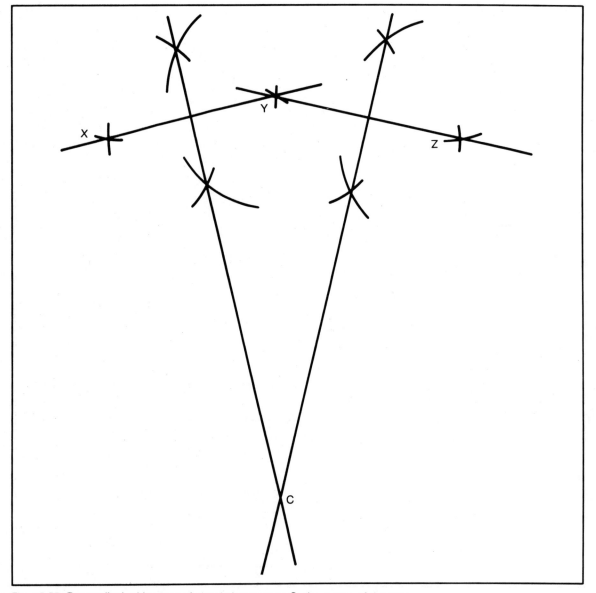

Fig. 15-55. Perpendicular bisectors of chords intersect at C, the center of the arc.

technique is almost the same as that used to find the center of an arc or circle.

IRREGULAR CURVES

More than three points not in a straight line can also be smoothly connected using a flexible straight edge or batten. Figure 15-5 showed soft solder used for this purpose.

TRIANGULATION

It is sometimes difficult to tell just how long a line or an edge is from the usual two or three views on a blueprint. Take, for example, the corner of the flared tank or box shown in Fig. 15-57. This is because the standard views show slanted corners as shorter than they actually are. The true length of such edges can be quickly found using a framing square and a process called triangulation.

To find the true length of a slanted corner, draw a right angle with a framing square as shown in Fig. 15-58. Mark off the *apparent* length d of the edge as it appears in the plan view on one side of the right angle. Now mark off the height h of the tank on the other side of the right angle. The true length to mark out on the metal for cutting is the slanted side of the triangle shown in Fig. 15-58. The layout for this particular tank is shown in Fig. 15-59. (A triangle as in Fig. 15-58 can be used as a check of the layout.) It is important to check that the edges that must go together at the corners are of the same length.

The triangulation technique can be used to find the true length of any slanted edge. It is only necessary to know the two sides of the right angle in the layout triangle. The third side is the true length of the slanted edge.

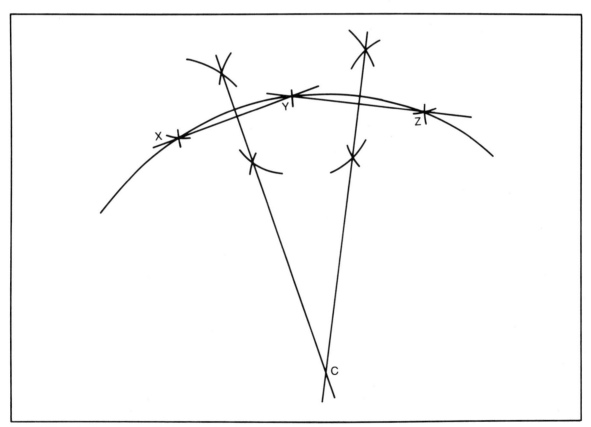

Fig. 15-56. Swinging an arc with C as center through the 3 points.

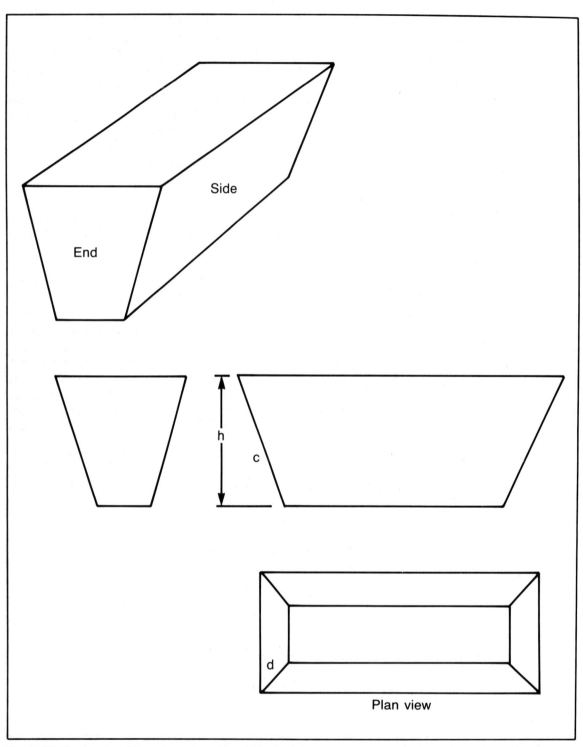

Side

End

h

c

d

Plan view

Fig. 15-57. The lengths of the edges of an object with slanted surfaces are not obvious from the standard views.

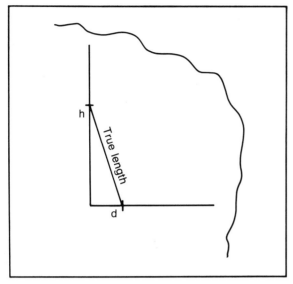

Fig. 15-58. Using the apparent length d and the height h to find the true length of an edge.

PATTERNS AND TEMPLATES

Now and then a little research and development to test an idea is needed before beginning any layout. A taped-together cardboard mock-up of a proposed project often saves time and trouble. This model, either full-or part-scale, gives a sense of the proper proportions and the assembly procedures for the project. Once fabrication is underway, a cardboard pattern of a subassembly might help you to cut the real parts only once.

A full-size pattern is often called a template. These permit transferring layout points if placed directly over the material to be cut or formed. Layout lines are either scratched around the template outline and/or punch marks made at key locations. A template can be saved for repeated use.

CAPACITIES

The home welder might need to fabricate a tank or similar vessel of a certain capacity. The easy way to do this is to consult a table of volumes like the one in Appendix B. When making a tank, it is important to first know if any outside dimensions are critical.

Flat-sided tanks (or boxes) are the easiest for weld fabrication. If possible, try to assemble the panels with a corner-to-corner fit as was shown in Chapters 9 through 11. Layout of such structures is simplified by working with the inside dimensions. If baffles or other internal features are needed, they should be laid out, inserted, and welded before the assembly proceeds very far. This not only makes the welding itself easier, but also permits proper postweld cleanup. Any cutting or drilling on a closed tank should also be done before assembly to enable proper removal of slag or chips.

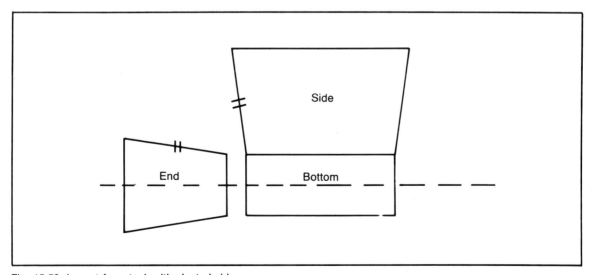

Fig. 15-59. Layout for a tank with slanted sides.

Chapter 16

How to Cut and Form Metal

CUTTING AND FORMING ARE BASIC TO FABRI-cating new projects and for maintenance and repair work. Flame cutting heavier plate and structural shapes in the home shop is mainly done with an oxyacetylene torch, although other fuel gases can be used. The heat of the electric arc with carbon electrodes and compressed air is an effective way to "gouge" or remove metal quickly.

Small, precise cuts on light metal are made with the familiar hacksaw or portable power saws. Sheet stock can be cut with shears or snips. Thin grinding wheels or discs can also be used to cut metal in certain applications. Drills and hole saws cut metal to make holes. Threads are cut using taps and dies.

Metal can be cold formed or hot formed into different shapes. Heat forming is sometimes done to shape a new part and sometimes used to correct welding distortion. Some metal shaping is done with files and grinders.

CUTTING STEEL

What follows is a more or less complete listing of industrial methods for cutting steel. Only some of these are practical for the home workshop. However, it is worth knowing about all of them because you might want to have something done "outside."

Thermal Methods

- Flame cutting of oxyfuel "burning" (Fig. 16-1)
- Thermally cut by "stick" cutting electrodes
- Thermally cut by Air Carbon Arc Cutting (AAC), also known as "gouging" or "scarfing"—widely used by industry
- Thermally cut by *Plasma Arc Cutting* (PAC) (Fig. 16-2)

Sawing

- Manual sawing—hacksaw and coping saw
- Power sawing—"sabre" saw and Skilsaw for light gauges. Also power hacksaw and bandsaw
- Abrasive ("hot" saws) and thin abrasive discs fitted to a sander/grinder

Fig. 16-1. Flame cutting, or burning with an oxyacetylene torch, remains one of the most popular methods to cut steel.

Other Methods

- Shearing operations that include punching and notching
  - Chipping and chiseling
  - Using high-speed burring tools
  - Drilling and knocking out
  - Filing and grinding

- More exotic and elaborate forms like Water Jet and laser (Fig. 16-3)

CUTTING METAL IN THE HOME SHOP

You sometimes have a choice of cutting processes. Which method you select is determined partly by

Fig. 16-2. Plasma Arc Cutting (PAC) will cut most metals such as this aluminum plate.

Fig. 16-3. While not exactly practical for the home shop, the very high-pressure water jet cutting system made by Flow Systems can cut steel over 4 inches thick with no heat input. (Courtesy of Flow Systems.)

cost and partly by physics. The practicality of a given cutting process in your workplace must be considered. For example, Air Carbon Arc Cutting (AAC) is efficient but needs compressed air or nitrogen and really throws the slag. Moreover, it is noisy. Such factors might well save you any further thoughts about using AAC in the home shop.

CONSIDER THE KERF

The slot or groove made by a saw or cutting torch is called a *kerf*. Metal cut by any means other than with a shearing operation involves some loss as either sawdust or slag. This loss is called *kerf allowance* and must be taken into account or you will end up short of material.

Shearing is quick and clean and causes no metal loss, so it is widely used in industry. But, plate and bar shears are not normally practical for the home workshop. Sheet stock up to 16 gauge (.062 or 1/16 inch) can be sheared for shorter cuts with a *throatless bench shear* commonly called a "Beverly shear" (Fig. 16-4). Some home welders make their own bar shears modelled after a commercial design.

The most common ways to cut metal in the home shop are flame cutting, sawing, abrasive cutting, and arc cutting. Each of these is discussed in the following sections.

OXYACETYLENE
CUTTING, THE OLD RELIABLE

Oxyacetylene "burning" is suitable for both fairly light and heavier steel. Acetylene fuel gas is still the industry favorite for cutting heavy plate. Flame cutting makes little noise and the slag is readily contained. There is another oxyacetylene advantage—you get a cutting torch with your combination oxyacetylene welding outfit.

Flame Cutting Equipment

Before you start flame cutting, review the safety procedures in Chapter 9.

A cutting attachment is installed on the welding torch handle in place of a welding tip. Use the correct size cutting tip for your torch. There is no industry standardization with cutting tip number

Fig. 16-4. A throatless bench or Beverly shear can be used to cut irregular shapes in sheet metal as heavy as 16 gauge.

sizes, so they vary somewhat. If you have it, follow the manufacturer's advice to match tip size to the metal thickness being cut. Otherwise use a size 0 or 1 tip for material from 11-gauge (1/8 inch) to 1/2 inch thick. Avoid overtightening the tip retaining nut; you might stretch the threads.

A cutting tip contains four or six preheat flame orifices (Fig. 16-5). Preheat flames bring the material to a kindling temperature of about 1600 degrees Fahrenheit. Depressing the oxygen cut lever on the torch admits a jet of pure oxygen from the center orifice. Pure oxygen slices easily through the heated plate. The torch is moved at a uniform rate to produce the desired shape. The path of the torch is usually directed by hand, sometimes with a fence or guide.

After the equipment is properly assembled and pressures set, the torch can be lit. Multiple preheat flames on a cutting torch are adjusted the same way as the single flame of a welding tip.

Lighting and Adjusting a Cutting Torch

Use the following steps to light and adjust the cutting torch attachment:

- Make sure that the oxygen preheat adjusting valve on the cutting attachment is closed.
- Open wide the oxygen valve on the torch handle.
- Purge the lines by opening the acetylene valve on the torch handle about 1/2 turn for one second, then close the valve. Do the same for the oxygen preheat adjusting valve on the cutting attachment.
- Light and adjust the flames to a neutral condition.
- Depress the oxygen cutting lever (Fig. 16-6) and check the condition of the cutting stream of oxygen (Fig. 16-7); if necessary, shut off the torch and clean the tip as explained in the following section.

- When not in use, hang the torch in a place with free air movement; never place a connected torch (even if you think it is not pressurized) in a confined area where gases could collect.

The importance of using cutting tips that are in good condition cannot be overstated. Do *not* use bent, smashed, clogged, or otherwise damaged tips. Tips must also be kept clean of slag and other obstructions. A jet of water is widely scattered by a small obstacle at the point of discharge. A speck of slag over a tip orifice has a similar effect. Slag scatters both the preheat flames and the oxygen cutting jet. This disperses the preheat and prevents cutting oxygen from making a straight and clean cut. Always test the cut stream as shown in Fig. 16-7.

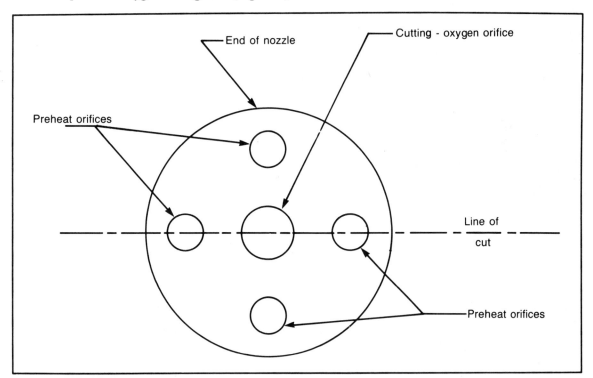

Fig. 16-5. The nozzle end of a cutting tip shows 4 or 6 orifices for the preheat flames and a larger orifice to admit the cutting oxygen. Most flame cutting is done with the orifices aligned to the cut line as shown.

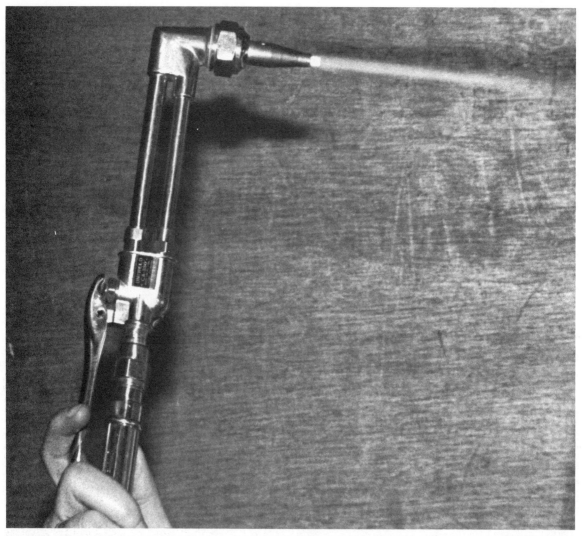

Fig. 16-6. Depressing the oxygen cutting lever on the cutting attachment to check the condition of the preheat flames and the cut stream.

Cleaning the Tip

Tip cleaners are sold as sets of serrated wires and often include a shoddy, almost adequate file (Fig. 16-8). It is easy to damage tips by overcleaning. Removing metal (not slag) from a tip orifice causes it to become bell-mouthed. The gas then tumbles instead of flowing in a smooth stream causing ragged preheat flames and rough cuts.

Open the preheat oxygen valve very slightly to keep contaminants from entering the tip orifices. File squarely across the end of the tip. Use a tip cleaning wire small enough to avoid oversizing the orifices and pass it in and out only a time or two. Depress the oxygen lever to clear the oxygen orifice. Close the oxygen preheat valve when you finish the cleaning.

Protect cutting tips in storage. The sealing surfaces must not be wire brushed or polished. If a tip leaks at its seating surface, replace it. If cleaning fails to restore a proper flame and cutting stream, replace the tip. Do not modify tips or any other oxyacetylene apparatus.

The Cutting Surface

The cut passes right through the plate, so some thought should be given to what is underneath. There should be no barrier to the free passage of the cut stream and slag. Burning tables consisting of flat bars on edge are often used. The cut should be made parallel with and between the bars as seen in Fig. 16-1.

Like using a saw, try to avoid flame cutting through metal that is supported so that it collapses into the cut. When possible, support the metal so that the cut section can freely drop from the rest of the plate (Fig. 16-9).

If cutting on an assembly, plan where the heat and slag will go and provide shielding as required. Fiberglass cloth or sheets of metal are sometimes used. Never burn on an unvented container or system. Also avoid containers that have held flammables or cleaning solvents. Fumes from burned paint and metallic coatings are dangerous; use plenty of ventilation.

Of course, all flammables must be removed from the area, and provision is needed to catch the slag. (Hot slag dipped on a lawn effectively removes grass for quite some time.) Cutting should never be done directly on or near a concrete surface; remaining moisture boils, causing a violent "spalling" or outward thrust of particles. Hot slag easily damages asphalt surfaces. A sheet of light metal makes a good shield from flame cutting slag. However, avoid using painted or coated metal because the heat will generate fumes.

Firebricks are sometimes used to support the work to be flame cut. Never use ordinary red brick or even concrete (cinder) blocks. These will crack and spall.

A simple way to contain cutting slag and sparks is to use a clean drum with vent holes near the bottom to prevent accumulation of unburned gases. Though unlikely, collected gases could ignite with a dangerous upward blast. A little water in the bottom of the drum cools hot slag quickly. (Some peo-

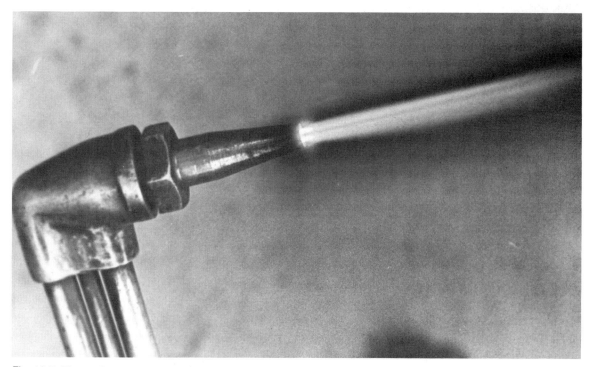

Fig. 16-7. The cutting stream should be clearly defined for 3 or 4 inches inside the flame pattern, and the preheat flames should remain neutral.

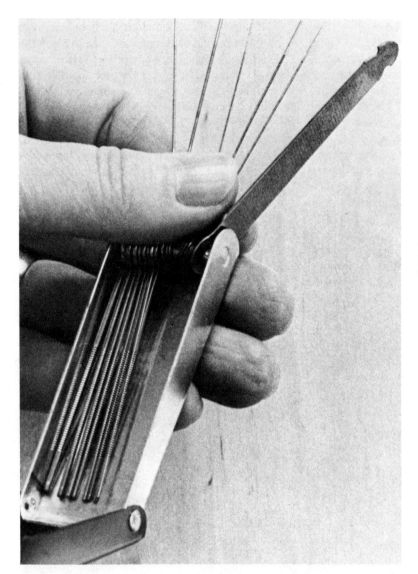

Fig. 16-8. A set of tip cleaners. Use them sparingly because tips can be easily bell-mouthed.

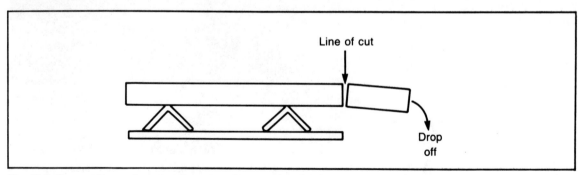

Fig. 16-9. Support the work so that the cut section can drop freely.

ple collect slag. They say that it is good for their camellias and hydrangeas.)

Marking out the Cut Line

Use a sharpened piece of soapstone to mark a cutting line (Fig. 16-10). Soapstone lines on steel look like chalk but hold up under the torch heat. Chalk burns up and blows away. Use a center punch to mark the end points of a cut and even perhaps every inch or so to help with following the line. Mark the excess side of the cut line with "XS."

If cutting out a number of closely "nested" parts, identify each part on the plate before starting to cut. This helps in sorting out parts from scrap.

Cutting in a Straight Line

Straight line free hand cutting requires skill. You will get better at this with practice. Try to avoid making cuts any longer than you find comfortable. As you feel stretched out, stop and reposition yourself. A stop and start can produce a slight burn scar, but this is preferred to a nonstop, full-length ragged cut. Small scars can be easily "picked up" or filled with weld metal.

Before lighting the torch, make a "dry run" with it through the cutting path to discover any obstacle that might prevent a smooth motion. Like shooting pool, one hand should be steady and serve to guide the other. Notice how, in making a straight cut (not an arc), it is necessary to slide the torch fore and aft as well as sideways.

Obtain some plate of the type and thickness that you will be using the most. Mark out several straight lines with soapstone. Following the safety procedures outlined in Chapter 9, light the torch and adjust for a neutral flame. Depress the oxygen lever and readjust if necessary. Start the cut at the plate edge by preheating just the top corner (Fig. 16-11).

Hold the torch so that the flame cones are 1/16 to 1/18 inch off the plate. As the plate reaches the red kindling temperature, begin the cut by depressing the cutting lever. Point the torch slightly ahead in the direction of travel. Maintain a uniform travel speed and torch-to-plate distance throughout the

Fig. 16-10. A soapstone stick sharpened against a grinding wheel will produce a distinct line.

cut. If you travel too fast, the cut stream will not have time to completely penetrate the plate. Travelling too slow causes excessive melting and allows slag at the bottom of the cut to fuse back together. Best results for cutting a straight line are achieved if the holes in the tip are aligned as shown in Fig. 16-5.

Piercing

To *pierce* or start a cut in the middle of a plate, pre-

heat a small area to the kindling temperature by moving the torch in tight circles. Just before adding cutting oxygen, lift and angle the torch slightly to permit the slag to escape to one side as seen in Fig. 16-12. This is important or slag might splash back onto the tip.

If you make a small nick in the plate surface

Fig. 16-11. Preheating the top edge at the start of the cut line, the cutting oxygen lever is not yet depressed.

Fig. 16-12. Holding the torch at an angle when piercing allows the slag to escape to the side, thereby sparing the tip.

with a punch or chisel, the raised burr reaches the kindling temperature fast. This reduces the required preheat. You might even prefer to drill a hole to start the cut. Once the plate is pierced, the burning procedure is then the same as starting a cut at an edge.

Intermittent Cutting

It might be necessary to flame cut a long strip. (In fact, the collected impurities at the "roll edge" of hot rolled steel should be discarded and not included in a weld joint.) If not restrained, the application of high heat causes metal along the cut to expand. Upon cooling, it contracts. The tendency is for the strip to shrink more along one edge, resulting in a curved shape that may be useless. Such shrinkage can be limited with a strong enough restraint.

A way to burn off a straight strip is to leave sections of the cut line attached (Fig. 16-13). Cutting such intermittent segments is often called "skip burning." It allows surrounding metal to restrain expansion or contraction of the hot edge. Intermittent or "skip burning" can be done as follows:

● Pierce or drill starting holes along the cutting line for the cuts every four to six inches.

● Mark "stop cut" points with soapstone about 1/2 inch before each starting hole.

● Light and adjust the torch; cut one segment at a time from a starting hole to a stop line.

● Avoid excessive heat input to any one area of the plate—cut segments at different areas to avoid overheating.

● Allow the plate to cool, then cut apart the small sections that were left attached.

Fig. 16-13. Leaving segments of the cut line attached helps prevent distortion.

Cutting Light Metal

Lighter plate is easily warped and melted rather than burned apart by a cutting torch. To cut steel plate 11 gauge (1/8 inch) or lighter, use a severe torch angle and a very high travel speed. Cutting guides to slide the torch help; a length of angle bar works quite well.

Use the smallest standard tip or a special "stepped" tip to cut light metal. The special tips are designed to be dragged on the plate for a very rapid travel speed.

Vertical Cuts

Begin a vertical cut at the bottom and travel upward. By moving away from the slag, the tip and cut zone remain clear.

Cutting Round Shapes

It is often necessary to cut out a round shape. With practice, you should be able to make a fairly respectable hole by guiding the torch freehand. After piercing or otherwise starting the cut, keep the tip square to the plate. It helps to have the hole outline marked with a center punch.

Where more accuracy is required, use a circle cutting attachment (Fig. 16-14). These are basically an adjustable centering (pivot) point for the torch. The centering point is placed in a healthy center punch mark on the plate and the torch is moved around the point. Good circle cutting devices can be home made.

Cutting a round shape is easier if the plate can be rotated and you do not have to swing the torch. Simple turntables can be devised with worn-out but intact heavy-duty bearings. (A heavy equipment repair shop is a good place to find them.) Figure 16-15 shows a rotating cutting table.

To avoid slicing across the bars of a cutting table, many welders use a short length of large diameter pipe as a base for cutting circles. This provides another advantage in that the *burnout* piece is free to drop out as the cut is completed.

Shutting Down

When finished with the oxyacetylene equipment,

Fig. 16-14. A circle cutting attachment made from an electrical cable splicing bolt.

close the cylinders, and *one at a time*, bleed both gas lines and then close the torch valves. Back out the regulator adjusting screws and hang the torch in an unconfined place.

OTHER FUEL GASES

The welding industry uses other fuel gases besides acetylene for flame cutting. Propane is generally less expensive and easier to handle. Acetylene cylinder pressures drop rapidly with cold temperatures, to the point where gas cannot be drawn from a full cylinder. Propane and other fuel gases do not present this problem. Natural gas is also widely used, as is a patented blend distributed by Airco called MAPP gas (Fig. 16-16). Other fuel gases al-

Fig. 16-15. A rotating table makes circle cutting easier. This table rotates on top of an old ball bearing.

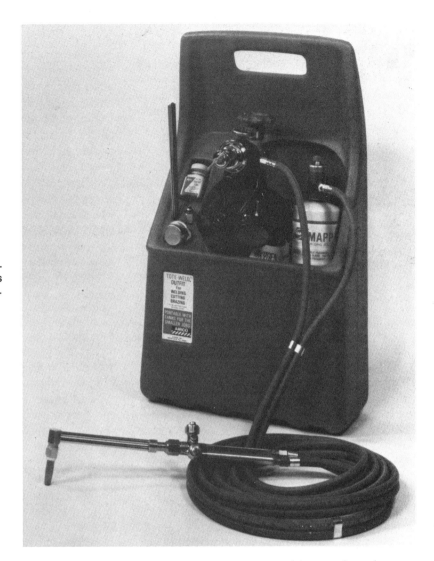

Fig. 16-16. A portable cutting out-fit containing MAPP fuel gas. This could also be used for brazing. (Courtesy of Airco.)

low high pressures for long hose runs as they do not have the inherent instability of pressurized acetylene. They also burn cleaner and produce no smoke when lighting up. Different regulators and torches (or tips) are used for these fuels.

Please, something good about acetylene? Gas welding requires acetylene, and it is the hottest fuel gas. In your home shop, it is not likely that you will work in sub-zero temperatures; nor are you apt to need hose runs several hundred feet long. The equipment needed for other fuel gases means more expense. So unless you have a very large amount of flame cutting to do, acetylene is still the best fuel.

If you still want to cut with natural gas, be sure to include your local gas company in your plans. Do not attempt to jury rig your own hookup. Gas pressures and even gas compositions vary widely. Besides needing a special siphoning type of torch, special shutoff valves and regulators are needed to safely use natural gas. You could blow up your water heater without the right installation. Involve an expert.

SAWING

A hand-held torch is difficult to move fast enough across thin steel to avoid warpage. If shearing is not available, sawing might be the best avenue to cut such thicknesses (Fig. 16-17). While large sheets of the thinner gauges are not usually arc welded, you will sooner or later find yourself working with thin-wall tubing and other formed shapes. The use of a fine-toothed saw is often the most practical way to cut these materials.

There are a few facts to keep in mind when using saws:

- The cut takes place with the blade moving in one direction only. For example, a correctly-installed hacksaw blade cuts on the forward, push stroke only.
- At least three saw teeth should always contact the material being cut. For thin sections this means using fine-toothed blades and slanting the saw to the work. A scrap wooden back-up provides extra support to the saw blade (Fig. 16-18).
- Friction from sawing produces heat which can quickly draw the hardness from a saw blade. Using a reasonable cutting speed and possibly a lubricant can keep you cutting better and longer.
- Both the material being cut and the saw must be secured from bouncing about. Remember this and you will avoid snapping off blades and/or producing ragged saw cuts.
- A saw cut edge is usually sharp. The burr is best removed as soon as possible to prevent injury and for more accurate fitup. A burr along a weld joint can also produce an inferior weld by contributing to cold lapping.
- Wear safety glasses when sawing and maybe ear plugs as well. Never operate electrical equipment in a wet environment. When changing blades be sure that the saw is first unplugged.

HOLE SAWS

Hole saws are driven by drill motors to make holes in lighter metal. They should be operated according to the manufacturer's advice with the proper speeds and lubrication. If overheated, hole saws dull quickly. A pilot hole is required, so the disc cut out by a hole saw might not be useful.

SNIPS AND PIPE CUTTERS

Snips have an advantage over saws in that less support of the work is required, and there is no kerf. Short cuts in 18-gauge and lighter sheet metal are easily made with snips. Two varieties are shown in Fig. 16-19. Tin snips should be used with one jaw resting on the bench. Aviation snips can be used to cut out circles once a starting hole has been made with a drill, punch, or chisel.

Pipe cutters and their miniaturized versions, tubing cutters (Fig. 16-20), consist of a hard sharp wheel that scores an ever deeper line as it is rotated about the pipe. These cutters are little rotating shears that make square cuts. The inside burr must be removed if the pipe or tube is used for fluids.

DRILLS

The familiar drill bit makes holes in metal of different thicknesses. Punching is preferred for sheet metal lighter than 18 or 20 gauge. It is generally a good idea to use a lubricant to keep the drill bit from overheating.

Drills are driven mostly by a hand-held motor of 3/8- or 1/2-inch capacity. The three-jaw Jacobs chuck that holds the drill clamps best if it is tightened at all three locations. Small drill presses for home use are also becoming more popular. Provide a pilot or center mark to start the drill, or it can "walk" over the plate. To ensure an accurate start, turn the drill on the work around once by hand before using power. On metal less than 18 gauge, use a wood backup to save snagging as the drill passes through the plate.

When drilling metals other than cast iron, a continuous chip should form, indicating adequate feed pressure (Fig. 16-21). With a drill press, the chip can get dangerously long. Interrupting feed pressure breaks it. Never touch a moving chip. A drill overheats with too much or too little feed pressure. Keep drill bits sharp and use lubrication. Cast iron is drilled dry, cool it with compressed air.

A "man saver" or "dead man" is a second-class lever, used to gain extra force to feed a drill into the work. These can be a simple chain fastened to the plate at one end and to a wooden stick at the

Fig. 16-17. A sabre saw can be used to cut irregular shapes.

Fig. 16-18. Backing up sheet metal with a piece of wood permits cutting with a hacksaw.

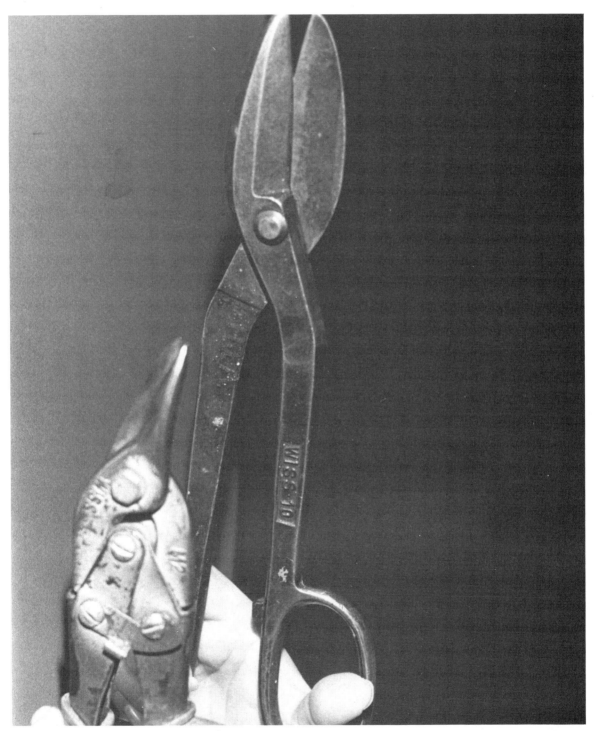

Fig. 16-19. The larger ones are tin snips and the smaller aviation snips. These are made in several sizes and varieties, with some designed for cutting curves.

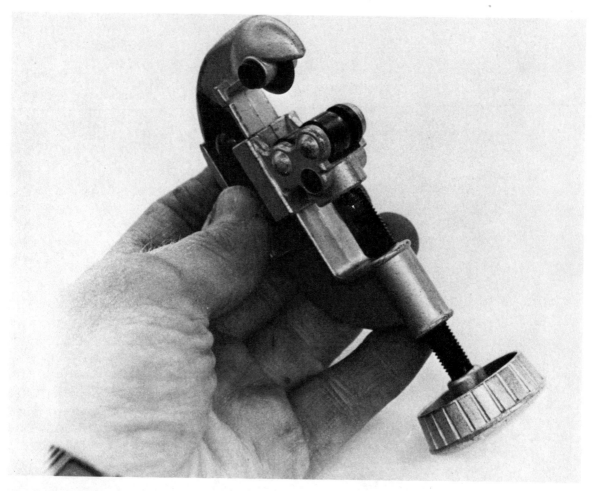

Fig. 16-20. A tubing cutter is a miniature pipe cutter.

other. The stick is slipped through the drill motor handle. A more elaborate pivot can be devised to attach to the motor.

Drill guides are helpful for keeping the drill square to the plate. As with drilling wood, making straight holes in thick plate can be a problem. Several types of guides are available.

THREAD CUTTING TOOLS

Taps cut internal threads in holes and *dies* cut external threads on round bars (Fig. 16-22). Threads are also repaired with these tools. *Die nuts* (Fig. 16-23) are used to restore external threads. Follow the instructions with threading tools and be gener-

ous with cutting oil. Most thread diameters come in a coarse and a fine *pitch*. Pitch is the number of threads per inch.

A tapped hole must first be drilled to the right size. If too large, there would not be adequate metal left for cutting the thread. If too small, the hard-but-brittle tap might seize and break off. Appendix B has a tap and drill table.

CHISELS AND PUNCHES

Chisels and punches are impact tools that are used with eye protection. Remember that a mushroom (Fig. 14-2) tends to form on the driven end of chisels and punches. These should be ground off as they

Fig. 16-21. A continuous chip indicates adequate feed (speed) pressure when drilling steel.

accumulate. High-speed chips from mushroomed heads can cause blindness and other serious injury.

Chisels and punches, made of high-carbon tool steel, are heat treated for a careful balance of hardness and toughness. Most chisels are the *cold* variety which means they are intended for use on cold material. A *hot* chisel is more expensive and can be used to cut off red-hot metal. Keep chisels sharp. When sharpening on a grinding wheel, cool the ground edge in water to avoid overheating and drawing out the hardness.

Arch hollow punches are useful for cutting holes in thin metal and some other materials. The thin cutting edge should be protected from nicks. Place the sheet to be punched on a firm backup material like hardwood, end grains of softer woods, or a lead block. The punched-out material is driven into the backup.

A solid punch for driving is also used to punch out holes in thin sheet. As with an arch hollow punch, use a firm backup.

ABRASIVE (FRICTIONAL) CUTTING

Cutting metal with a thin abrasive disc, often called "hot sawing," is a flame cutting alternative (Fig. 16-24). Reasonably priced "chop saws" are suitable for home use. An abrasive disc can also be fitted to a circular or "Skilsaw" saw for slicing thin metal. This method is quite noisy and has its hazards. Full face protection is essential to guard against the possibility of a disintegrating blade. Hot sawing produces a sharply burred edge but a straight, distortion-free cut. Recall that a disc sander/grinder can also be used for cutting.

Thin abrasive discs equipped with mounting hubs are also available for use with disc sander/grinders. Again, full face and hearing protection is necessary when cutting with a disc.

FILES

Files are effective cutting tools that are not very interesting so, with strong marketing efforts for power tools, tend to be overlooked. Filing affords good control of the location and amount of metal removed and the surface finish.

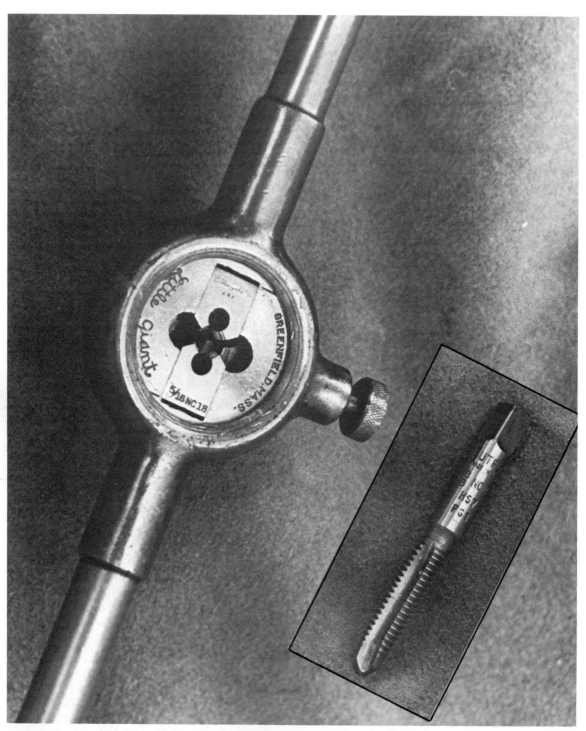

Fig. 16-22. A die, here held in a diestock (handle), is used to cut an external thread on a round bar. A tap (insert) is used to cut a thread in a hole that is first drilled to a certain size.

Fig. 16-23. A die nut is used to "chase" or recut a damaged thread.

Types of files vary according to their cross-sections, the space between teeth, and the nature of the cut. There are too many to name here; machining handbooks tell all about files. In your shop you will need a "mill file," a round or "rat-tail" file, a half-round file, and possibly a triangular file. Check the spacing of the teeth, there are at least seven varieties.

Files are hard but brittle, avoid damage by storing them separately and transport in wooden or plastic containers. Always use handles to avoid injury from the sharp tang of a file.

A clogged file, "loaded up" with chips, cannot cut smoothly and can damage a smooth surface. Use a file card often to remove chips; avoid rapping to remove a buildup of chips. Scrub soapstone across your files from time to time to keep chips from sticking between the teeth. Do not use deep cut *Vixen* files on hard metals, the teeth will chip. Some files are designed especially to resist clogging when used on "sticky" metal like aluminum and lead.

There are many file types and varieties, so find an informed tool dealer. Do not overlook the value of your files. As you reach for that grinder, question if the job could not be done better with a file.

AIR CARBON ARC CUTTING (AAC)

Air Carbon Arc Cutting (AAC) is mainly an "industrial strength" item (Fig. 16-25). Called "scarfing" or "gouging," AAC is a *metal removal* process that can also cut metal, although it leaves a wide slot (kerf). AAC is often used to "gouge" or carve out a groove in thicker metal to prepare for welding. It is a very effective way to remove an undesired weld without greatly affecting the parent metal (Fig. 16-26).

AAC requires a power source, compressed air, and a torch to use the special copperclad carbon gouging electrodes. For home use, the small 5/32- and 3/16-inch electrodes are the most practical. Besides a hood and other protective arc gear, the operator should wear safety glasses and ear plugs.

If the noise and slag can be tolerated, there are smaller AAC torches suitable for home use. Among these are L-TEC's "Groovey" and Arcair's Arky II which use small volumes of compressed air.

An air compressor is desirable but not essential for the use of AAC. Compressed air or nitrogen in cylinders work well for the occasional job. Oxygen, however, must never be used with the carbon electrodes.

347

Fig. 16-24. A small, easily moved abrasive hot saw is ideal for making straight cuts on thin wall tubing.

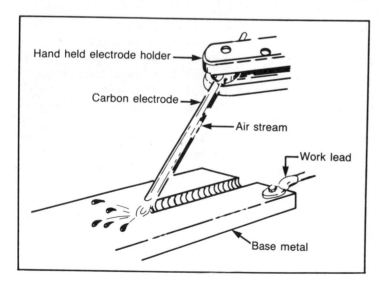

Hand held electrode holder

Carbon electrode

Air stream

Work lead

Base metal

Fig. 16-25. Air Carbon Arc Cutting (AAC) is a rapid way to remove metal involving little heat input.

Fig. 16-26. The carbon cutting electrode is extended about 3 inches from the special AAC torch.

SPECIAL CUTTING ELECTRODES

There are special coated stick electrodes that do not deposit metal, but rather are used for cutting. These are sometimes called *chamfering* electrodes. A chamfer is a bevel that does not include the entire edge of a surface. Cutting electrodes were dis-

cussed in Chapter 10.

In an emergency, a 1/8- inch E 6011 electrode and high amps can be used to cut metal, but not to gouge. If you try this, start at an edge and work back towards the center of the cut with a "gnawing" technique.

FORMING METAL

Building welded projects also involves forming. Home resources for metal forming are usually limited; forming equipment is expensive heavy-duty machinery. Sometimes, it is best to buy metal forming services, such as rolling, from a professional. But, even with limited equipment, some operations can be done at home.

Bending is often a part of metal fabrication. Considerable technology is used with bending and forming to close tolerances. The home craftsman can use a simple trial-and-error approach.

Several factors are involved to form a bend. In Fig. 16-27, an *applied force* forms the wire around the *dies* formed by the thumbs. The index fingers serve as *stops*. In similar way, a piece of wire clamped in a vise is restrained by the vise and formed around the shape of the vise jaw.

The center point of the bend falls somewhere between the far end of the die and the point of applied bending force—in this case the other hand. The exact location of the center point of the bend depends on the location of the applied force. The bend can be made "hard" with a sharp corner or "gentle" with a rounded corner. This is done by changing the shape of the die and/or the location of the applied force.

A simple bend in thicker material involves the same factors as bending the wire with an added fact that metal stretches on the outside of a bend and is compacted on the inside (Fig. 16-28). This means that you might have to consider *bend allowances* when planning your projects. Despite an impressive body of engineering data describing the mechanics of bending metal, time and metal are saved with simple practice bends. Use narrow strips of the same thickness as the metal to be used for your project.

When forming metal, wear eye protection because the mill scale on hot rolled steel shatters and flies outward. While unlikely, defective steel can contain hard areas that could shatter under forming stresses.

Figures 16-29 and 16-30 show a simple bending jig made with (die) pieces of round bar welded to a plate. Note how the straight section at the end

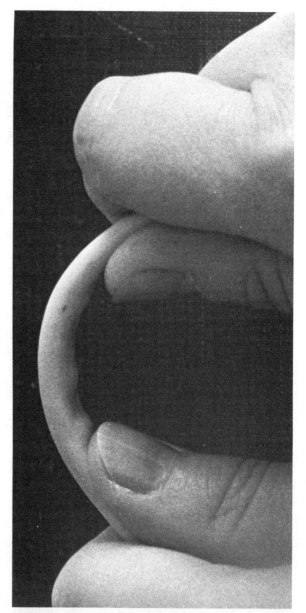

Fig. 16-27. Wire is bent as it is forced over a die. Here the thumbs act as dies and the index fingers act as stops.

of the formed bar rests against the angle bar stop. If heat is used, forming is simplified because the metal springback is greatly reduced.

Round and Irregular Bends

With pipe or other shapes as a die, you can form

350

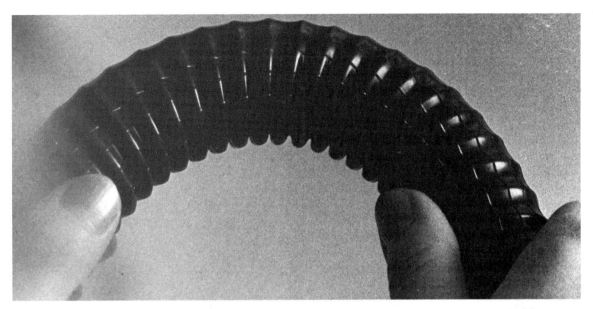

Fig. 16-28. During a bend, the outside of the curve is stretched while the inside is compacted. (The length of the center-line remains "neutral.")

round bends in your home shop. Oftentimes a vise can hold both the pipe die form and serve as a stop for the metal. An eye on the end of a piece of bar stock is easily formed in this manner (Figs. 16-31 and 16-32).

Irregular curves like scrolls can be "free formed" with a bending fork or bending jig (Fig. 16-33). It usually helps to have a full-sized pattern. If many identical scrolls are desired, flame cut a jig to conform to the shape.

Twisting Metal

Bar stock can be twisted by clamping one end in a vise and rotating the other with a wrench (Fig. 16-34). Slipping a pipe over longer bars helps to maintain straightness. A sample should be made to determine the proper number of rotations and the amount of material "lost" in the twists. Avoid using heat. It is difficult to make uniform twists with uneven temperatures at different locations on the bar.

Bending with Heat

Heat (from a torch) lowers the yield strength of

metal and makes bending easier. With the limited forming equipment of most home shops, heating metal is often the only practical way to make sharp bends. Apply heat evenly and keep the torch moving to avoid melting the steel.

After heating and forming, you might wish to cool the part rapidly. There is seldom a problem with quenching mild steel in water. However, if a heat-treated article such as a wrench handle is to be heat formed, it should first be *annealed*. This involves first heating to a critical temperature (dull red) and then allowing it to cool very slowly. (Burying in a bed of fine ash works well to retain heat.) Once annealed, the hardenable item can be more easily formed. It will then be necessary to reharden and temper the object to restore its original condition.

Heat Forming and Flame Straightening

You might recall how flame cutting a strip of steel from a plate produces a curved strip. The forces that distort metal can also form it to shape.

If just one side of a bar is heated and then allowed to cool, it will shrink on the heated side,

Fig. 16-29. Heating 1-inch round-bar lowers its yield strength to make forming around the die pins easier.

Fig. 16-30. The completed bend made to an irregular curve. Notice how the angle bar clip served as a stop.

Fig. 16-31. Starting to form flatbar over a pipe while clamped in a vise.

Fig. 16-32. The vise is alternately opened and clamped to allow forming the flat bar into nearly a full circle or "eye."

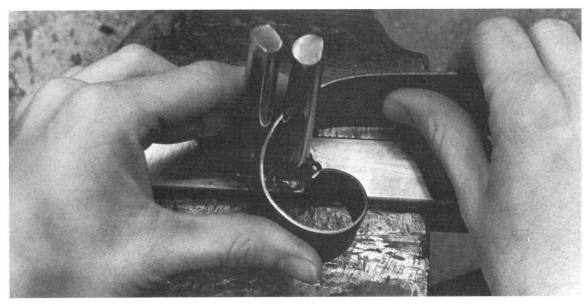

Fig. 16-33. Scrolls and other irregular bends can be formed in a bending fork.

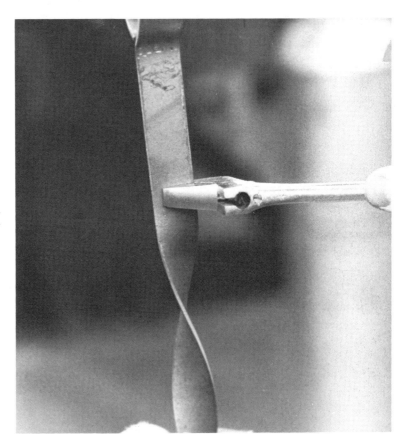

Fig. 16-34. Twisting bar stock is sometimes done for appearance and sometimes for function.

forming a curved shape. This effect is explored in the next chapter. The trick is to keep heat from reaching the other side of the bar. The work should be positioned so that the area to be heated is up to take advantage of the fact that heat naturally rises. It is also important to avoid overheating the metal. This same technique works on structural shapes. For example, one leg of an angle bar can be heated to put a curve in the bar.

Structural distortion from the forces of weld shrinkage can often be corrected by introducing counter forces. Such methods, called *flame straightening*, are often used to straighten items such as car frames and bent shafts. Heat forming at home can avoid the expense of having parts formed in a professional shop.

The distorting effects of heat are at work during a welding operation. How to cope with the effects of welding heat is discussed in the next chapter.

Chapter 17

Fitup and Distortion Control

A HIGH POINT OF BUILDING WELDED PROJ-
ects occurs when you begin putting together
parts of the basic structure. A sense of accomplishment develops along with the shape of the project.
Up to this point in the construction hurdles have
already been crossed, but a few yet remain. There
are forces at work that, if you let them, have a nasty
way of changing the wonderful project you planned
into something you want to hide. This chapter includes methods to help ensure that your final
welded project will be the same one that you
planned.

In the welding industry, the term project *fitup*
often includes all of the fabrication activities prior
to the final welding. The fitup operations of laying
out, cutting, forming, and finishing are discussed
elsewhere.

An important difference between the construction of a wood and a welded metal project is the
potential for destruction with the final joining
process. While it is possible that a light wood project could be destroyed if an attempt were made to
nail it together with oversized spikes, this seldom

happens because splitting wood is readily noticed.
However, a precise assembly of metal parts can be
seriously affected in rather subtle ways as soon as
the welding begins. The heat of welding produces
extreme but gradual forces that affect not only the
dimensions of a project, but can even distort its basic shape. Developing the means to anticipate and
control these forces is the challenge (and glory) of
welding fabrication. It is a game, with the final outcome uncertain. With the rules known and under-
stood, this game is seldom lost.

EFFECTS OF HEAT

In the last chapter, it was stated that the heat of
a torch can be used to form a curve in a length of
straight bar stock. This is due to the basic princi-
ple that metal expands when heated and contracts
upon cooling. It is worthwhile for home welders to
experiment with the effects of heat. You will gradually develop a feel for what happens to a plate with
the heat of a welding operation.

Place a bar of mild steel in a vise as shown in
Fig. 17-1. The exact size of the bar is not critical—

357

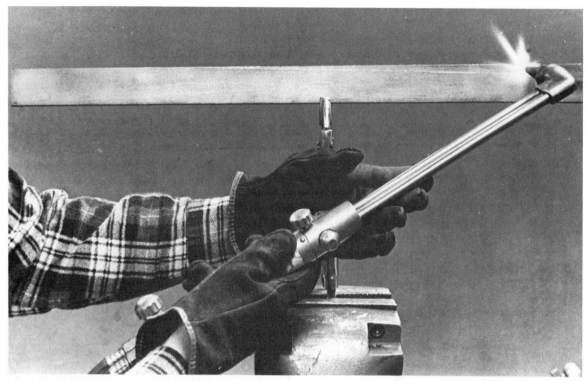

Fig. 17-1. A straight bar heated along one edge will change dimension at just that edge, causing distortion.

about 1/8 inch thick by 1 or 2 inches wide, and maybe 18 inches long should do just fine. Using an oxyacetylene flame, heat just the top edge of the bar. Note how, while being heated, that edge first expands to form a curve. What is not as readily apparent is why, upon cooling, the bar does not simply return to the original straight shape. In fact, once cooled, the bar undergoes *distortion*. This means that it takes on a permanent change of shape—in this particular instance, a curved shape opposite to that when it was fully heated (Fig. 17-2).

RESTRAINTS

The reason that a heated bar distorts is due to colder metal inside the bar acting as a *restraint* to prevent full (outward) expansion along the heated edge. However, upon cooling, there is less restraint to prevent contraction. Hence, the bar finally ends up with one edge more contracted and therefore shorter than the other. Where did the metal go that

originally made up the straight edge? Once hot enough, it flowed like soft plastic from the squeeze of contraction to make the bar slightly thicker along the affected edge. This thicker area is said to be *upset*.

An Upsetting Experience

A graphic demonstration of the expansion, plasticity, and contraction of metal is often made by placing a short length of bar stock endwise in a strong vise (Fig. 17-3). The vise jaws act as a restraint to prevent any increase in length. The bar should be held snugly without overtightening the vise. The heat of a torch is applied until the bar is red hot. (Do not apply heat directly to the ends or the vise could be damaged.) Remove the heat and allow the bar to cool until it nearly reaches room temperature. It should be quite easy to dislodge without loosening the vise; it might just drop out by itself. The bar has become shorter. Where did the metal

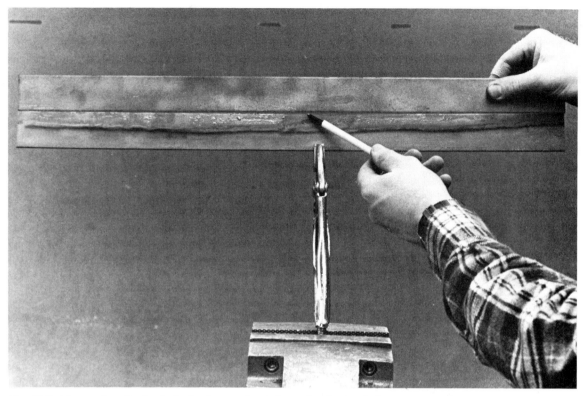

Fig. 17-2. After cooling, the heated edge has contracted, causing the bar to become curved.

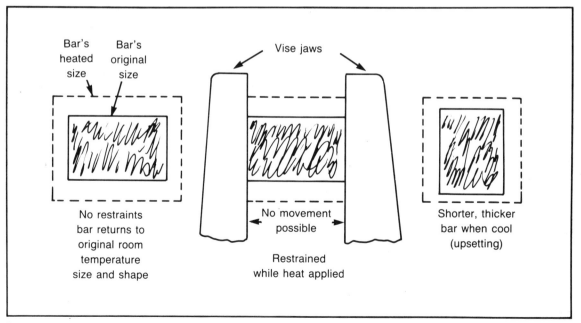

Bar's heated size   Bar's original size

Vise jaws

No restraints
bar returns to
original room
temperature
size and shape

No movement
possible

Restrained
while heat applied

Shorter, thicker
bar when cool
(upsetting)

Fig. 17-3. Effects of heat with a restraint.

go? It moved plastically within the hot bar. This plastic flow causes the bar to become slightly thicker or upset. You would have to subtract two micrometer readings, one taken before and one after the test, to see just how much thicker.

None of the interesting shape changes just discussed would occur if a piece of metal were heated and then cooled slowly and evenly in something like an oven. There would be an even expansion and contraction without the distorting effects of any colder metal restraints. Upon cooling, the piece of metal would contract to its original size and shape. Such uniform heating and cooling conditions do not exist for most welding operations.

WELDING AND DISTORTION

It is the essence of weld metal to be hot and totally expanded when it is first stuck onto the plate. The metal that surrounds the very hot weld zone is considerably colder and therefore acts as a restraint. Remember that where there is a restraint, there are also forces to cause distortion. This is why welding done on a flat plate causes the plate to warp.

When all parts of a welding joint are free to move, the weld metal will simply carry the parts of the joint along as it cools. Figure 17-4 shows distortion to a lap joint. Unrestrained expansion and contraction of a groove weld made on a butt joint, also cause *angular distortion* to the lineup of the plates (Fig. 17-5). The root opening, or gap, between the plates also changes with the heat of welding. Ideally, parts to be welded would always be arranged so that, with welding, they would naturally assume the desired angles and positions. This condition seldom exists. Unless you first experiment with some samples, it is difficult to predict exactly how much movement will occur. In actual practice, parts are usually secured to minimize plate movements with ample tack welds and/or bracing.

Fig. 17-4. The contracting weld deposit caused angular distortion between the lap joint members.

STRESS

If you were to completely restrain joint members from moving with clamps or rigid bracing, no apparent distortion would take place. However, even though the parts are not able to move, a contractive force, or *stress*, still develops as the weld cools.

It might seem that welds cooling in a highly restrained joint ought to crack. Sometimes they do, but most of the time weld metal simply tries only so hard to contract, then gives up and *yields* (or

Fig. 17-5. Angular distortion to a welded butt joint.

stretches) to relieve the stress. The yield strength is the loading value at which the weld metal begins to permanently stretch. This wonderful ability is possible only because welding filler metal is made of exactly the right blend of ingredients and is also very pure. Welds are largely stress relieved by the time they cool. However, the weld zone usually contains some left-over or *residual* stresses.

The postweld process of stress relieving is used to remove residual stresses when the weldment will be subject to impact loads or vibrations. Stress relieving prevents welds from cracking on items such as motor mounts and lifting frames. In the home shop, stress relieving of small assemblies can be done by slow heating with a torch to about 1200 degrees Fahrenheit. This temperature can be guessed at when the steel is seen to be red hot in the sunlight. The project should be "soaked" at this temperature for as long as is reasonably practical. The welded object can then be buried in ashes to cool slowly over several hours. More precise stress relief is one of those activities that you might want to have done on the outside.

CONTROLLING DISTORTION

Welding distortion, like taxes, is something we must learn to live with. And, like taxes, if you play the game right you might actually be able to make things work in your favor. Distortion is often controlled, but seldom eliminated. Several practical ways to control distortion in the home shop include:

- Temporary restraints
- Permanent restraints
- Limiting the heat input
- Balancing weld forces
- Preheating
- Presetting and Prebending

Temporary Restraints

Clamps are used to rigidly hold parts from moving while welding takes place. For example, a butt or lap joint clamped to a table can be released once the weld cools. In this case, the heavy plate of the table is a temporary restraint. If clamps had not been practical, the plates could even be tacked directly to the table. After the weld has cooled the tacks would be cut.

A slightly more involved way to hold plates in alignment can be done with *dogs* (clips) and wedges. The plates are moved by wedges driven against dogs welded to one member (Fig. 17-6). The wedge is driven from the welded side of the dogs. If you do not see why this is important, try driving a wedge the other way. One advantage with using dogs and wedges is that the plates are held down but can also slide laterally, thus allowing some stress relief. Wedges like that shown in Fig. 17-7 are easy to make.

A *U assembly clip* can also be used to align plates for either lap joints or butt joints (Fig. 17-8). A scrap piece of square or rectangular tubing can be split to form the U.

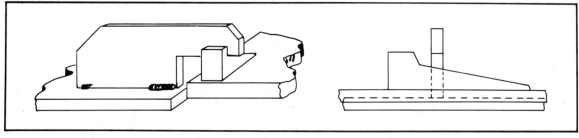

Fig. 17-6. Dogs and wedges produce a great force for moving plates.

To make two plates even is to make them *fair* (Fig. 17-9). Plates which are not fair are said to be "out-of-fair." A *fairing bar* (Fig. 17-10) is sometimes useful in lining up plates. The bar is used to align plates for tack welding by starting at one end of a joint and working towards the other end. The bar is tacked just enough to permit pulling the high plate down. After the joint is tacked at that point, the fairing bar is easily "hinged off" by prying in the opposite direction. It can then be moved to the next location along the joint.

Clips and bolts can be used to pull the members of a T joint together (Figs. 17-11 through 17-14). These, and other temporary restraints that involve welding to parts of the project, are practical if clamps cannot be used or are unavailable.

Strongbacks are temporary heavy stiffeners most often used across butt joints to keep the plates fair and to prevent angular movement from the heat of welding. They need to be drawn tightly against the plates they are restraining. One popular way to do this is with the use of *saddles* (yokes) and

Fig. 17-7. A wedge is useful for fitup. Notice how the large end has surfaces to allow driving either way.

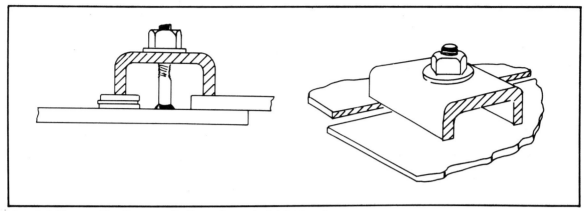

Fig. 17-8. **U** assembly clip can pull either a lap or butt joint together.

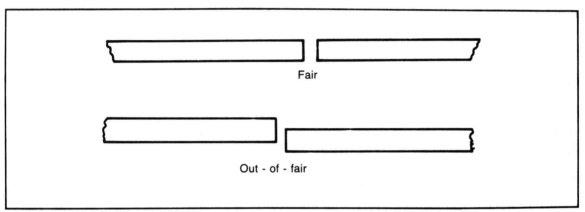

Fair

Out - of - fair

Fig. 17-9. Fair and out-of-fair.

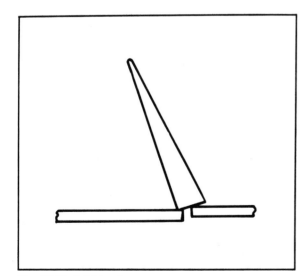

Fig. 17-10. A fairing bar.

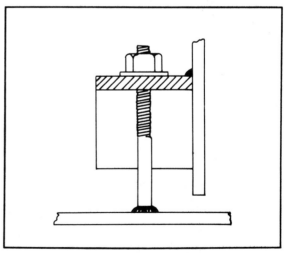

Fig. 17-11. A puller clip and bolt if correctly applied draws an edge to a surface as the bolt is tightened.

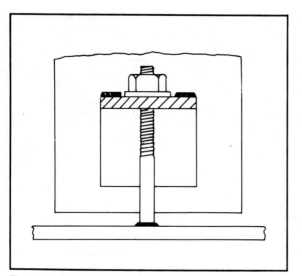

Fig. 17-12. The puller clip will collapse if improperly attached.

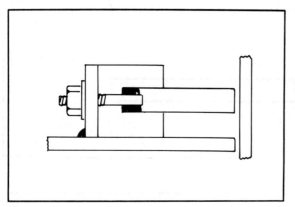

Fig. 17-13. Lateral pulling can be done with a puller clip and bolt.

wedges (Fig. 17-15). The strongback is drawn against the low plate starting at one end and working towards the other. This progression is important to avoid developing a buckle (wrinkle) in the plates.

Once the strongback is pulled tight, it is sometimes tack welded (on one side only to permit easy removal). The wedge can then be knocked out and the saddle hinged off (Fig. 17-16). The saddle is moved progressively to the next location along the strongback or, if you have enough saddles and wedges, they can simply be left in place. If the strongback is tacked to only one of the plates, the other plate will be free to slide laterally as the weld cools. This is not always possible, but allows for considerable stress relief while holding the joint flat.

On larger weldments like commercial tank work, a strongback is often notched with "ratholes" at the butt joint (Fig. 17-17). This is done to permit continuous welding of the joint. Without the rathole, the weld would need to be stopped at each side of the strongback. The unwelded segment would then have to be picked up (welded) after the strongback is removed. After the joint is welded, the strongback can be hinged off along with any remaining saddles or dogs. Grind off any tack weld scars from the saddles and strongbacks remaining on the plate.

A *brace* is a steel supporting member tacked or clamped in place long enough to permit the assembly, final welding, and sometimes even the trans-

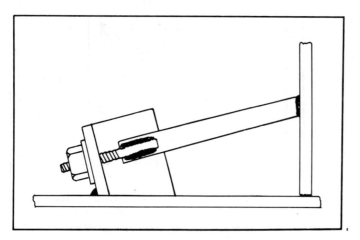

Fig. 17-14. Placed at an angle, the puller clip and bolt can pull both over and down.

Fig. 17-15. A plate and stiffener can be pulled together with a saddle and wedge.

portation of a welded project. A brace can be placed across the corner of an angle-bar frame prior to welding. A T joint can also be held from moving by using braces. Braces need enough tack welding to stop them from snapping as weld forces develop. Be sure, however, to place your tacks on one side only to make later removal easy.

Permanent Restraints

A permanent restraint is an integral part of a welded structure that deliberately or coincidentally prevents plate movement and distortion. *Stiffeners* are bars that provide rigidity to plate, similar to the way framing studs give rigidity to sheathing for the wall of a house. Stiffeners are often pulled tightly against the plate they are to strengthen with the use of saddles or dogs and wedges.

Most of the time a stiffener does not require a continuous weld. Instead, *intermittent* or "skip" welds are used to reduce distortion and for economy. Skip welds are often placed on both sides of stiffeners. During the fitup and assembly, place tack welds so that they will be incorporated by the later skip welds. This makes for a much neater assembly. As a refinement of this, tack welds should even be placed at the *ends* of the skips to provide extra metal for helping to fill the craters. It is important to clearly mark the location for skip welds prior to making the tack welds so that precise placement is easy.

The same general assembly methods used for installing stiffeners also apply to working with *gussets*. A gusset (Fig. 17-18) is a type of permanent corner brace providing diagonal strength at a T joint. When making gussets, the sharp corners

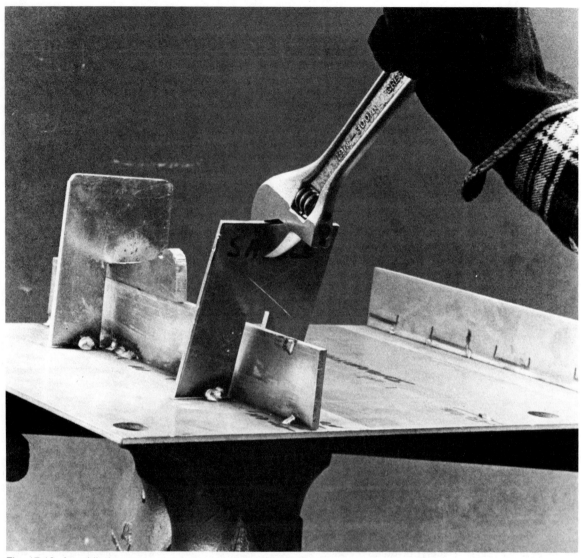

Fig. 17-16. A saddle is easy to hinge off on the side with the weld.

should be cut off. Such cuts are referred to as "snipes." The snipe at the right angle corner allows a weld to be run continuously along the joint under the gusset. The other two snipes provide adequate material to stand up to the heat of welding without burning away.

Limiting the Heat Input

A very effective way to control distortion is to limit the amount of welding heat put into a structure.

This can be done by controlling the size and location of the welds.

A fillet weld is measured by the length of its legs. Most do-it yourself welders tend to make welds too large. A weld should normally be flat or slightly convex across the face, unless there is some special reason for making it concave (Fig. 17-19). As a general rule, the size of a fillet should not be larger than the material thickness. In fact, if both sides of a T joint are to be welded, the size of each

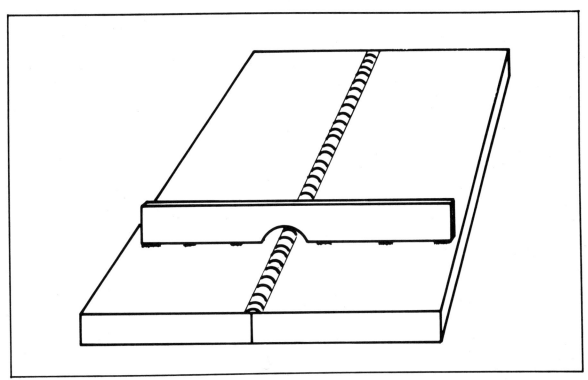

Fig. 17-17. A strongback with a rathole permits complete welding of the joint.

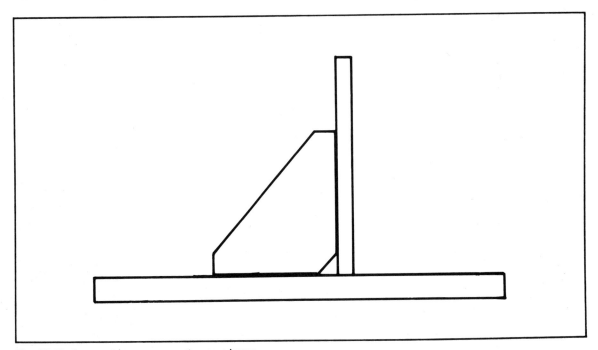

Fig. 17-18. A gusset is a permanent corner brace.

367

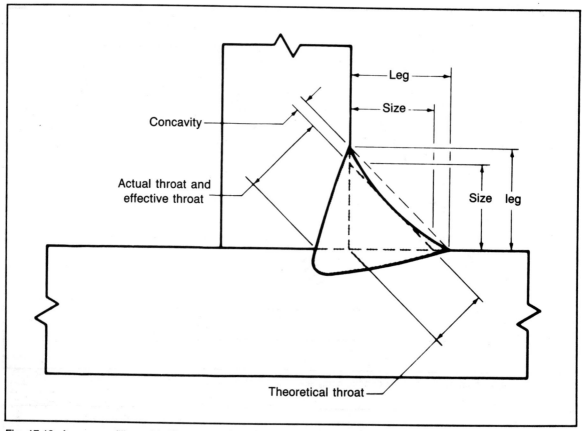

Concavity

Actual throat and
effective throat

Leg

Size

Size   leg

Theoretical throat

Fig. 17-19. A concave fillet weld. (Courtesy of American Welding Society.)

of the fillets only needs to be equal to 1/2 the plate thickness.

If two different medium metal thicknesses are combined, the rule is to size the weld according to the lighter member. A T joint made up of 11-gauge (1/8-inch) and 1/4-inch plate would require total welds measuring 1/8 inch. According to the rule in the last paragraph, each weld needs to measure only 1/16 inch. However, it is not practical to try to make welds smaller than 1/8 inch. Unequal leg fillets with more weld energy and metal directed towards the heavier plate are used with thick-to-thin assemblies.

Limit the heat input to a butt joint by keeping root size (gap) to the minimum needed for access to the back of the joint to obtain adequate penetration. Space the plates of butt joints about a wire diameter apart or, with stick electrodes, 1/32 to

1/16 inch. As always, test your welds on scrap before working on the real thing. Butt joints should not be covered with excessive reinforcement. Also avoid a wide side-to-side weave pattern. (Fig. 17-20).

Intermittent (skip) welds have already been shown as a way to avoid distortion. Skip welds are used wherever a joint does not require the strength of a continuous weld or a leakproof seal. Besides involving less total weld footage, skip welds also keep heat from accumulating in any one area. Skip welds are located either back to back or staggered. Back-to-back skips (also known as chain intermittent welds) concentrate the welding heat, so are suitable mainly for heavier metal thicknesses. With thin plate, staggered skip welds keep distortion to a minimum.

To avoid an excessive heat buildup when welding a long joint, a *backstepping* technique is sometimes used (Fig. 17-21). Whether or not you use this will have to be weighed against the disadvantages of extra starts and stops.

Weld *sequence* is the order in which parts of an assembly are welded. A sequence is followed for two reasons, both to control distortion. One reason is to balance weld forces, and the other is to avoid a significant buildup of heat in any one spot.

A general rule is: less welding time, less heat input. In your home shop you can often increase welding amperage along with your travel speed. The net result is less distortion.

Balancing Weld Forces

A T joint welded on only one side is subject to cracking if the root of the weld is placed in tension. (You have probably seen how easily a T joint welded on one side only can be "hinged" apart.) To withstand stresses from both sides, many T joints need to be welded on both sides. If both sides

could be welded at exactly the same time with exactly the same heat input, angular movement (distortion) to the standing joint member would not take place. In such a situation the forces that cause distortion are in balance and cancel each other. It is not likely that, in the home shop, you can simultaneously weld both sides of a joint. You need another way to balance the weld forces.

Most welders jump back and forth when welding a T joint. First, weld about 1/3 of what is to be done on one side of the joint. *Allow this welding to cool.* Second, do all of the welding on the second side. Third, weld the rest of the first side. Whatever distortion took place with the first welding should have been balanced (and then some) by the second phase. The final welding should even things back out.

Preheating

Recall that distortion results from the action of a restraint upon heated metal. Colder metal surround-

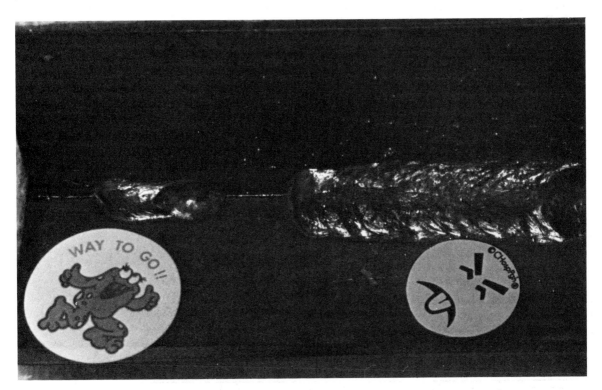

Fig. 17-20. Avoid overwelding with a wide weaving pattern.

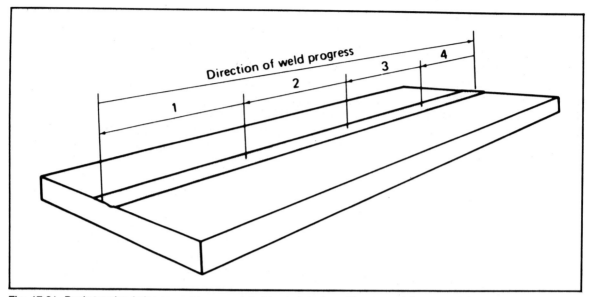

Fig. 17-21. Backstepping helps to avoid accumulated heat distortion. (Courtesy of American Welding Society.)

ing welds is a restraint. Limiting temperature difference controls distortion. It is sometimes advisable to preheat a joint so the entire weldment moves as a unit. Preheating eliminates condensation, reduces cracking, dries the plate, and improves fusion on heavier plate. It is difficult to preheat larger projects in the home shop. The heat required is usually high, perhaps 1000 degrees Fahrenheit (red hot) or more. This is often not practical with an oxyacetylene outfit. Also, preheat should be applied evenly and done gradually over a period of time.

For preheating larger objects, rent a propane outfit. As soon as you are finished working, cover the object with heat retaining blankets such as fiberglass. A slow, even cooling period helps to avoid distortion.

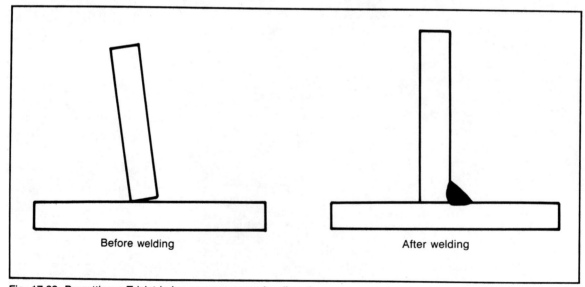

Before welding

After welding

Fig. 17-22. Presetting a T joint helps overcome angular distortion.

370

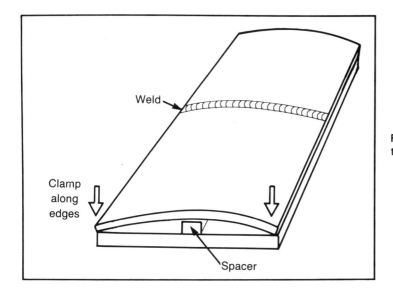

Fig. 17-23. Prebending compensates for the contractive forces of welding.

Weld

Clamp along edges

Spacer

Presetting and Prebending

If you can predict the forces of distortion, you can take steps to minimize it. An example is the leg of a T joint that, after welding, leans towards the welded side. If the leg is *preset* and tacked the opposite way, the final result should be square (Fig. 17-22). If welding is likely to cause warpage, plates can sometimes be *prebent* (Fig. 17-23).

FITTING TOGETHER T AND BUTT JOINTS

Some methods for assembly of a T joint have been seen earlier in this chapter with the use of clips and bolts. Temporary braces (Fig. 17-24) are also sometimes used to hold a T joint in alignment. The members of a tall T joint can be drawn together with

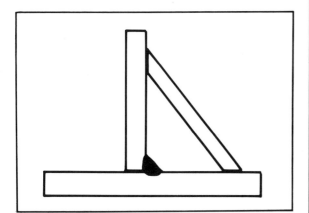

Fig. 17-24. A temporary brace should be placed in compression by being located on the same side of the joint as the weld.

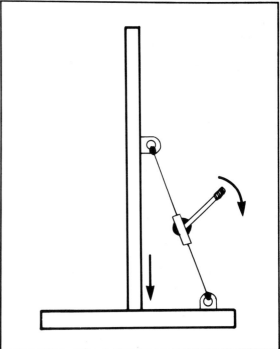

Fig. 17-25. A comealong connected to padeye (plates) can be used to draw a tall T joint together.

pads and comealongs (Fig. 17-25). The installation of a strongback across a butt joint as was shown in Fig. 17-17 prevents angular distortion.

Figure 17-26 is a T joint assembly with gussets. Try to tack gussets before welding the joint they strengthen. Joints should be assembled from one end, working towards the other.

A butt joint welded from one side is likely to distort. Yet, it might not be practical to fasten the member plates to a flat surface. Other ways to control distortion with a butt joint include U clips, strongbacks, and presetting.

As a rule, metal lighter than 1/4 inch does not require any joint preparation (bevelling, etc.). If you bevel the edges of the member plates to achieve adequate penetration, try to avoid leaving a sharp feather edge. A grinder or file can be used to remove a feather edge after the bevelling cut has been made. It is very easy to burn through a joint with feather edges unless a permanent backup bar is used. Backup bars are seldom practical.

If a butt joint can be placed for welding in the flat position, a dissimilar backup bar of aluminum, copper, or special ceramic tape can be used. These materials permit fast one-sided welding. These materials will not stick to the weld. Figure 10-57 showed a steel butt with a dissimilar aluminum backup.

A joint requiring a continuous weld can be properly fit and tacked and still have start and stop defects. Starting a weld on cold metal before the arc gets established tends to cause cold laps and inclusions. At the end of a weld, the heat buildup is often so considerable that cratering is excessive. The use of run-on and run-off tabs (small temporary platforms) can overcome these problems. The start and stop defects are then located on the tabs, which are easy to hinge off and discard.

An ordinary spirit level can be used to check your fitup as you progress. Several compact and convenient designs of levels and angular measuring devices are equipped with magnetic bases.

JIGS OR FIXTURES

A jig is a device to line up parts in the right arrangement for welding. So is a fixture. The terms mean about the same, but a C clamp or other non-dedicated tool used for holding plates is usually called a fixture. Sometimes it is more practical to use a jig only for tacking. The assembled part can then be moved about for further welding. A tacking jig does not have to be as strong as a welding jig. Wood is sometimes used to build a tacking jig.

A frequent operation in the home shop is to assemble pieces of bar stock or pipe into square corners. A universal corner assembly jig can be fabricated using bars with adjustable slots. Any suitable type of clamping arrangement can be used for holding the parts against the bars. If joint members are placed so the joint is located over the edge of a table, the corner can be completely welded out while still securely clamped in place. This same idea can be used to line up parts at any angle. This type of jigging also works well for assembling wooden structures.

An angle plate is easy to make and can be used to assemble parts in a variety of ways. Figure 17-27 shows the web (gusset) of the angle plate clamped in a vise. This positions the square surfaces at a more comfortable working height.

The V shape cradle in the web of angle bar can be used to line up pieces of round stock. Figure 17-28 shows a welding jig made from a locking welding clamp and angle bar. It is used to clamp drill bits to lengths of round bar. This simple unit can save money by enabling you to turn a standard length drill bit into an expensive long-shank drill. Notice how the heel of the angle bar was cut away to gain access for welding all the way around without needing to release the clamp. Use stainless steel electrode to weld a drill bit to an extension rod.

Circular bends are used to make rings, U bolts, and eye bolts. The trick is to do it without kinks. Circular bends can be made with flat or round bar by forcing the bar over a rigid round shape as was shown in Figs. 16-31 and 16-32. This round form could be a smooth piece of pipe, a fitting, or any other object of a suitable diameter. The scrap bin at your local machine shop might have just what you need. Experiment to see how much the metal springs back after you have made one complete bend. A torch applied to the bar makes it easier to

Fig. 17-26. A welded assembly with gussets.

Fig. 17-27. Clamped by the web, the angle plate can be used to fit outside corners with ease.

form and greatly reduces the springback. If you use a hammer, hold the free end of the bar with a gloved hand to prevent it from vibrating and possibly assuming an unwanted shape as it whips.

An easy way to make a number of small rings is to clamp or even weld the ring stock to a rigid round form. The stock is forced around the form to make a continuous spiral. The spiral is then cut, forming the individual rings. Next, the ends of each ring section are clamped together and welded. This

Fig. 17-28. A welding jig for joining round bars, including drill bit extensions.

method involves flattening each ring after it has been welded. If the ring is to be used in an application where maximum strength is needed, be sure to bevel the ends of the ring before welding.

To build identical assemblies, it is often best to build the first one very carefully and then use it as a jig for the rest. It might help to tack temporary stops to the original assembly to keep parts from sliding.

A simple forming fixture or die, for example to make an offset bracket, can be made as shown in Figs. 17-29 and 17-30. Such a die can be forced together in a vise.

A pattern plate and transfer punch are often a blessing for duplicating precise layout points—such as center marks for holes to be drilled or burned. A transfer punch is not tapered and has a center locating point. The pattern plate is first carefully laid out and then drilled with a special size of drill. This size matches the diameter of the transfer punch. The pattern plate is clamped on a project part and center punch marks are transferred at each hole location.

Welding jigs are usually made of metal, but do not overlook the possibility of using wood. The strength of wood might not be adequate for a true welding jig, but often tacking operations can be set up fast on a wooden surface. A sturdy old flat entry door can be picked up for a few dollars at a demolition salvage outlet. Make sure that you pick a flat one. Place the door on sawhorses or stands at a convenient working height. With the use of a screw gun and self-tapping drywall screws, jig-building becomes a high-speed activity. Attach angle bar clips to the door with screws. Parts for tacking can be clamped to the angle clips. A wood door

Fig. 17-29. A simple die for clamping in a vise.

is light enough to be taken down once it is no longer needed. A similar steel surface could not be moved around with the same ease.

TACK WELDS

Often underrated, tack welds are very important. A first impression is that if they hold together long enough to get the main welding done, they are good tacks. Tack welding is a part of fitup and tack quality has an effect on the overall quality and appearance of the finished project. A tack needs to be strong but it must not be so large as to form a conspicuous lump in the final weld. If the correct procedures are followed, there should be few signs of tack welds on your completed projects.

Generally, tack welds should be made hot and small. When using a wire feed, turn up the voltage and, for any work other than short circuit transfer on light material, increase the wire feed speed as well. With the stick electrodes, use E6011 for tacking. Also use an extra 10 to 20 amps or select an electrode one size smaller. With experience you will soon learn how small a tack can be and still have sufficient strength.

One of life's more unpleasant sounds, heard during final welding, is the sharp "pop" of a breaking tack weld. If a tack should break, do not weld through it with the idea of healing the crack. The extra cold metal from the tack can absorb enough welding energy to leave the crack buried in the weld. With load impacts or vibrations these cracks often begin to migrate. Stop welding and check for any serious misalignment. Use a grinder or chisel to remove the broken tack. It might be necessary to use clamps or other devices and retack the joint.

More rework might be needed if the member plates have moved too far.

Place tack welds so they will be inconspicuous on the finished project. Use your welding hood to its advantage as outlined in Chapter 10. A careless arc strike (Fig. 10-20) is a poor signature to have on your work. If you make a bad tack, replace it.

Tack welds are made quickly on cold plate, so there is a tendency to form a cold lap at the start of the tack. As you later weld through these tacks you stand a strong chance of merely covering over a weak spot in the joint. To minimize this tendency, "feather out" any large tack welds with a disc grinder or a diamond point or gouge (round nose) chisel. Always remove the unbonded cold area of a tack.

FOUR BASIC FABRICATIONS

Certain representative fitup operations can be seen in the assembly of some useful projects. Four are shown: a box shape, an angle frame, a pair of trammel points, and the fabrication of a welding table. You can probably find use for each of these.

Fabrication of a Welded Box

The box shape is basic to many objects besides an actual box. Welded tanks, cabinets, and other familiar items such as wood burning stoves are essentially box fabrications. In the example, exact dimensions of the box are omitted for simplicity. Always check the dimensional accuracy of the parts before starting to tack them together. Diagonal distances should be equal for square and rectangular shapes.

The corner joint is the basic joint of a box shape. A full-open "corner-to-corner" fitup is shown in Fig. 17-31. With this fitup, the panels are cut to the inside dimensions of the box. Assemble

Fig. 17-30. An offset formed in flat bar with the die.

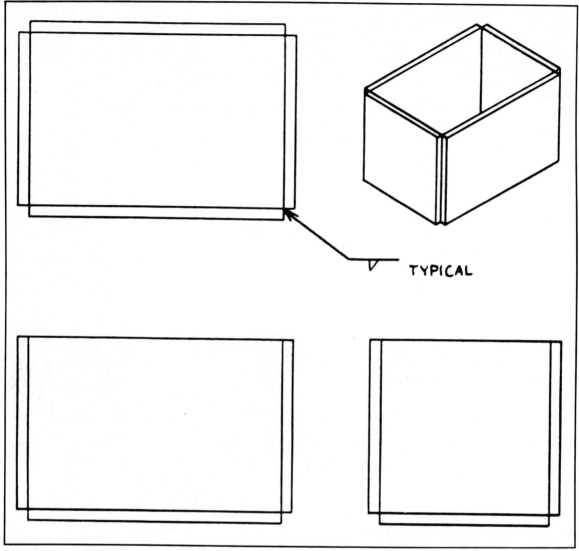

Fig. 17-31. A box with corner-to-corner fitup is easy to both assemble and weld.

a full-open corner joint for tacking with an aluminum shim, spacing one member the proper distance from a flat surface. Figure 17-32 shows a side assembled to the bottom of a box. The tacks are placed on the inside with this method. Aluminum shims will not adhere to the steel in case your tacks burn through (which can easily happen). After the sides are tacked to the bottom, they are tacked to each other on the outside (Fig. 17-33). At this stage, again check diagonals (Fig. 17-34) to ensure that

everything is square. If things look right, tack corner braces to hold the square shape (Fig. 17-35).

A furniture (pipe) clamp can be helpful to align the sides of a box shape. Open boxes should be braced at the corners before welding. If a tank is equipped with anti-sloshing baffles, tack them in as early as possible to provide extra rigidity.

Welding an outside corner joint with thin or narrow members tends to cause distortion (Fig. 17-36). If the box sides are not high enough to pro-

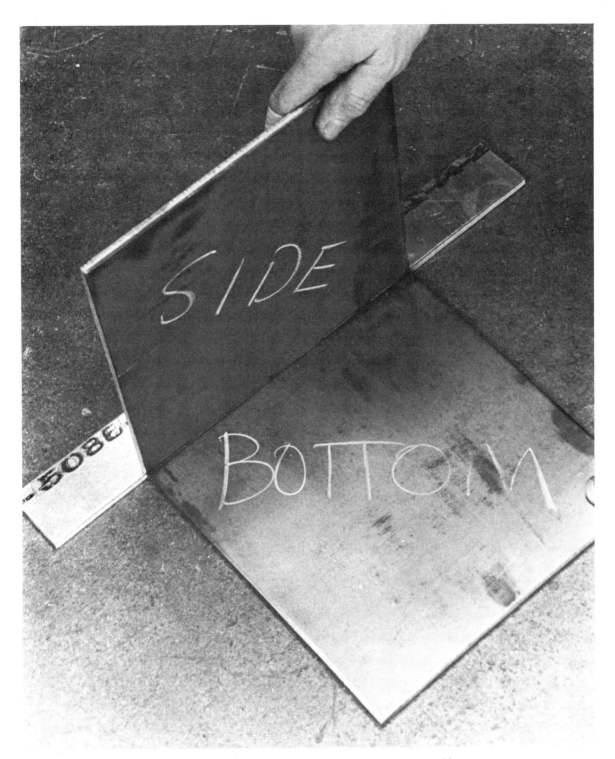

Fig. 17-32. Assembling the side and bottom. Note the use of an aluminum spacer or shim.

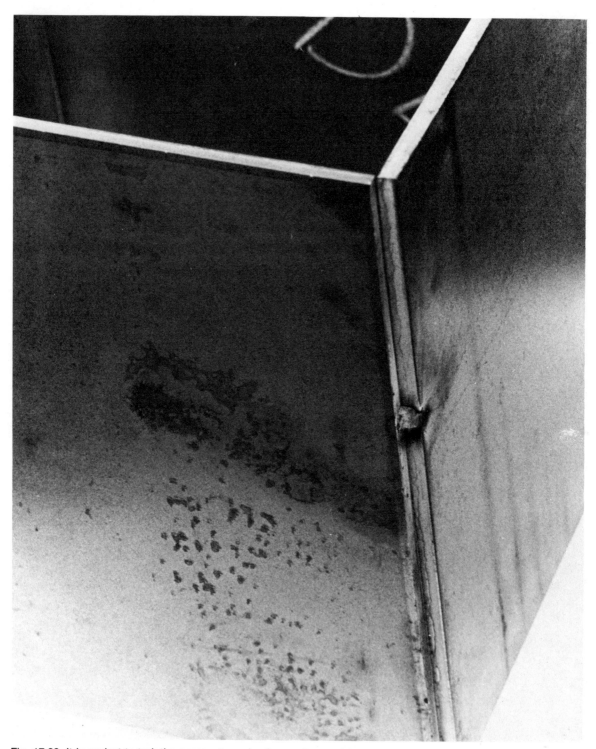

Fig. 17-33. It is easiest to tack the corners to each other on the outside.

Fig. 17-34. Comparing the diagonal dimensions ensures squareness.

Fig. 17-35. Corner braces hold the box square during welding.

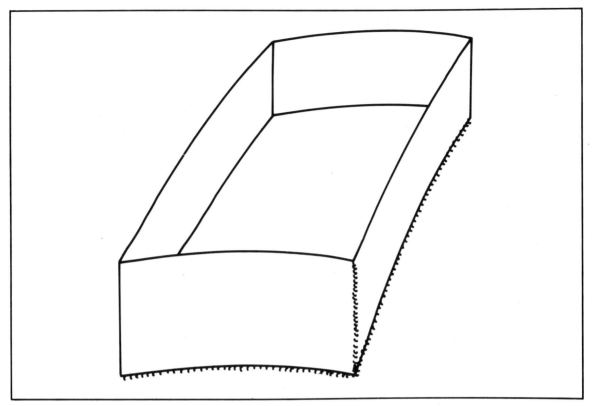

Fig. 17-36. Welding tends to distort a flat box.

vide a healthy restraint, the bottom can be bowed. Such distortion can be minimized by tacking or clamping the box upside down to a heavy plate before the welding begins. Avoid a localized heat buildup with a backstepping weld sequence.

Attaching the top or final corner of a closed container can be tricky because it tends to drop inside. Use strips of thin metal across the corners as shown in Fig. 17-37 to avoid this problem. Remove each strip as you tack that corner. Also, provide a vent to prevent a buildup of hot gases inside the container. A pipe coupling can be attached and later closed with a plug.

A box shape can develop some impressive distortion if the heat input is not controlled. If possible, slightly incline the corner joints so that they can be welded with a fast travel speed in a downhill direction.

Sometimes the inside corners of a box must be smooth and free from crevices or rough areas that might trap its contents. For example, containers for foodstuffs and chemicals must be easy to empty and flush out. In these cases, a concave fillet on the inside of the container is needed. The outside of such corner joints do not require a full fillet because some of the strength for the joint will come from the inside weld. A *partial lap corner* fitup (Fig. 17-38) is often used to save on welding and to achieve good appearance. When welding the inside of a box, be sure to provide positive ventilation for yourself.

A box shape welded on the inside tends to distort inward. To keep the sides from "sucking in" use wooden *shores* to hold them out. In fact, it is often advisable to slightly prebend the sides with oversize shores. After they are removed, the sides will spring back flat.

Fabrication of an Angle Frame

There seems to be no end to the items that can be made from standard structural shapes such as

Fig. 17-37. Strips at the corners keep the last panel from falling in.

square tubing, round tubing, or angle bar. A representative construction sequence is now included for building an angle bar frame. Such frames are used as table or bench tops, stair landings, frames for heavy duty cabinets, etc.

An angle frame is built with the flanges either in or out. Most of the time, flanges are directed inward as in this example. This gives an outside cor-

ner to cover plates like bench tops. With the flanges outward, a convenient mounting surface for bolting is formed. An example might be a base for a box or tank that must be kept from moving about on a vehicle.

There are two widely used ways to fit angle bar corners together. Both are shown in Fig. 17-39. It is often just a matter of personal preference as to

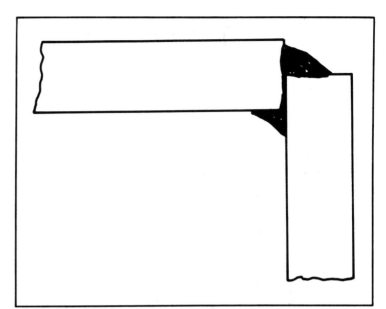

Fig. 17-38. A partial lap corner is usually welded on both sides.

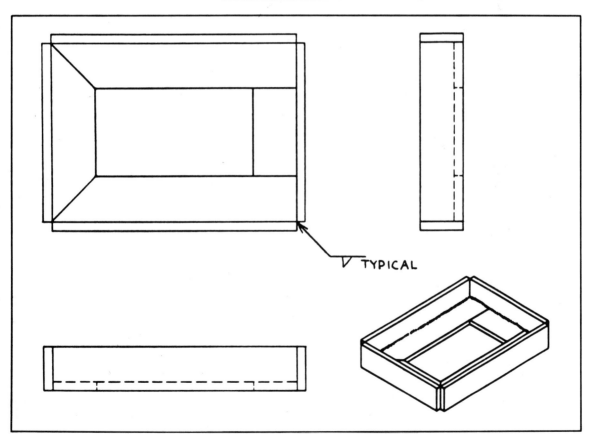

∇ TYPICAL

Fig. 17-39. An angle bar frame can have either coped or mitered corners.

which one you will use. One way is to *cope* or remove a part of one flange in order to insert a flange of an adjoining member. Figure 17-40 shows the outside of a corner with a coped construction. The cutting must be accurate to provide a decent joint for welding.

Another, perhaps more familiar, method of joining angle bars at corners is with a *miter*. This requires making a 45-degree cut on the end of each frame member. A mitered fit corner is seen in Fig. 17-41. Notice that for both methods, the flanges of the angle bars that meet in a corner joint were left in the full open configuration. It is important to be aware of this before you cut the bars to length. This type of fitup avoids distortion from welds made on the inside of the frame.

There are a few measures that enhance the accuracy of a fabricated angle frame. To build to a rigid dimension, first build a simple jig by tacking angle bar clips to the bench top or other plate. A frame built inside such a jig is held both square and to the proper dimensions.

Corner braces (Fig. 17-42) keep an angle frame from pulling out of shape from welding. Try to locate the frame so the outside corners are welded in the vertical down position. Monitor the shape with frequent diagonal checks. Weld the (outside) corner joints first, then the butt joints. Figures

Fig. 17-40. A coped corner joint ready for welding.

Fig. 17-41. A mitered corner joint ready for tacking.

Fig. 17-42. Bracing the corners permits moving the frame without breaking the tacks, and holds squareness while welding.

Fig. 17-43. Aluminum backup prevents burn-thru.

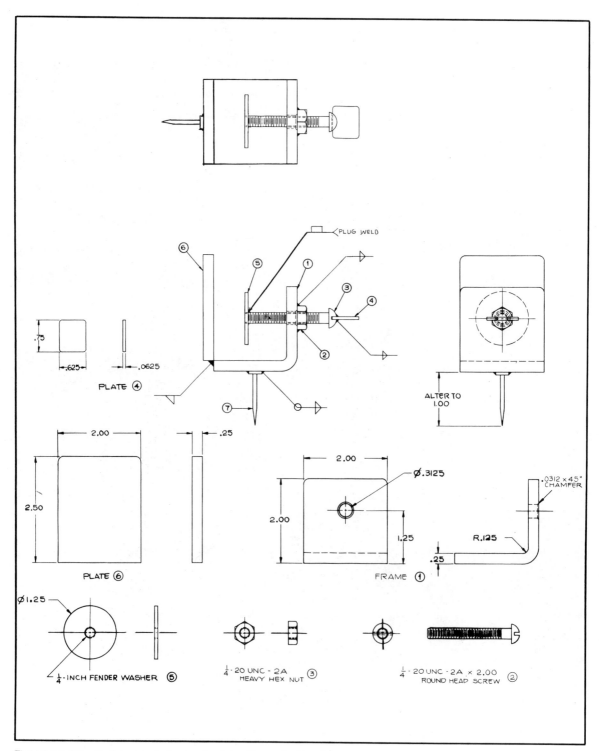

Fig. 17-44. Plan for homemade trammel points. (Drawn by Bao Nguyen.)

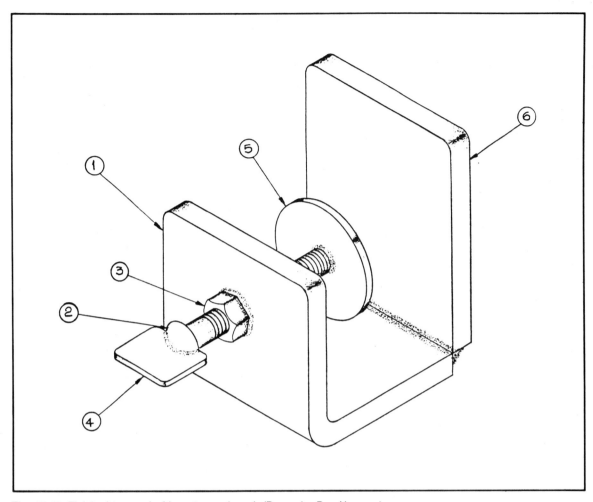

Fig. 17-45. Finished trammel with parts numbered. (Drawn by Bao Nguyen.)

Table 17-1. Parts List for Trammel Points.

ITEM	QUAN.	DESCRIPTION	SIZE	MAT.
7	1	#6d nail (alter)	Approx. 1″ lg × .25 ϕHD	Steel
6	1	plate	2.00 × .25 × 2.50	Steel
5	1	washer	1/4-20 unc - 2A	
4	1	plate	.625 × .0625 × .75	
3	1	round head screw	1/4-20 unc - 2A × 2.00	
2	1	hex nut	1/4-20 unc - 2A	
1	1	frame	2.00 × .25 × 2.00	Steel

Parts List by Seattle Central Community College, Trade and Industrial Department

17-40 and 17-41 also show how a grinder can prepare the butts for better weld access.

Sometimes it is important to avoid all weld burn-thru to the inside of a frame. Figure 17-43 shows a piece of aluminum angle bar used for a corner backup. It is smooth and will not stick to the weld. It also acts as a "chill plate" to absorb some heat that otherwise would contribute to distortion.

Fabrication of a Pair of Trammel Points

Trammel points span distances greater than the range of ordinary dividers. They can be used to transfer dimensions or for swinging larger arcs. Unclamped from their supporting beam, they store easily. The trammels shown in Fig. 17-44 can be clamped to any bar shape. The opening has been made large enough to accept a standard 2-inch piece of wood.

What follows is a rather detailed construction narrative of a typical small welding fabrication. Several key construction ideas are worth noting:

- By using nuts, the use of drills and taps can often be avoided without sacrifice to the quality of the final project.
- The small welds are representative of more delicate fabrications. For example, a weld made all around the nut would be likely to distort or otherwise damage the nut.
- The sequence of assembly is such that alignment of parts is easy and protection of the threads from spatter is natural.
- As with the fabrication of the box and the angle frame, a corner-to-corner fit is used both for strength and appearance.

Figure 17-44 is a blueprint of the overall project in a more or less standard format. This drawing is more elaborate than what is needed for most home weld projects (where communication among a large group is not a major concern). A list of necessary parts is given in Table 17-1. Figure 17-45 is a drawing of a finished trammel, with parts numbered.

The steps for fabrication of trammel points are as follows:

1. Cut and form the parts.
2. Drill or burn the 5/16-inch hole in frame (1), make sure that the screw can travel freely through the hole (Fig. 17-46).
3. Assemble nut (2) to screw (3), protect threads from spatter with tape (Fig. 17-47).
4. Center the screw and nut in the hole.
5. Weld (but do not overweld) the nut to frame (1) (Fig. 17-48).
6. Back out screw and place washer (5) as shown in Fig. 17-49.
7. Clamp and weld washer (5) to screw (Fig. 17-50).
8. Clamp back plate (6) to frame (1) and weld (Fig. 17-51).
9. Center masonry nail (7) on bottom of frame (1) and weld (Fig. 17-52).
10. Weld plate (4) to screw to form thumbscrew (Fig. 17-53).
11. Clean all welds and finish as desired.

The finished trammels are seen in Fig. 17-54. These were spray painted with the threads masked off. A light coating of oil was later applied to the threads.

Fabrication of a Welding Table

A plain steel plate is a handy surface upon which to assemble parts for welding. Jigging those parts, however, often means welding locating clips and stops directly to the surface. After each job, weld scars then must be ground away. You might find the next project, a slotted-top welding table, interesting (Fig. 17-55). The concept for this table has been around for a long time and many variations exist; you might wish to change the design to better suit your needs. Versatility is the claim to fame of this table. It can be an "adjustable jig" or a flat surface. A bench vise or other fixture can be fastened to the top, as needed, with the hold-down bolts.

The construction uses operations already seen—like fabrication of an angle frame. The top consists of channel bars (2 × 1 and 1/8 or 3/16 inches thick). The frames and legs were made from 1 1/4- × -2- × -1/8-inch angle bar. This was a bargain

Fig. 17-46. The hole (either drilled or burned as shown here) must permit free passage of the screw.

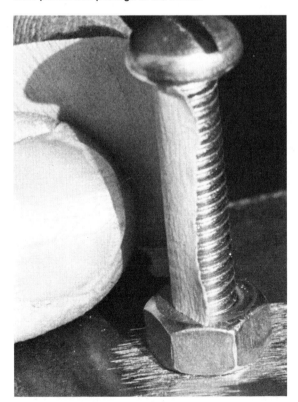

Fig. 17-47. Masking tape lasts long enough to effectively protect the threads from spatter.

Fig. 17-48. Only small welds are needed to join the nut to the frame.

oddball size. The dimensions of the table are not critical. This one measures 24 inches square and stands 36 inches high with the adjusting leg bolts.

The main steps in the fabrication of the welding table are:

1. Lay out the outside dimensions for the top frame and fasten angle clips to the working surface (Fig. 17-56).

2. Assemble the top frame members and clamp to the clips. (Use either coped or mitered corners on the angle frame.)

3. Check for squareness by comparing the diagonal dimensions (Fig. 17-57).

4. Grind bevels on the flatbar spacers (upon

Fig. 17-49. Locate the screw about flush with the washer for welding. Connect the ground cable directly to the screw to prevent damage from the welding energy arcing as it passes through the threads.

Fig. 17-50. The washer welded to the screw.

Fig. 17-51. The back plate welded to the frame.

which the table top channels will rest) to clear the inside corner web of the angle frame (Fig. 17-58).

5. Tack the spacer to the top frame members (Fig. 17-59).

6. Use shims to gauge the location of the table top channels from the side of the top angle frame and from each other; provide 5/8 inch between the channels (Fig. 17-60).

7. Weld (using only short 1/4-inch long beads)

each top channel to the spacer bars as it is inserted.

8. Turn the assembly over to gain access for welding the top channels to the *edge* of the spacer bars and the spacer bars to the angle frame. To avoid distortion, use only small welds as shown in Fig. 17-61.

9. Build a second (lower) angle frame for the leg braces; this can be made on top of the top frame.

10. Weld the corner joints (copes or miters) of

Fig. 17-52. The (hard) masonry nail ready for welding.

Fig. 17-53. The completed thumbscrew.

the top frame. Be sure to allow enough room for the legs; grind away some weld if it interferes (Fig. 17-62).

11. Space the top end of the legs 1/4 inch below the top of the top frame (Fig. 17-63).

12. Check the legs for squareness (both ways) and tack weld to the top frame.

13. Measure and mark an equal distance from the bottom of each leg for the location of the (lower) leg brace frame.

14. Clamp in place, locating stops for the lower frame.

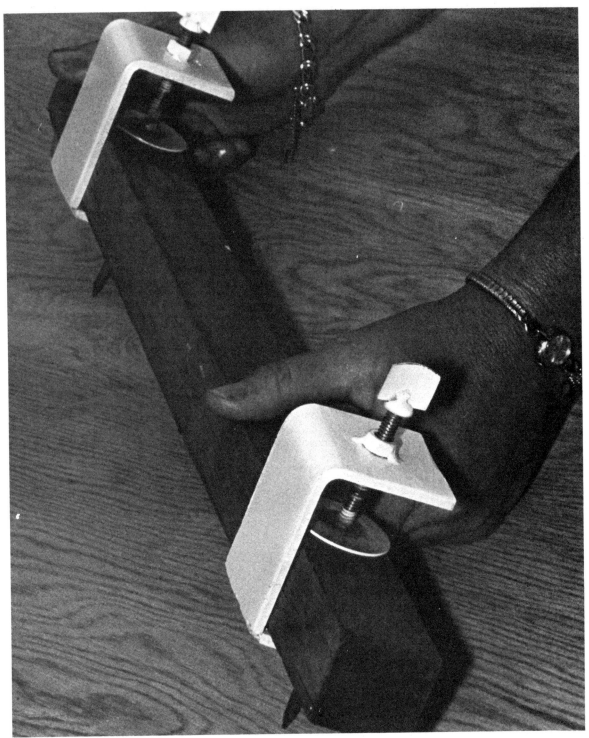

Fig. 17-54. The trammels clamped to a 2 × 3, ready to swing large circles or arcs.

Fig. 17-55. A small slotted-top fabrication table.

15. Install and tack the lower frame to the legs as shown in Fig. 17-64.

16. Check the diagonals on all openings for squareness (Fig. 17-65).

17. Remove the stops and reposition the table assembly for welding of the legs to both the top and bottom frames (Fig. 17-66).

18. Burn 5/8-inch-wide holes in suitably sized 1/4-inch plates for the 1/2-inch leg (levelling) bolts (Fig. 17-67).

19. Assemble the bolts and nuts and weld the inside nuts to the plates for all four legs. Figure 17-68 shows the assembly of a bolt, nuts, and plate about to be placed on the end of a leg angle bar. The lower (inside) nut in the figure should be tacked to the plate.

20. For each sliding clamp bolt, cut a flat washer and bend up 1/2-inch-wide tabs as shown in Fig. 17-69. These tabs slide in the 5/8-inch-wide slots between the top channels and prevent the clamp bolts from turning as setups are secured to the table top. (Like T bolts on machine tools.)

21. Weld (small tacks are sufficient) the modified flat washers to the 1/2- x -5-inch long (coarse thread) clamp bolts (Fig. 17-70).

22. Burn adjusting slots in either bar stock or heavy angle bar for the adjustable locating bars.

If the table is to be subjected to the elements, cover it with plastic sheeting and also be sure to paint it. All threads should be protected with a light application of grease.

Fig. 17-56. Tacking the angle (locating) clips directly to the working surface.

Fig. 17-57. Check for squareness.

Fig. 17-58. These bars serve as spacers for the top channels. Note the bevel for clearing the web of the angle.

Fig. 17-59. Tacking the spacer to both flanges of the top frame members.

FITTING ROUND
AND CYLINDRICAL SHAPES

Steel pipes and other cylindrical shapes are often used in *structural* welded fabrications. Structural means the pipes support members of a project rather than transport fluids. A piping system to carry fluids must be streamlined and requires additional techniques beyond the scope of this book. Before taking on the fitup of cylindrical shapes, it might be worthwhile to review the "Round Lay-

Fig. 17-60. Tacking one of the top channels to the spacer bar. Note the use of shims to gauge the distance between the channels. (3 pieces of 3/16 aluminum are being used here to provide 9/16-inch slots.)

Fig. 17-61. The use of small (hot) welds avoids distortion.

Fig. 17-62. Checking the corners of the frames for clearance inside the legs.

Fig. 17-63. Spacing the top of the legs below the top of the frame to ensure a flat top surface.

Fig. 17-64. Inserting the lower frame.

408

Fig. 17-65. The final check for squareness before the final welding.

Fig. 17-66. Welding the legs to the frames.

Fig. 17-67. The bolt must pass freely through the hole.

outs" material in Chapter 15.

Some interesting intersections are possible with pipes. The examples here involve a simple 90-degree intersection and a 90-degree corner. These are both common. In both cases, all pipes are the same diameter and there are no offsets (all pipes are centered). Use empty bathroom tissue or paper towel rolls to practice the layout and fitup before trying it with steel pipe. Refer to one of the many excellent pipe fitting or sheet metal layout manuals for more complex pipe intersections.

Dividing a Circumference with a Wrapper

A *wrapper* is a pattern with square corners made

of paper (or other material) that can be wrapped around a cylindrical shape. Figure 17-71 shows a wrapper used to divide the outside of a pipe into equal divisions. A layout like this might be used to locate spokes at the hub (and rim) of a wheel.

Figure 17-72 depicts a cylinder divided into eight equal parts. To the left of this view is the wrapper pattern. The first step for making the wrapper is to draw the circumference or distance

Fig. 17-68. The levelling bolt assembly for one corner.

411

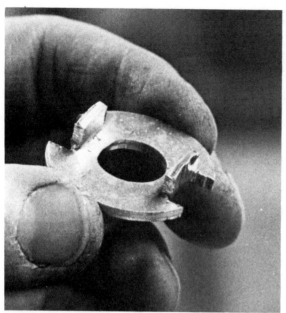

Fig. 17-69. The washer tabs bent up.

height lines that correspond to the eight divisions.

Transfer the distance of the height lines in the end view to the height lines in the stretchout. (Line 1 is emphasized in Fig. 17-74). This is done for each of the height lines. Notice how the same heights are shared by more than one line. When you have finished transferring all of the heights, sketch through the points with a smooth line. This will be the cut line for the branch pipe. You can use this stretchout directly for your wrapper or transfer it to other paper. Use a center punch to transfer the curved cut line to the pipe (Fig. 17-75). The finished cut on the branch pipe should look something like Fig. 17-76. Usually some grinding or filing is necessary to obtain a proper fit. Figure 17-77 shows the branch pipe fitted to the main pipe ready for tack welding. Use braces to keep the weld from pulling the branch out of square.

90-degree Corner Formed by Two Pipes

Where pipes join to turn a corner is called an *el-*

Fig. 17-70. A completed sliding clamp bolt.

around the pipe. The base line on the wrapper equals the circumference and is called a "stretchout line." Next, the stretchout line is divided into equal segments. (Line division is covered in Chapter 15.) *Height lines* (perpendiculars) are drawn from each of these divisions and numbered. In this example, there are 8 equal divisions and, therefore 8 height lines.

With the wrapper placed on the cylinder so that both ends of the stretchout line are brought together, the equal divisions are then transferred to the cylinder. This is done by making center punch marks through the paper pattern for each of the divisions.

90-degree Intersection of Two Pipes

Where two pipes intersect, the branch pipe straddles the main pipe (Fig. 17-73). It will, therefore, be necessary to cut some material from the end of the branch pipe. Begin by dividing a cylinder into 8 equal parts as just outlined.

Figure 17-74 shows the pipe intersection with two views. The end view shows the pipe divided into eight equal parts, each is numbered. To the right of the end view is a stretchout with eight

Fig. 17-71. A wrapper for dividing a pipe into equal parts.

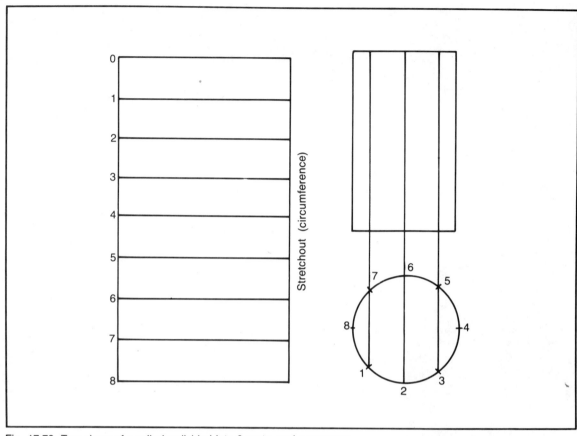

Fig. 17-72. Two views of a cylinder divided into 8 parts, and a wrapper.

bow. A technique to form a two-piece, 90-degree elbow follows. The process involves dividing a circumference and making a stretchout as just outlined. Figure 17-78 shows the wrapper layout.

Welding an elbow is interesting. The outside *heel* of the fitting is a sort of butt joint, and the inside *throat* is a **T** joint. The transition from one type of joint to the other is gradual as the welding progresses around the intersection. When you are fitting the corner together, grind a preparation for the weld on the heel. Weld the heel first to hold the throat fillet from distorting. When possible, locate the elbow so that the welding can be performed in a slight vertical down position.

A more gradual elbow requires more sections. This type of layout is not difficult, but is rather lengthy. A good sheet metal layout book covers layouts for multiple-piece elbows.

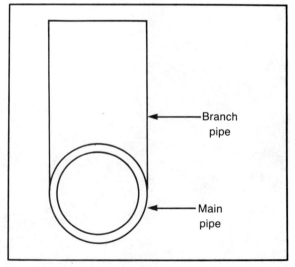

Fig. 17-73. A 90-degree pipe intersection showing the main pipe and the branch stub.

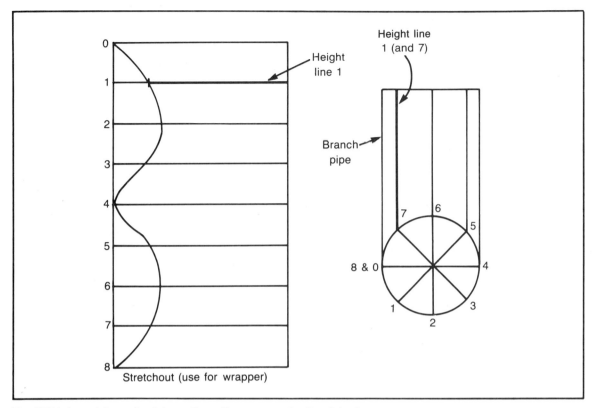

Fig. 17-74. Layout for a pipe intersection with a wrapper for the stub pipe.

Rapid Intersections

Although it may not produce the best-looking joint, Fig. 17-79 shows a "quick and dirty" method to make intersections with pipes less than 2 inches. It involves flattening the branch pipe and butting it against the main pipe. This is fast and often more than adequate. Many manufactured items are made in this way.

If you use galvanized or coated pipes for your projects, remember to have adequate ventilation. Zinc fumes are produced from flame cutting, grinding, and welding. Welding is easier and better if the zinc in the weld zone is first ground away.

USING HAMMERS AND WRENCHES

The fitup techniques in this chapter have involved the use of a hammer and wrenches. In Chapter 14 some hammers and wrenches were discussed. It is worthwhile to consider the use of these tools.

If a tool does its intended job without any damage to itself, the workpiece, or the user, it was probably used correctly. Hammers are not very complex. They ought to be large enough to do the job with only a few blows. In fact, a hammer should be used sparingly. Every time metal is forcibly struck with a hammer, it is marked, deformed or work hardened. As mentioned, a 16-ounce ball peen hammer can be used for much of your weld fabrication, but something larger is also required.

Before using a hammer, look at the handle. If it is cracked, replace it. If the head is loose, drive it tightly onto the handle and drive a wedge into the end of the handle. Also, consider the arc through which a hammer (or any pounding tool) must travel. Avoid swinging a hammer in line with any person. Wear eye protection to guard against possible high-speed particles, and do not forget to protect your hearing.

Many times a large hammer can be used

Fig. 17-75. Transferring the cut line on the wrapper to the pipe by center punching along the line.

Fig. 17-76. The freshly cut stub pipe. This shape is sometimes called a "fish mouth."

backup for a simple forming operation; bent bars can be straightened using two hammers. Hammering the edge of a bar or leg of an angle bar will stretch that edge imparting a curved shape. Anticipate the effect of every hammer blow. When straightening or otherwise shaping a bar, step back and check your results frequently. It is easy to go too far.

To avoid hammer marks, the hammer head must strike the work squarely. The hammer face should also be free from nicks, etc. These act like a marking punch, leaving their stamp with each blow. If a hammer face has hardened so that a file can no longer remove nicks, first use a fine grinding wheel and then finish with emery cloth. Avoid excessive grinding of a hammer face.

If a great deal of force is needed, it might be difficult to avoid leaving marks. Place a sacrificial metal plate between the hammer and the work. Light plate such as aluminum absorbs the marks and still transfers most of the hammer's energy.

Wrenches are also widely used in fitup. Besides turning nuts and bolts, they also increase your levering strength for hinging off pieces of metal. With an adjustable wrench, be sure to pull in the correct direction as seen in Fig. 17-16. If you pull the other way, the wrench is likely to slip. Avoid excessively long "cheaters" (pipe or tubing slipped over a wrench handle to gain more leverage). To gain more leverage, weld together a hinge-off bar

Fig. 17-77. This fitup of the stub to the main pipe is close enough for most applications and is ready for tack welding.

like shown in Fig. 17-80.

If you want to modify (bend) a wrench for a special purpose, avoid overheating it. Most wrenches are drop forged and then heat treated for maximum strength. Welding on a wrench is always a risk because of the heat input and matching with the right filler metal. There are expensive specialty filler products for welding on tools, but stainless steel electrodes in the 300 series work.

MAKING vs HAVING FITS

Project fitup is a high point in welding fabrication. If you check the alignment of project parts at each step, the effects of heat will seldom get out of hand. Do not underestimate the shrinkage caused by that very last weld. If the project is correctly aligned before the final weld, there is a good chance that it will have moved after it. Place welds when and where the heat effects will be the least destructive.

When possible, sight along the edge of a straight member to check straightness. Use a straightedge when that is more practical. Step back and look for parallels and perpendiculars in the outlines of your project. And while you are back there, be sure to take a moment to enjoy what you have made.

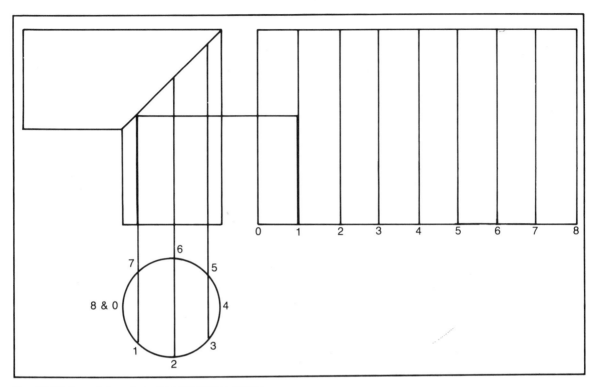

Fig. 17-78. Layout for a two-piece 90-degree elbow.

Fig. 17-79. A "rapid" pipe or tubing intersection for structural applications can be made by flattening the end of the branch member.

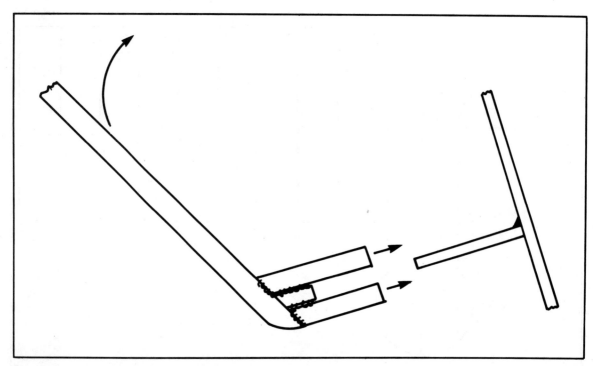

Fig. 17-80. A hinge-off bar can be made to any practical size.

Chapter 18

Are We Having Fun Yet?

DO-IT-YOURSELF ACTIVITIES ARE MANY things to many people. But most of all, they ought to be fun. Alas, this is not always true. Are you welding more but enjoying it less? There are several ways to get turned off (and to get turned back on).

TOO MUCH TOO SOON

It sometimes happens that a person begins home welding with an interest in building just one big project. With minimum skill and maximum enthusiasm, work on the project starts off like a desperate race against time. Super projects tend to go sour and become a chore to finish. Nothing succeeds like success; we need successes even with recreation.

If you find yourself in this situation, you might set the big project aside and work on a simpler one. A more modest proposal is especially encouraging if it involves aspects of the larger project. As a way out of a ponderous megaproject, consider the information in Chapter 20 about outside services.

TOO COMPLEX

An overly complex project can be a welding impossibility. Joints must be reasonably accessible for a proper weld to be made. Struggling to weld joints that are nearly impossible to reach does little to instill enthusiasm for a project. While a certain challenge might be associated with playing the part of a welding contortionist, difficult welds are often of such a poor quality that they cancel out any pleasurable sense of accomplishment.

If you have such complexity in a project design, chalk it up to experience. Take it apart and rebuild it more simply.

TOO MANY PARTS

If your project involves a great number of parts, it is possible to lose track of them. Mark parts with a name or number and size as you make them, otherwise you could mistake a part for excess material. Cutting up a good part to make other parts is demoralizing. Try welding or at least tacking subassemblies, or components of the larger project, as you make them. This reduces the number of loose parts.

TOO BIG OR HEAVY

Every welder appreciates the ability to locate the workpiece in the best position, usually the flat or horizontal. A large or heavy project can get too unwieldy to permit easy movement. Where practical, one solution is to strengthen the framework of your shop and install a jib crane. Even a simple block and tackle arrangement might help. A portable engine hoist (an interesting project) can also be used to position objects up to a certain size and shape.

Larger projects might be better managed in smaller subassemblies. If having a crane is out of the question, a simple angle frame (or two) equipped with casters can serve as a rolling platform during the fabrication.

TOO EXPENSIVE

It happens to the biggest and the best. They run out of money. Hopefully, the cost of a project is realized before any substantial waste occurs. Postponing a project might be viewed as a setback, but it need not be totally demoralizing.

As mentioned earlier, a more modest proposal can be undertaken or even subassemblies built over a longer period of time. Remember to get at least one other opinion on your selection of materials before starting your project. As Chapter 7 points out, metal prices can vary with the popularity of the size and shape. Substituting a different size or alternative shape might save the day.

TOO DEMANDING

Welded fabrications are always subject to the forces of expansion and contraction. Enter distortion and shrinkage. Most professional weld shops do not seek to hold dimensions any closer than plus-or-minus 1/32 inch from the specified value. This gives a tolerance of 1/16 inch (Fig. 18-1). To maintain even these generous dimensional tolerances might require heavy jigs and fixtures unavailable in the home shop.

Those inexperienced with welded structures

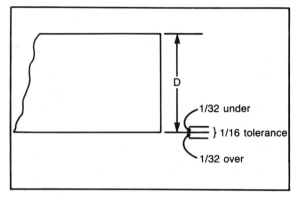

Fig. 18-1. Tolerance is the permissible variation from a specified dimension.

might not be prepared for the dimensional effects of heat on metal. For example, a welded frame could end up too small to accept what is supposed to fit inside of it. Doing fabrication operations in a certain order or *sequence* can compensate for the forces of metal shrinkage. For example, it is often best to drill matching holes and other critical operations involving locations after the welding has been completed.

If something is not dimensionally correct or parts are misaligned so that you are unhappy with the situation, change it. Disassembly of a welded structure should not mean the same as destroying it. However, starting over might be best if rework is impractical. There is little comfort in the adage: "always enough time to do it over but seldom enough time to do it right." Yet, rework is a part of all worthwhile activities, especially those involving fabricating the first or just one item. When possible, the use of AAC greatly speeds the disassembly of a welded structure, with little damage to the parts.

Try to avoid welding on a thin or delicate assembly. Such situations arise with attempts to repair small valuable items. It is too easy to cause a delicate item to literally go up in smoke if the heat input is not controlled. Home welding shops seldom have the more sophisticated welding tools for extremely fine work. (Even heavier items can be damaged if the heat input is not anticipated.) Always test your heat by making practice welds on scrap material.

Threads and machined surfaces need protection from the effects of welding. Spatter sticks to screw threads with special tenacity. Attempting to repair spatter damage can just cause more. Check your scrap bin for a pipe or sleeve to place over external threads. Even paper masking tape (Fig. 17-47) lasts long enough to shield surfaces from the spatter of brief welding operations. Anti-spatter compounds can be very effective.

UNFAMILIAR MATERIAL

One of the basic rules of welding is to first identify the base metal. This is essential for selection of the proper filler. A metal's identity is not always apparent. It might be painted. Most metal that you encounter will be plain carbon steel, but there are many others. Bolts and other round shapes are often made of heat-treatable steels that are not practical to weld by ordinary means.

Using the wrong filler metal can be catastrophic. For example, mild steel electrodes used on cast iron produce a hard deposit that cracks. Also, repairing a broken braze joint by welding with steel filler metal forms a brittle structure that might appear to be sound, but usually fails in service. Avoiding some pitfalls of weld repair is covered in Chapter 21.

THE FAILED WELD SYNDROME

One understandable reason for wanting to throw in the towel with home welding is the disappointment if your welds fall apart. This frustration may be rooted in the fact that permanence is just assumed with strong weld deposits. If the right filler

Fig. 18-2. A weld shop on wheels has some advantages.

Fig. 18-3. The exhaust pipe from the motor-generator power source is routed down and out.

metal and welding method are used, failure on medium-to-heavier thicknesses are usually caused by a lack of fusion. Failures with thinner metal are mainly associated with excessive heat.

To avoid weld failures, the practice of making and testing sample welds must become second nature. Most of us are in a hurry to be finished, especially with small jobs. Preparation time is often much longer than actual welding time, but you can't rush quality. The payoff comes in having confidence that your weld will do its job.

POOR WELD APPEARANCE

Another cause for disappointment could be the appearance of your welds. Ugly welds do not evoke good feelings or confidence. With practice, the ap-

pearance of your welds should improve. If you are unhappy with the look of your welds, check with an expert. The problem might be with your equipment or its adjustments. In general, a weld must be adequate to its intended purpose. While a weld's appearance can be something less than ideal, the ability of the weld to perform in service should not be left to chance.

TOO MUCH NOISE

Noise pollution has at last been recognized as a powerful threat to the safety and welfare of those exposed to it. Noise jars the nerves of those generating it and can also elicit some hostile activity from its bystander victims. As mentioned in Chapter 3, the practical workplace should not be burdensome

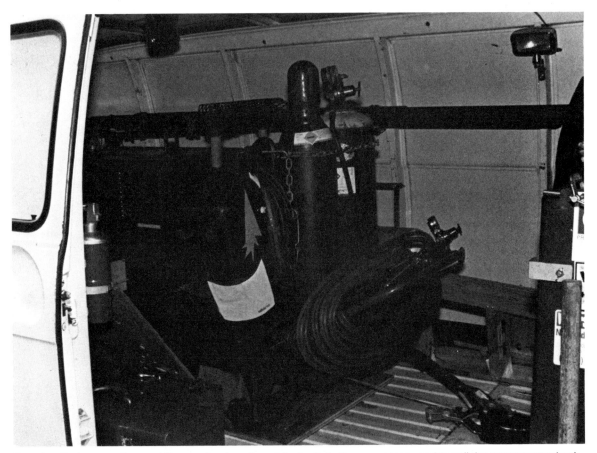

Fig. 18-4. Both gas and arc welding can be done from this rig. Note the comealong used to pull the power source back up the loading ramp.

to either yourself or to others. Noise will wear you and other victims down. Use ear plugs at the proper times and also consider the neighbors when hammering or using power tools.

Some hobbyists have successfully "buttered up" their neighbors with an offer to repair something or with a little project of wood or metal. This tactic may not always be advisable, but a neighbor with an interest in your shop activities is less likely to be irritated with a little noise.

TOO MUCH TIME PREPARING

A problem many home craftsmen reluctantly accept is the time spent setting up and putting away equipment in a multi-purpose work area. For some of us, a permanent, dedicated shop area remains an elusive dream—yet there are solutions. An innovative do-it-yourselfer who likes to get together with other hobbyists on weekends has located his shop in a Chevy van (Figs. 18-2 through 18-4). With this vehicle under a carport, fabrication takes place in nearly any weather.

Efficiencies of time and motion are realized with items like portable work benches and vise stands shown elsewhere in the book. These are good first projects. Their convenience is often appreciated at the start of the next work session.

Welding fabrication ought to be creative fun like woodworking and other do-it-yourself activities. If you get stuck in a doldrum of inactivity, consider some of the ideas brought out in this chapter and also contact some other home welders. Enthusiasm and good ideas are, fortunately, contagious.

Chapter 19
Finishing a Steel Project

W OOD PROJECTS ARE SANDED AND SHARP corners are removed to enhance appearance, prevent injury, and to preserve them. Welded metal projects involve similar finishing touches.

CLEANING AND FINISHING

It is necessary to clean and finish a project following the fitting and welding operations. "Cleaning" metal projects means more than removing dirt. A cleaned project has no sharp edges. Slag, spatter, and smoke dust have been removed, and oil, grease, and rust eliminated. "Finish" in this sense means a final surface treatment that might involve some added minor cutting and/or application of a coating material. Coatings decorate or preserve, some do both. Certain coatings involve special equipment and processes requiring outside services. These are discussed in the next chapter. However, there are several finishing operations done by the home welder.

The satisfaction that ought to follow a creative effort can be undone in an instant with an injury caused from an overlooked sharp edge or a bit of welding residue (Fig. 19-1). Before you consider a project finished, pass a "snag rag" over all surfaces and edges to locate sharp spots. This takes extra time and effort, but it pays off in quality appearance and a safe product.

As soon as you are finished with welding, remove all slag and spatter. (It is much easier to clean a weld zone that has first been protected with an anti-spatter compound.) Use a slag hammer with care to avoid leaving marks on the work. Also recall that a small chisel and hammer offer more control in removing slag and spatter from delicate items.

FINISH GRINDING

When possible, avoid grinding a surface just to clean it. Hand held grinding, which can remove a lot of metal fast, is always potentially destructive. Using a grinder to remove a bit of spatter is like using a roto-tiller to remove a dandelion; you get rid of the offending object, but face a considerable

427

Fig. 19-1. A metal fabrication left with sharp edges or burrs is not yet finished.

scar. Notice the irregularities in the surface of a work table that has been ground off many times. Imagine trying to paint such a rough surface. The coarse grinder lines would always show through.

To grind a surface that needs to be fairly smooth, do it in stages. First, take off heavy deposits with a solid reinforced disc (Fig. 14-16). Then, depending on the final polish desired, use sanding discs with backup pads. Figure 19-2 shows the various items used on a sander/grinder to obtain a sanded finish. Figure 19-3 shows a die grinder fitted with a spot sanding disc. Sanding discs pro-

duce smoother finishes than is possible with a hard wheel.

Notice how grinder lines form arcs on the surface of the work. A random pattern is avoided by paying attention to the direction that the disc passes across the surface. Experiment with the use of a disc on some scrap metal to see if you can produce both a "draw-filed" look and a "cross-filed" look on edges and corners (Fig. 19-4).

Stickwax applied to sanding discs makes cutting more efficient, reduces heat buildup, and the discs last longer. The roughest grit you might need

Fig. 19-2. Items used to mount a sanding disc on a disc sander/grinder. The extra sanding disc helps prevent ripping of the outer sanding sheet.

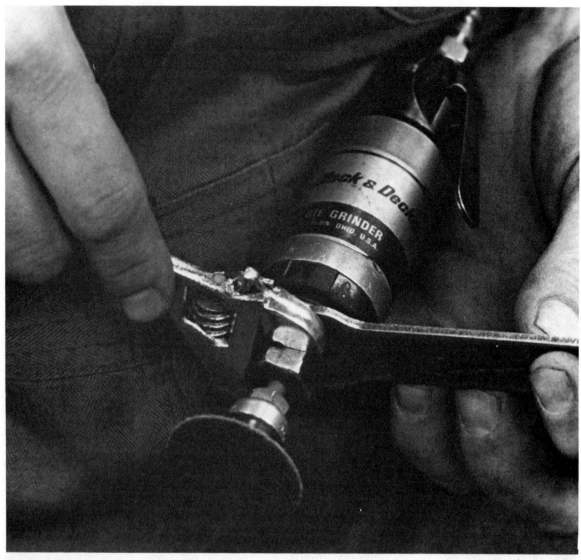

Fig. 19-3. A spot sanding disc is useful for breaking sharp edges.

is 36, and 100 about the finest. A pair of cautions: a sander/grinder used on thin metal can easily go all the way through. Also, the frictional heat produced by a sanding disc is high enough to cause warpage to the workpiece.

FILING

An ordinary file is very useful for removing snags and burrs at corners. In fact, even a worn-out file can be made into a good scraper to remove spat-ter. Like grinding discs, files can cut either in line with an edge surface or across it. *Drawfiling* in line produces a very smooth edge by planing down high spots, yet sharp corners along that edge are created. These are removed by more drawfiling strokes with the file gradually lifted with each stroke. The fine scratches or finish lines of drawfiling run along the length of an edge and impart a high quality, finished look to your work. *Crossfiling* concentrates the cutting action to a small area. It produces file lines

Fig. 19-4. A grinder can produce a drawfiled look with the grinder lines running in line with the edge (left), or a crossfiled look (right).

431

across the edge which, if highly visible, detract from an otherwise completed appearance.

A folded sheet of emery cloth can often be used like a file. Like sanding wood, finish lines should mainly run the length of, rather than across, a surface. Emery cloth torn into strips and wrapped around or held onto a file, produces a very smooth finish—depending on the grit used. Keep several on hand. If a little light machine oil is used along with the emery operation, the polish is very high.

A thread file is a special knife-like file. These are sometimes used to repair a male thread if a die or a die nut (Chapter 16) are not available.

A *countersink* is a rotary cutting tool that produces a bevelled or chamfered edge at the top or bottom of a hole. For example, the seat for a flat-head screw is produced by a countersink. A countersink fitted to a file handle is handy for breaking the sharp edges of small drilled or punched holes. Lacking this, you can do a respectable job with an ordinary flat file to scrape (break) hole edges.

Keep your files clean and store them carefully. See Chapter 16 for more about the general care and use of files.

PAINT AND OTHER SURFACE APPLICATIONS

There are many paints for use on metal surfaces. Do not use water-based paint designed for wood. With paint, you often do get what you pay for. A spray-painted or dipped finish usually looks much better on a metal object than a brush application. Home welding projects tend to be small, so spray painting them poses few problems. You can probably improvise an effective spray booth with a cardboard box on top of the garbage can in the backyard. With larger projects that require a spray gun, overspray can be a problem. You might consider taking your larger projects to a professional painter.

The preparation for painting a welded project is very important to the quality of the finish. Even the best metallic paints do not hold up on an improperly prepared surface. Weld spatter often contains entrapped slag residue that reacts with paint. This is another reason for removing spatter. Certain antispatter compounds are claimed to be compatible with most paints.

The directions on practically any can of paint say that the surface must be clean and dry. An easy way to ensure this is to wipe everything off with a volatile solvent like lacquer wash thinner. You must have adequate ventilation and a location away from possible ignition. Using a rag to wipe down your welded project also helps to locate snags and burrs.

The ideal paint preparation involves a light and uniform roughening of the surface to provide more area for paint bonding. Sand blasting and similar processes have this effect but are not often available or practical in the home shop. Special etching primers or metal "preps" do an excellent job of imparting many fine pits to the surface. If you wire brush a surface, be advised that brushes often contain oil to prevent rust in storage and shipment. A lacquer thinner wash removes any oil.

Be sure that the thinner on the metal has evaporated and that containers and rags have been removed from the area. Only then, use a torch to heat the metal to about 125 degrees. Heating removes any residue of thinner and oil, along with moisture. Warming the plate also allows the paint to dry quickly, thus avoiding runs and sags.

With aerosol paint, place the can in a container of warm (not hot) tap water long enough to warm the contents. (Never apply any source of heat directly to the can.) This also helps to ensure that the paint will dry quickly. Follow the directions regarding shaking. Some paints contain a primer element that must be thoroughly distributed to adhere properly. As you apply the paint, do like the pros and keep the nozzle moving all the time to avoid heavy buildups and runs.

Besides painting, there are other more complicated ways of protecting or enhancing the surface of a metal project. Some of these are discussed in Chapter 20.

You can apply plastic coatings like those found on purchased hand tools by using one of the colorful compounds designed for dipping. These are useful for handles and gripping surfaces, and also provide a soft, non-marring surface for racks and hooks. Even a liquid plastic type of finish such as clear Varathane can be used as a dip for smaller projects.

While true galvanizing (coating with zinc) is an outside service that the home welder cannot practically do him/herself, some reasonable facsimiles are available. The application of "hot stick" is quite effective in preventing rust. This is applied to *clean* metal that has first been heated with a torch. The low-melting stick is smeared over the heated metal and then scrubbed with a wire brush. While not practical for larger surfaces, this product is used by the welding industry for touch-up of a galvanized product damaged by cutting and welding heat. It is also suitable for coating smaller items if all of the surface areas are accessible.

There are "cold" galvanizing compounds that are like a heavy paint. Because of their heavy consistency, applying these seems more like squirting a mushy banana than painting. These are usually available in both aerosol and brush-on forms at marine supply houses. They are rather expensive but, if the directions are followed, do a good job of preventing rust. Certain types of cold galvanizing compounds permit later color coating with other paints more easily than others.

STORING UNCOATED STEEL

Sometimes it is necessary to store steel with areas shiny from grinding or filing. This often happens with parts of a partially completed project. Unless your home shop is a very dry environment, rust soon forms. It is tempting to apply a coating of machine oil over the exposed areas. While this measure effectively prevents rust, oil can also work its way into the crevices between weld joint members. Even if you wipe off the plate with a solvent, some oil will remain trapped. As you begin to weld, this oil boils out and elements of it, particularly hydrogen, enter into the weld deposit. Recall that trapped hydrogen causes underbead cracking.

If you could heat oiled parts to a high temperature (maybe 800 degrees Fahrenheit) and long enough (maybe 5 hours), the oil would be driven out. However, adequately heating projects of any size is not practical in most home shops; a special oven is required. Therefore, avoid splashing oil over your projects.

There are some silicon-based protective sprays that do not contaminate the weld, but applying them can be difficult and expensive. When possible, just cover and enclose your unfinished projects. Smaller articles can be placed in plastic bags; trash bags work for larger items, and plastic sheeting taped along the edges with duct tape can be used to cover yet larger areas. If moisture is a major problem in your shop, you can even enclose a supply of a dessicant (dryer) in with the project. One of the best ways to escape rust formation is to avoid exposing your unprotected projects to extreme changes in temperature.

Chapter 20

When to Use Outside Services

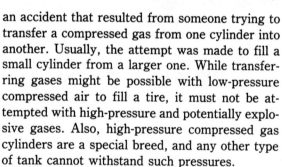

THE DO-IT-YOURSELFER IS NATURALLY THE self-sufficient sort who does not always take kindly to having someone else do part of his or her projects. However, certain things must be done by an outside expert due to safety considerations or because of the specialized equipment and techniques required.

WHEN SAFETY IS INVOLVED

It has been mentioned that regulators and other gas apparatus are not items for tinkering and modification. Even if you think you know what you are doing with that equipment, the stakes are too high for any risks. A number of grim accident reports document this. If you have problems or doubts about gas equipment, call your supplier at once. Most have 24-hour emergency numbers.

The inside of power sources and wire feeders are strictly off-limits to all but trained electricians. The only owner-serviced parts of most power sources are fuses and connecting cables.

From time to time there comes the report of

an accident that resulted from someone trying to transfer a compressed gas from one cylinder into another. Usually, the attempt was made to fill a small cylinder from a larger one. While transferring gases might be possible with low-pressure compressed air to fill a tire, it must not be attempted with high-pressure and potentially explosive gases. Also, high-pressure compressed gas cylinders are a special breed, and any other type of tank cannot withstand such pressures.

Explosions have happened while gas cylinders were being transported in a car trunk. It is worthwhile to consider having compressed gases delivered, especially oxygen and acetylene.

WHEN SPECIAL EQUIPMENT IS NEEDED

Certain fabrications require the use of specialty equipment and, even if offered, rental for just one use might not be practical. It could be wiser to send it out.

Surface Treatments

Steam cleaning or sandblasting are excellent ways

to clean dirty surfaces before (or even after) welding. It is possible to rent portable equipment. However, the main problems with using either steam cleaning or blasting in your backyard is that they require space, involve considerable noise, and make a mess. (Of course, they annoy people too.) Someone in the steam cleaning or sandblasting business has solutions for these problems. Whenever possible, transport your work to their place.

If a steel or iron object to be welded has an accumulation of paint and/or grease, a hot bath of caustic chemical solution will strip it to the bare metal. Most automotive machine shops offer such "hot tanking." In light of the harsh chemicals used, the home welder should be glad to farm out this service. One problem with hot tanking is that the cleansed part begins to rust as soon as it dries off. Wire brushing is often necessary to remove the rust.

Certain coatings can only be applied with specialty equipment. Figure 20-1 shows part of an electroplating operation. The variety of metals deposited as plating include: copper, brass, nickel, chromium, and even gold and silver. To keep the costs down, find out from the plating service what preparations you can perform—like removing slag, spatter, grease, dirt, or paint.

Galvanizing

Hot-dip galvanizing with a thin coat of protective zinc is a surface treatment to preserve weldments exposed to the elements. This is not the sort of thing you can do at home however. Figure 20-2 shows a part of the process.

It is essential that items to be galvanized contain no dangerous air pockets that might cause molten zinc to erupt. Also, drain holes are needed to allow the liquid zinc to escape as the project is withdrawn from the zinc tank. If necessary, cut or drill temporary vents and drains that can later be welded shut and touched up. It is also important that all slag be removed from welds and that welds be free from defects like excessive porosity or pinholes.

There are special paints that adhere to zinc, but preparation is necessary. The zinc should be properly etched or roughened before painting. Tell your galvanizing service operators if you intend to paint the articles. After coating, they will flash dip your parts in hot acid and rinse them for a proper etch. You can approximate this at home by painting vinegar on a galvanized surface and leaving it for a half hour or so before rinsing.

It is not generally realized that blistering or flaking of paint and surface coatings applied over welds can be caused by the slow release of hydrogen entrapped in the welds. It seems incredible, but minute hydrogen bubbles can migrate to the surface of a weld even days after the weld was made. One way around this is to simply allow a few weeks to elapse before applying any coating to a weldment.

Flame spraying (Fig. 20-3) is a way to apply a thin coating of metal, usually for functional purposes. For example, the inside of a steel tank can be coated with aluminum to reduce corrosion. (Normal plating methods do not permit this.) Another popular use for flame spraying is to impart a wear-resistant metal over a softer one. The bond is so good that machining of the deposit is possible. Flame spray advantages over conventional hardfacing include precise control of the deposit thickness and the avoidance of distorting heat. Before a flame spray finish is applied, the outside of the workpiece needs to be sandblasted to provide a larger surface area for the deposit.

Heat Treating Services

Heat treating involves several operations that make metal products more useful. These include:

● Stress relieving at a temperature in the 800-degree range to remove locked-in welding stresses. This prevents cracking in rough service.

● Annealing to remove all stresses from metal, usually before it is assembled into a fabrication. With steel, the metal is heated to slightly above 1250 degrees Fahrenheit, held for a "soaking" period, and very gradually cooled back down. This also removes any hardness.

● *Normalizing*, which is almost identical to annealing except a rapid cooling time somewhat compromises the final stress relief.

Fig. 20-1. Chrome and nickel plating require special tanks equipped for electrolytically transferring the coating metal.

Fig. 20-2. Immersing fabrications into a bath of molten zinc can be hazardous. If the parts should contain any pockets of trapped air or water, an explosion can take place.

Fig. 20-3. A mild steel shaft being coated with a layer of stainless steel with flame spraying.

● *Hardening* whereby the heat-treatable steel is heated above 1250 degrees Fahrenheit, held for a short time, and rapidly quenched in either icy brine, water, oil, a salt bath, or even with an air blast. The more rapid the quench, the harder the final product. Brittleness under impact often accompanies a hardened condition.

● Tempering (drawing) which involves reheating a hardened article to a temperature below that used for hardening. Cooling is usually slow and removes some hardness and brittleness but makes the article tougher under impact.

Heat treating involves precise operations with equipment designed for exact temperatures, and accurate heating and cooling rates. Automated ovens

are used to stress-relieve larger articles. Figure 20-4 shows a medium-sized oven. The controlled atmosphere inside many heat treat furnaces is free from oxygen and nitrogen, and so the work emerges without any millscale.

If you wish to weld heat-treated parts, a professional heat-treat plant can first remove the hardness with an annealing operation. After welding, stress relieving might be desirable, as might hardening and tempering.

Cutting and Forming Operations

If you know the exact sizes of plate needed for a project, consider having it sheared to size where you buy it. Sheared steel has nice straight edges to work from, and also has no heat damage from flame cutting. Besides straight line shearing, other cutting is often available in the form of notching and punching.

A power brake is a machine to bend metal at a specified angle with a certain radius. Plate is often bent in hydraulically operated brakes. Arcs in structural shapes, are often rolled to an exact radius as shown in Fig. 20-5.

Machining

Machining operations include lathe work, drilling, boring, precision milling, and grinding. When extreme accuracy is required, take your problems to a machine job shop. You will save time and end up with a more useful project. Machining would be desirable if you wished to cut and reweld a drive line. Facing of the members produces a uniform

Fig. 20-4. This heat treating furnace is fired by natural gas.

Fig. 20-5. Forming rolls provide an even radius to shapes, free of kinks.

joint that helps to hold the shaft straight. If you need a hole of an exact diameter, you could first burn it about 1/4 inch undersize and then have a machine shop bore it to the final dimension.

Equipment Repair

Faulty welding equipment can cause damage and injury. Have your gas equipment serviced only by the trained experts at a torch and regulator shop. These shops must, by force of law, go through an extensive check list for each piece of equipment serviced.

Metal Indentification

If matching the base metal with the right electrode is difficult and the job warrants it, take a metal sample to a testing lab for analysis. It is rather expensive, but exact. The correct filler and weld procedure (like preheat, etc.) can then be determined. It could be that the type of metal identified should be welded by an outside expert.

OUTSIDE WELDING SERVICES

Certain welding jobs, because of a strict procedure or potential liability, are best left to a certified expert. Some other welding jobs demand a high degree of a specialized skill.

Welding Certification

Many municipalities require that welding done on a structure concerned with the public be performed by a certified welder. This has become the norm because of problems in the past with street signs and the like falling on people. You might even be liable if someone got hurt as a result of a structural weld failure on your house. That is why building departments often specify that structural welding of such commonplace items as handrails and security gates be performed by a certified welder.

Weld repair and modification to pressure piping, like a hot water heating system, require excellent welds. Again, the potential for injury and destruction has prompted some building codes to restrict welding of this sort to those who hold a certification.

There is nothing to prevent you from becoming a certified welder other than a high degree of skill and an application and renewal fee. Also, some assurance must be made to the licensing body that you will "keep your hand" after the initial certification. For renewals, endorsements by work supervisors are required, usually for each quarter of the year. This might be a catch unless you do a great deal of welding at home.

Repair of Jewelry and Precious Metalwork

Jewelry is delicate. It is sickening to see something you thought you could repair either go up in smoke or melt all over the bench top. Regard the trip to a jeweler as a learning activity and make the best of it. Watching tiny torches in deft hands can be very pleasurable.

Heat Exchangers

Heat exchangers are things like radiators, condensers, and evaporators used in heating and cooling systems. You probably have 3 or 4 on your car, a few in the refrigerator, and some in the heating and cooling system for the house. If the item is irreplaceable, take it to an expert for repair or modification. Reconcile this to the do-it-yourself ethic of self-reliance by viewing the referral as another learning experience. It might look like simple soldering and/or brazing but it is difficult to work on only a section of an exchanger (which was probably furnace brazed initially) without popping a great deal of it apart.

Weld Repair of Large Castings

Because of the need for high preheat temperatures and outsized welding torches, it is difficult to properly repair a large casting at home. This is a welding specialty and professional welders often refer such work to an expert. The repair of many smaller iron castings is not terribly difficult and is discussed in the next chapter.

You might be able to assist preparing the casting by removing paint and grease, maybe bevelling the repair area (get an OK for this first), and removing hardware such as bolts, plugs and attached parts.

Lifting and Hauling

As sections are joined, sometimes a project gets too heavy or unwieldy to be moved by hand. Enlist the services of a crane and maybe a transport service as well. Trying to wrestle with large steel assemblies is a loser's game. The steel always wins.

Chapter 21
Fix It? Maybe

A SOURCE OF SPECIAL DELIGHT FOR THE home welder is restoring a broken and useless item with a competent repair. After a few of these repairs, one might begin to regard metal objects with a certain mastery. Then it happens. Enter The Unfixable. It is a wise person who knows when to quit, and this chapter will hopefully provide some guidance as to the best time for this. When someone asks you about fixing their broken whatchamacallit, respond with: "Maybe."

One of those first rules is to identify the base metal within at least a broad category, because you cannot fool Mother Nature. For example, even with great sincerity you cannot weld cast iron the same way as mild steel (and get by with it.)

METAL IDENTIFICATION

There are ways to identify metal at home that at least give you a good clue of what you are dealing with. Among these are:

- Color

- Magnetic test
- Sound
- Weight
- The spark test
- The chip test
- The file test for hardness
- Reaction to chemicals

A general strategy should be adopted before looking at these techniques. A collection of known samples is most valuable. Comparing the unknown to a known is the best way to identify metals in the home shop. After awhile, you will acquire a general idea of which metals are used for which products. Not every technique listed is possible or valid in every case.

Color Test

The color test is a bit overrated, most metals are a sort of light gray. However, look at a penny, that is the color of copper (old pennies are mainly copper), brass is generally yellow—the color of low-fuming brazing rod, bronze is darker than brass. Zinc is a duller gray than ordinary solder. Stainless steel has a clean look that is sometimes hard to tell from cold-rolled (mild) steel. Unbroken cast iron is difficult to tell from cast steel.

Of course, the color test is more difficult if the metal is painted or otherwise coated. Coatings should be removed carefully or the metal may be polished and smeared.

Magnetic Test

The magnetic test is very simple: a magnet sticks to steel and iron. Most stainless steels used in fabrication are non-magnetic and weldable. However, if the spark test shows that a metal is stainless and yet a magnet sticks to it, avoid welding on that metal. It is probably a 400-series stainless, which is difficult at best to keep from cracking after welding.

Sound Test

The sound test distinguishes a hard from a not-so-hard metal. When struck, the hard metal rings, the softer metal makes a duller sound. You can test this by dropping a piece of mild steel onto a concrete floor and comparing the sound to that of an old drill bit or piece of spring steel.

Weight Test

Comparing weights can be useful. Magnesium, for example, is lighter than even aluminum. Cast iron, stainless steel, and mild steel all weigh about the same (cast iron is the lightest, mild steel the heaviest). A precision scale and other instruments are needed to perform an accurate metal weight identification.

Spark Test

Many metals give off a characteristic spark when ground. The spark can be compared to that given off by a known metal. This is a very handy test. Aluminum and copper alloys do not give off sparks (and, in fact, are not normally ground at all). Tests are best with a hard wheel on a pedestal grinder,

6 to 8 inches in diameter with a medium coarse grit. Even the sparks given off by a hand-held disc grinder often have enough distinction to be of use.

Compare sparks given off by grinding mild steel, cast iron, stainless steel, and the edge of an old file. The differences are quite pronounced.

Chip Test

It might not be practical to grind a metal to analyze the spark, yet you know from the magnetic response that it is either cast steel or iron. The chip test involves attempting to chisel a continuous chip from an edge of a sample. An iron chip breaks, steel can be made continuous. Recall how a long chip forms when drilling steel with a steady feed pressure on a drill press.

File Test for Hardness

The comparative difficulty with which a sharp file cuts into a block of metal can gauge its hardness. (As with the other identification methods, it is essential to have an established basis for comparison.) Hard metals are difficult to weld because they do not have flexibility to allow expansion and contraction. Moreover, hardness is likely to be lost with the heat of welding.

Chemical Reaction Test

Some metals have distinct reactions to chemicals, which helps in identification. Two common tests are described.

Mild steel can be distinguished from stainless steel by applying one drop of *Nital*. The mild steel will be etched right away, but the stainless will be unaffected. (Nital is a mixture of 1 to 5 percent nitric acid and 95 to 99 percent methyl alcohol.)

Aluminum and magnesium look about the same. One drop of either silver nitrate or zinc chloride turns magnesium dark, but does not affect aluminum. The reaction continues, so the test solution should be washed off.

Aluminum or Pot Metal

One frequent identification error is to mistake zinc

die cast material ("pot metal") for aluminum. This book does not deal directly with aluminum, but weld repair of zinc die castings is sometimes successful. These metals can be distinguished by placing a *small* torch with a neutral flame at an expendable edge of the object in question. If it melts quickly away, it is zinc die cast (it might even fume like overheated brazing rod). Aluminum is a super conductor of heat, so it transfers the torch energy to the rest of the plate before it melts.

SOME BASIC REPAIR TECHNIQUES

If you intend to fix a broken object, the question "why did it break?" must be answered. Also, if it wore out, what can (or should) be done to prevent future wear? Finding the answers to these questions is well worth the effort.

Examine the face of the fracture (break). Is it clean (Fig. 21-1) or is it partially filled with rust and grime? This gives a clue as to whether the fracture occurred suddenly or developed gradually. Is there a sign of bending or sudden impact that could have caused the break? These questions are important to avoid going through the motions of making a repair only to have the object break again right away. If you have an idea of the break cause, you can often strengthen or redesign the object so that it will be better than new.

If a weld repair involves a worn-out part or section of a part, try to ascertain why it wore out. Stronger or harder filler metal or even reinforcing members might prevent future wear. Keep in mind, however, that bearing assemblies are often designed with a sacrificial member. For example, a bearing insert or bushing is made of soft metal and supposed to wear, thereby protecting the more expensive shaft that turns on it. If an assembly with a failure is subject to vibration and/or impact, see if there is any way to reduce shock transmitted to the failure zone. Making a broken part stiffer might merely relocate the problem, not solve it.

WHAT NOT TO DO

There are endless possibilities for weld repairs; many are safe and easy. However, potential dangers lurk in some repairs.

Fig. 21-1. This heat-treated shaft broke in a brittle fashion. The smooth and clean break indicates a sudden fracture. A weld improperly made around the shaft while it was in the hardened state caused the break.

Do not weld on a casting near a pressed-in bushing. The expansive and contractive forces can eject the bushing with a violent force. If removing the bushing is not practical, secure it with a bolt and large flat washers.

Do not apply heat to a hollow item without proper venting. Avoid repairing containers that have held gasoline or cleaning solvents. Some gas and diesel engine valve stems have been filled with liquid sodium for cooling. Applying a hard face layer to the valve stem was the very last thing some people ever did.

Do not weld near a gas tank. If welding is required, first remove the contents and then the tank to a safe distance. Beware of a smouldering car fire starting in carpeting or upholstery.

A lead-acid storage battery produces hydrogen

gas. Do not make a spark near a battery. Also, avoid possible damage to electronic components of an automobile by disconnecting first the grounded and then the insulated battery cables before arc welding on the vehicle or a trailer.

Do not attempt weld repairs to wheels, steering, or suspension components of motor vehicles. Damaged components should be reconstructed or replaced.

Never allow chlorinated solvents to be touched by a flame or arc. Deadly phosgene gas is the product. It is best to avoid all use of chlorinated solvents. Their vapor and even contact with the skin is harmful.

Apply no heat to panels lined with styrofoam or polyurethane foam. This destroys the lining and produces a deadly gas.

Do not assume that welding heat will remove water, oil, grease, or paint from the base metal. More active precleaning is required. The only sure way to remove oil and grease from a casting is by heating to an almost red heat and holding this temperature for several hours. This removes carbon and hydrogen that would weaken any weld in the vicinity.

Avoid welding near a machined surface (including threads), glass, decorative panel, or plastic surface without first taking measures to prevent spark and spatter damage. Automotive finishes are especially susceptible to such damage. Use fiberglass blankets to shield irregularly shaped surfaces and objects.

WELD REPAIRS TO CAST IRON

Cast iron is a blanket term that includes four main types: gray cast, white cast, malleable, and ductile (nodular). All contain about 2 percent carbon (compared to the 0.16 percent of mild steel). These variations are attained mainly by the sequence of heat treating received in production.

Most cast iron is *gray cast iron* used for machine bases and housings, gear boxes, engine blocks, and many machine parts. Gray iron is weldable with qualifications. White iron is very hard, and welding it should be avoided. Malleable iron is weldable but must be heat treated afterward to avoid

forming a brittle structure alongside the weld. Ductile iron should be welded when it is hot (about 600 degrees Fahrenheit) and must cool very slowly.

Your main contact with iron is bound to be gray cast. It can be welded at home with oxyacetylene or SMAW. Gray iron can also be joined by oxyacetylene brazing and with GMAW using silicon bronze wire.

Compared to mild steel, cast iron is brittle and does not yield much before it breaks. The control of heat is especially critical when welding iron. Preheating and post heating are often used.

The selection of a welding or brazing method is often determined by the available equipment. Professional casting repair shops use furnaces to preheat and hold the castings at a red heat—even while they weld them! Oxyacetylene is the tried and true welding method, but for a larger casting this is not practical because the home shop has limited heating abilities and small torches. Brazing is often done because it involves less heat. When heating the entire casting is not a practical option, arc welding with special nickel stick electrodes works sometimes.

If the broken iron will be subjected to temperature changes of more than 100 degrees or so, it should be gas welded. Also, if patches or inserts are needed use iron, not steel. Steel and iron can be joined, but crack as the casting heats and cools. This is because steel and iron have different rates of expansion and contraction.

Cleaning and Preparing Cast Iron for Welding

Most welding done to cast iron is to repair a crack or broken-off section, so the most likely joint is a butt joint. Preparing a butt joint in cast iron requires opening it up, often to a single V, to allow fusion to the sides all the way through. A small 1/16-inch land at the root prevents weld metal from simply melting through and dropping out. Bevelling the joint sides also provides more surface for the weld deposit.

Cast iron (Fig. 21-2) forms a skin as it cools. You must remove this and any paint and surface contaminants such as oil. To get rid of all the oil, it is recommended that the iron be heated to 750

445

Fig. 21-2. A cast iron fracture reveals a characteristic grainy surface. Preparing the parts for welding is the largest part of the job.

degrees Fahrenheit for an hour or more. This "cooks out" oil trapped in the casting. That much heat for that long means improvising a furnace. It also means stripping the casting of all parts to avoid cracking from uneven expansion rates.

As iron is ground, the skin and surface graphite are smeared which prevents proper action with the filler metal. This is mostly a problem with braze welding. With fusion welding, the surface is melted and stirred about by the filler rod coated with flux. Use a sharp chisel to remove the smeared surface of the joint face.

Always preheat iron to remove moisture and to reduce the sudden shock of a rapid quench to the weld zone. A small oxyacetylene torch is impractical for heating large castings. Rent a weed burner. A temporary furnace made of stacked firebricks conserves heat and ensures even heating. (Remember, uneven heating causes more cracking.)

For oxyacetylene welding, use cast iron filler rod and cast iron welding flux. Use a large tip and keep the flame constantly moving. Molten iron does not flow out like steel, it must be "worked" more aggressively. Do not attempt to fill deep joints in one pass, use several. When finished welding, cover the casting with a fiberglass blanket or bury it in a box of ashes to cool over several hours. Figure 21-3 shows an antique car manifold welded with cast iron filler. After being in use for awhile, the repair will be hardly noticeable.

There are two popular stick electrodes for welding cast iron. Check with your supplier. One electrode produces a hard wear-resistant deposit, the other leaves a ductile deposit. Avoid using mild steel electrodes. These form a hard structure that cracks. *Keep the heat input low.* Cast iron repair electrodes must be run at the lowest possible amps. Use 3/32-inch electrodes if you can, and weld no more than an inch or two at a time. Stop and allow the deposit to cool until you can rest your bare hand on it before proceeding with the next short deposit.

With thin castings less than 3/8 of an inch, clean the weld area thoroughly and establish the extent of the crack. This can be done by first wiping kerosene into the crack and then drying the surface completely. Next, rub blackboard chalk over the area. In a few minutes the kerosene still trapped in the crack soaks into the chalk, leaving a line. Center punch and drill 1/4-inch holes completely through at both ends of the crack. These holes keep the crack from spreading. Use a countersink or larger drill to bevel the holes so you can get the weld deposit into them.

If you use air carbon arc cutting (AAC) for opening up a crack for cast iron repair, use electrodes made for ac operation. Use ac if you have it; if you have only dc, use DCEN (straight polarity). Recall that this polarity is opposite to that used with most stick electrodes.

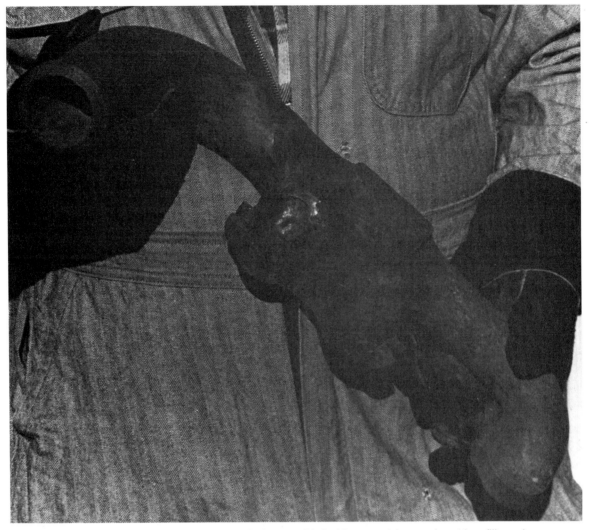

Fig. 21-3. This 1929 Model A Ford exhaust manifold was welded with oxyacetylene and cast iron filler rod.

Cast iron and steel can be joined. If mild steel electrodes are used, the joint will be especially hard and brittle on the iron side; use the softer nickel electrodes. It is often better to use braze welding with its lower heat and greater ductility to join iron and steel.

WELDED INSERTS

There is a proper way to close an opening in a plate with an insert. Avoid the temptation to simply tack it in and weld it. Once the insert has been tacked, it is welded only half the way around. Then a period of waiting is imposed while the weld cools and all the shrinkage takes place. The tack weld opposite the welded segment will probably break. If a weld is made nonstop all the way around, the cracking occurs in the second half of the weld. After cooling, the second half is welded. This technique should be used on all inserts—round, square, or rectangular.

PLUGGING A HOLE

It is sometimes necessary to fill a hole by welding. If it is threaded, remove the threads with an over-

447

size drill, a rattail file, or a small die grinder. This also removes oil that might be trapped in the threads.

If the diameter of the hole to be plugged is 1/2 inch or more, use an insert to avoid excessive weld metal shrinkage. For small holes, place an aluminum or copper bar at the bottom of the hole. This dissimilar metal backup keeps weld metal from running out of the bottom of the hole. E6011 is the best stick electrode for a clean deposit in the hole. Gas-shielded flux-core wire ("dual shield") is the best wire for plugging a hole. Use enough heat to avoid cold laps on the sides of the hole.

HARD FACING

Hard facing is the application of a hard and wear-resistant shell to a softer and more ductile base metal. This practice became widespread when it was discovered that oil well drills lasted longer with a hard facing.

In Chapter 20 a "cold" flame spray coating was said to give metal a hard outer shell. In the home shop, you can apply hard facing with a torch.

Use a carburizing flame with the acetylene feather about 3 times as long as the cone to introduce excess carbon to the surface of the steel. This lowers the melting temperature enough for a smooth application of the hard-facing rod. Steel appears to "sweat" under a carburizing flame. This indicates the proper temperature and surface condition for applying the filler. Do not melt the base metal and blend filler as with fusion welding.

A bronze overlay can also provide a wear-resistant surface to steel or iron. Clean the surface and, using flux, apply the bronze as you would low-fuming bronze filler rod.

Hard facing stick electrodes are usually accompanied by specific instructions as to amperage, polarity, and placement of the deposits. In general, it is not advisable to build up a solid pad. Rather, make low penetration deposits in individual stringer beads and keep the heat input to a minimum. These deposits are designed to crack at intervals across the line of the weld to form individual hard segments and avoid cumulative warpage to the base structure.

THREADED PIPE CONNECTIONS

Threaded steel pipes sometimes leak at their connection points. You might be tempted to run a weld bead around the connection to a pipe coupling or elbow to stop a leak. Don't do it. Also do not attempt to strengthen a connection with welding. Welding a threaded connection is a bad idea because:

● The thin area of a thread is easily undercut to the point where the connection strength is greatly weakened.

● Welding distorts the adjacent threads and shape of the pipe, upsetting the jam fit and destroying the seal.

● The extreme heat of welding easily destroys any sealing compounds or pipe thread tape.

Welded piping systems use connecting fittings with single V joints. No threads or sealants are involved. The assembly is lighter, faster, easier, and stronger than with threaded systems.

THAWING PIPES

You might hear something about a welding power source being used to thaw a frozen water pipe. This is not usually a good activity for the home craftsman for several reasons:

● A large heavy-duty constant current (CC) power source is required. (Constant voltage units have no built-in amperage limits.)

● There is a strong fire hazard from overheated pipes or connections. Melted pipes and additional flooding problems are also possible.

● Special extra-large cables and special monitoring equipment are required.

Unless you need to do a lot of pipe thawing, the added equipment expense is probably not advisable. Imagine trying to fight a fire started during a thawing attempt, with the water pipes still frozen.

Chapter 22

Joining Wood to Metal

THERE ARE MANY COMBINATIONS OF WOOD and metal in project construction for both aesthetic and structural reasons. Many larger commercial structures reflect this (Fig. 23-1). Wood and metal can also be combined on a more modest scale for home building projects.

WOOD AND METAL COMBINATIONS

Metal framing is sometimes the best solution for a housing carpentry problem. Stairways leading into areas of limited space might be better built from steel than wood. This is because a bulkier framework is necessary with lumber. However, if the stair treads and railings are made of wood or wood and metal, the warmth of a wooden structure can be retained. A spiral staircase made of steel is a stairway solution with an incredible economy of space. Knock-down kits are available from manufacturers with nationwide representation such as The Iron Shop.

Wood and metal are often combined in the construction of furniture, shelving, and other durable household items. A walk with a sketchpad through a display of office furnishings can provide ideas about both attractive and functional furniture. A sturdy table with an angle bar frame and a plywood top is very easy to make. This basic idea can be applied to produce many different table-like items such as aquarium stands, entertainment centers, and many more.

When you inspect wood and metal furniture closely, notice how often the finishes on these projects really set them off. Contrasts of color, tones, and texture can be especially pleasing to the eye. For example, chrome or brass and dark wood are often joined in elegant combinations. Such metal finishes are obtainable through plating and finishing services. Project parts that are to be plated should be made as smooth as possible.

MOISTURE

Structures of wood and metal can pose some minor durability problems that are best considered while a project is in the planning stage. Humidity

Fig. 22-1. This attractive lakeside wood pavilion is joined with bolted construction using fabricated steel connection brackets and plates. The assembly of the trusses and columns was simplified by using the prefabricated steel connectors as drill jigs. The steel is 3/8-inch thick with full-penetration welds used throughout.

can greatly affect unsealed wood. Variations in the moisture content of wood can cause curling and other problems due to expansion and contraction. Unless wood is sealed, restrictions to free movement can result in splitting or cracking.

Moisture also causes metal to oxidize or rust. It is necessary to seal the wood and protect the metal with appropriate products. Water-based wood paints cause rusting on bare steel, and varnishes tend to flake. Do not count on either the wood or the metal to shield the other from moisture. With changing temperatures and humidity, even plated or coated metal collects condensation that can deteriorate unsealed wood. There might be relative movement between the two mediums,

so it is advisable to finish the wood while disassembled from the metal to keep the finish from cracking at the interface.

Expansion and contraction movement also occurs in metals with temperature extremes. Such movement could cause rigidly attached wood to crack. (This is generally less of a problem with plywood and other wood laminates.) When joining wood to metal, you can allow for slight relative movement between the two mediums by making bolt holes slightly oversized, avoiding overtightening of fasteners, and by using a product such as H.B. Fuller's Max Bond—a cushioning rubber-like adhesive coating for application between wood and metal.

FASTENERS AND OTHER HARDWARE

There are many hardware items for joining wood to metal. Some items, like mounting brackets and clips, can be made in your shop from standard metal shapes at great savings. If you weld on plated hardware, avoid breathing the fumes.

Screws and Bolts

If you are unfamiliar with the items in the following list, go to the hardware store and look them over. Wood and metal are often joined with:

- Wood screws with various types of heads
- Sheet metal screws
- Machine screws
- Drywall screws
- Cap screws
- Self-tapping screws
- Carriage bolts
- Lag bolts
- Hex, square, wing, locking, and castle nuts
- Cut or flat, and various types of locking washers

Other fastening items to look for include:

- Eye bolts
- Cotter keys or pins
- Turnbuckles
- A clevis
- Shackles
- Brackets, plates, and clips such as Strong Ties
- Various hinges

When shopping for hardware, try to find hot-dipped galvanized products. Cadmium ("cad") plated bolts are more common these days but do not hold up as well. The zinc coating from hot dipping does not look as smooth as cadmium, so many hardware outlets have stopped offering hot-dipped fasteners. Once assembled, protect threads (and other areas from which the zinc may have been worn) with paint, a cold galvanizing compound, or perhaps even grease if no one is likely to bump into it.

Assembling with Fasteners

It is possible to develop very high compressive forces with fasteners, especially nut-and-bolt assemblies. The craftsman used to metal working might not be aware of this. He or she uses the standard practice for metal assemblies of tightening fasteners until they are very snug. The forces developed can easily crush wood to the point where it is likely to split or crack. Because of this, a limit is sometimes imposed to prevent overtightening fasteners for wooden assemblies. Compression sleeves that slip over the bolt can be purchased or just made from thin wall tubing to limit how far a bolt-and-nut can be tightened.

The handle or lever length of a wrench controls the ease of tightening fasteners. Short-length wrenches are usually adequate. If space permits, a "speed handle" is especially convenient. The use of long "breaker" or "cheater" bars can easily put too much squeeze on wooden members. Where fasteners directly contact wood, gnawing of the wood is avoided by using flat washers or plates to spread out the pressure.

WOOD AND WELDING

Try to complete all welding activities before bringing wood and metal together. It is, however, sometimes necessary to weld next to wood that can be easily burned. Exercise careful judgement when tackling such jobs. Be sure to use sheet metal or some other shielding to protect wood from the devastating effects of hot weld spatter and slag.

By the way, sometimes wood is deliberately burned on the surface with a torch and then lightly wire-brushed. This produces a rustic and dramatic effect that makes new wood look very weathered.

Woodworking with its inevitable chips, sawdust, adhesives, and finishes can present extreme fire hazards. Take extra efforts at fire prevention when combining both welding and woodworking.

Appendix A

Abbreviations and Terms

T HIS APPENDIX PROVIDES YOU WITH DEFINI-
tions of welding terms. They can be referred
to as needed or read completely through as you prefer. In your welding activity you will, no doubt,
come across terms not included here that require
definition; jot these terms down.

ABBREVIATIONS

AAC (Air Carbon Arc Cutting)—also called
Arcair, Scarfing, Gouging, this is a very useful
metal removal process that uses a copperclad
carbon electrode and compressed air. AAC is
suitable for use by the home craftsman using a
small AAC torch and carbons no larger than 3/16
inch. Special carbons are available for use with
ac power sources. See Chapter 16.

ac (alternating current)—The type of energy
supplied by electrical power companies. The
direction of current or electron flow alternates
back and forth in cycles (Fig. A-1). In North
America the alternation frequency is 60 cycles
per second (cps) also called Hertz (Hz). A power

source is plugged into an ac line connection. If
the power source is a simple transformer type,
its output for welding is also ac. See also phase.

AWS (American Welding Society)—A non-
profit technical society that sets procedural standards in materials and workmanship. The
homecraftsman can thank the AWS for ease in
obtaining high-quality welding equipment and
consumables. The AWS publishes *The Journal*,
a monthly periodical for the welding industry,
and provides technical handbooks and other
literature.

C-25—A popular shielding gas blend of 25 percent
carbon dioxide and 75 percent argon. See also
metal transfer, shielding gas, and Chapter 11.

CC (constant current)—Refers to a type of
power source output in which the maximum amperage produced by a power source is not excessive. A CC output curve is shown in Fig. 4-9.
This type of power is used mostly for the
SMAW and GTAW processes.

CRS (cold rolled steel)—This is low-carbon steel
that, after hot rolling, has been "pickled" or

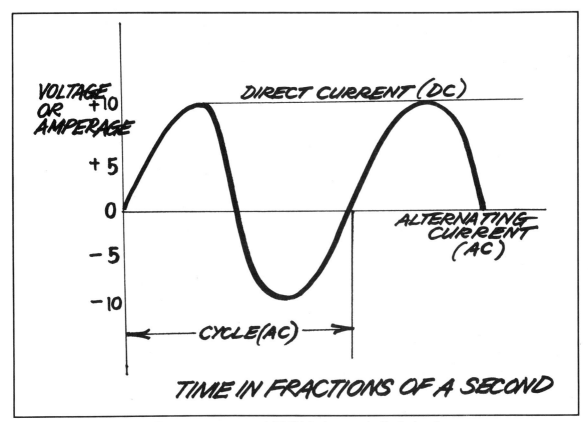

Fig. A-1. Alternating current. Sixty cycles per second (60 Hz) is the norm in North America.

cleaned in acid to remove all mill scale, and then rolled again. This second cold rolling produces a finer grain structure, more strength, exact dimensions, and a more expensive product. Lack of a mill scale necessitates an oil coating to prevent rust during storage. The home welder should use CRS only when absolutely required.

CFH (cubic feet per hour)—A measure of gas flow rate; a flow of so much volume in an hour past a point. In the home shop, CFH is used mainly to measure the flow rate of shielding gases used with certain of the wire feed processes. For safety, it is also important to know the gas flow rate of oxyacetylene torches in relation to the size of the acetylene cylinder. An AAC torch also has a compressed air flow rate requirement.

CP or CV (constant potential or constant voltage)—This describes a type of power source output. A CV/CP output curve is shown in Fig. 4-16. This type of power is used for the wire feed welding processes of GMAW and FCAW.

dc (direct current)—The type of electrical energy often used for SMAW and always used for the wire processes because it produces a smooth and easily controlled arc. A dc current consists of electron flow always in the same direction—from negative to positive. The dc current is obtained directly from batteries. For dc welding, ac is changed into dc by passing through a rectifier. Polarity refers to the direction of current flow in the welding circuit. See polarity.

FCAW (Flux-Cored Arc Welding)—A wire feed welding method in which a continuous, tubular, flux-filled wire electrode is fed through a torch or gun into the weld. FCAW is done with a constant voltage (CV), also called constant potential (CP), type of power source. Alloying elements can also be included with the powdered flux. Some flux-cored electrodes are designed for use without gas shielding, and some others require the use of shielding gas. These are generally termed self-shielded and gas-shielded respectively. An American Welding Society numbering system is employed which specifies gas or no gas, and possible welding positions for all tubular electrodes. An ongoing discussion persists as to the advantage of one over the other. The self-shielded electrodes are simpler to use because no gas apparatus is required. The claim of the self-shielded electrode producers is that they work best outside, in a moderate breeze. Operator appeal, however, is greater for the gas-shielded electrodes. The gas-shielded electrodes produce a more handsome weld with less smoke and spatter. Greater penetration is also achieved with the gas-shielded electrodes. Both varieties are suitable for use by the home craftsman. A popular slang term for the self-shielded flux-cored process is "Innershield" (Lincoln Electric); the gas-shielded mode is often called "Dual Shield" (Alloy Rods Corp.). The leading edge of filler metal development has been with the flux-cored wires. The diameter of self-shielded wire for the home shop is .045 inch. Gas-shielded wires for home use are .035 and .045 inch. Direct Current Electrode Negative (DCEN) is normally used for the self-shielded wires and Direct Current Electrode Positive (DCEP) for the gas-shielded wires.

GMAW (Gas Metal Arc Welding)—A continuously fed solid wire electrode and use of a shielding gas characterize GMAW. A CV (CP) power source is most often used; a wire drive mechanism is needed to propel the electrode at a constant speed into the weld pool. There are 3 variations of GMAW based upon 3 different modes of electrode metal transfer to the plate:

● Short circuit transfer, often called "short

arc," is an extremely low heat process in which the metal transfers by a rapid series of short circuit contacts to the plate.

● Globular transfer with relatively large droplets melting off the end of the wire and transferring across the arc to the plate.

● Spray transfer, in which minute droplets are melted off the wire and transferred across the arc to the plate.

There is yet another method called pulsed transfer which is something like a combination of short circuit and spray transfer.

HAZ (heat affected zone)—Part of the parent metal immediately adjacent to the weld metal deposit (Fig. A-2). See also Heat Affected Zone and Weld Metal.

HRS—Hot Rolled Steel—This is low-carbon steel as it comes from the mill. During the rolling operation, which imparts a specific shape to the metal, water sprays are used that cause a gray iron oxide or mill scale skin to form. Mill scale slows down the formation of rust on hot rolled steel. Mill scale inclusions are a detriment to weld quality; grinding is used on critical joints. Hot rolled steel is sometimes available without a mill scale. Called "pickle plate," this is easily confused with cold rolled steel. The scale is removed in an acid pickling solution and a light oil coating applied immediately to prevent rust formation during storage. The home welder works almost entirely with HRS.

Hz—Hertz, same as cycles per second. This is a term describing the frequency of an alternating current (ac). In North America, power companies supply 60 Hz ac. See also ac.

IC (integrated circuit)—The electronic marvel known as a "chip" that combine a number of separate components into a very small and standardized package. Welding equipment manufacturers started late in using IC's but they are now in full swing. As a result, newer welding equipment is smaller and lighter with greater capabilities.

454

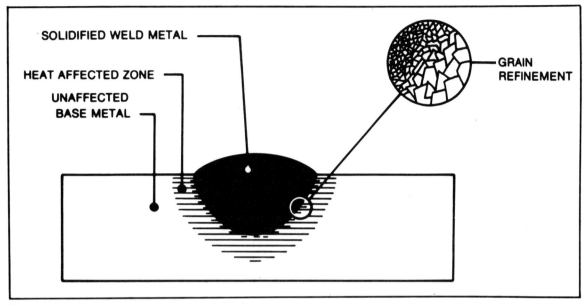

Fig. A-2. The heat affected zone (HAZ). (Courtesy of Alloy Rods.)

IPM (inches per minute)—1) the unit of travel speed of the welding heat source relative to the work. Bead width, height, and penetration are affected by the heat source travel speed. 2) the wire feed speed (WFS) of continuous wire feed electrode. The WFS in IPM is directly related to the welding amperage when using a wire feed process.

MIG (Metal Inert Gas welding)—An older term, now slang, used formerly to describe Gas Metal Arc Welding (GMAW). There is some ambiguity with its continued use. It is often applied to both GMAW and Flux-Cored Arc Welding (FCAW)—whether or not an inert gas or any gas at all is used as, for example, the terms "mig gun" and "mig machine." Generally speaking, MIG usually means GMAW.

NC (National Coarse)—A description of a thread series as found on nuts and bolts. Sometimes the older terms USS or UNC are used. The number of threads per inch (pitch) of coarse threads permits rapid assembly. Because coarse threads are deeply cut, they are often favored over fine threads to gain strength in threaded castings. In the home shop, threads are cut using taps and dies. See Chapter 16.

NF (National Fine)—A description of a thread series as found on nuts and bolts. Sometimes the older term SAE is used. Fine threads provide the maximum in thread surface area and, because they are not deeply cut, leave a great deal of shear strength in threaded rods etc.

OAW (Oxyacetylene Welding)—This welding method, which uses the heat of a properly adjusted flame, is used primarily in the home shop to join mild steel. Often the term is misused as a blanket to include other torch operations besides welding. OAW has drawbacks that are emphasized when compared to the more efficient arc welding processes. It is slow, and its use produces a great heat input to the plate. OAW should be limited to applications not suitable for arc welding. The term "gas welding" is slang for OAW.

OAC (Oxyacetylene cutting)—A very useful method for cutting steel. A specially designed cutting torch uses an oxyacetylene flame to preheat the steel to its kindling temperature. The torch operator then pulls a lever, which causes

455

a jet of pure oxygen to cut through the plate. Sometimes OAC is called "burning."

OCV (open circuit voltage)—The output voltage of a power source before the arc is struck. Power sources designed for SMAW use have an OCV of 77 volts; wire feed units have an OCV of 50 volts or less. The OCV value decreases rapidly, once the arc is struck, to that of the welding or arc voltage.

PC (printed circuit)—Refers to the method of mounting electronic components on standardized boards made of an insulating material, but coated with a thin copper cladding. During production, the copper paths used for connecting the components are masked off with a tar-like resist. This is usually done with a printing process. When the board is immersed in acid, all of the copper except for the protected paths is etched away. The boards are then cleaned and assembled—that is, the components soldered in place. There is usually no repair done to a "board;" if defective, they are usually replaced. Welding equipment is often full of PC boards.

PSI (pounds per square inch)—The familiar unit of gas pressure and also the unit of stress as applied to a welded object (weldment). The strength of steel and other metals is expressed in PSI of tensile (pull) strength. Filler rods and electrodes are also classed by their tensile strengths.

SCR (silicon control rectifier)—A semiconductor electronic component used to control the output of many power sources. Other familiar semiconductor devices include transistors and diodes. The SCR (Fig. A-3) is capable of rectification of ac into dc like a diode and additionally, for control, can be turned on and off (to conduct or not to conduct) very rapidly by a signal sent to its gate connection. This differs from a transistor in that the SCR gate signal is simply an on-or-off and not a variable up-or-down control. See also diode, solid state.

SMAW (Shielded Metal Arc Welding)—This is frequently called "stick" welding because of

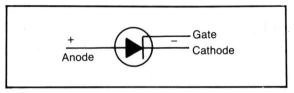

Fig. A-3. Symbol for a silicon-controlled rectifier (SCR).

the coated electrodes used. The coating is needed to provide arc control, to cleanse the weld zone, and for several other reasons. The equipment required for SMAW welding is simple and relatively inexpensive. Power sources are of the constant current (CC) design. A variety of metals can be joined with stick electrodes. The most convenient electrode diameters are 3/32 and 1/8 inch. While SMAW remains a useful home and industrial welding method, its popularity is declining due to the increased use of the wire feed processes. See Chapter 10.

SS (Stainless Steel)—Stainless steel is basically iron, very little carbon, at least 11 percent chromium, and usually some nickel. A very popular alloy is 304, which has 0.08 percent maximum carbon, 18-20 percent chromium, and 8-12 percent nickel. Stainless is rather expensive, costing about 4 to 5 times as much as mild steel. Stainless is used wherever conditions are harsh enough to require corrosion resistance, and for sanitary and decorative purposes. This is not an easy family of metals for the beginning home welder. Stainless can be welded with the home arc welding methods, although not all alloys are weldable. Stainless can be readily joined to mild steel using stainless electrodes and by carefully following a procedure.

TIG (Tungsten Inert Gas welding)—This is an old term, now slang along with "heliarc," used to describe Gas Tungsten Arc Welding (GTAW). The inert gas helium was initially used for welding aluminum and magnesium with a nonconsumed tungsten electrode.

TERMS

acetylene—A popular fuel gas derived from the

combination of calcium carbide and water. The flame is uniquely suitable for both welding and cutting. Acetylene (C_2H_2) is dangerously unstable at pressures over 15 psi; it can be stored at higher pressures only if dissolved in acetone.

alloy—A mixture of metals or an ingredient added to that mixture which imparts a special characteristic such as hardness, corrosion resistance, etc.

aluminum—The second most widely welded metal after mild steel. Aluminum is about 1/3 the weight of steel. It is a good conductor of heat and electricity and does not become brittle under cold conditions. Aluminum does "rust" or oxidize, but usually at a slow rate. It is rolled into plates and sheets as well as *extruded* or squeezed through dies into many shapes. Aluminum is available in several alloy groups, but some of these are impractical to weld. In the home shop aluminum can be welded by the GMAW wire process using an argon gas shield. Welding aluminum with SMAW or OAW is tricky and best avoided.

amp—or ampere, the unit of electrical (electron) current named after Andre Marie Ampere (d.1836). According to Ohm's law, 1 amp is the amount of current that flows in a circuit with 1 volt force and 1 ohm of resistance. An amp represents a large amount of electrons, in fact 1 coulomb's worth, flowing past a given point in one second. A coulomb is 6.25×10^{15}. Practically speaking, amperage is regarded as welding "heat." With SMAW welding, amps are limited by the setting at the power source; the welder has some control while welding by changing the arc length. With the wire feed processes, amps are controlled by the wire feed speed setting. The line draw, or input power, of a power source is also measured in amps and must be within the capacity of the household electrical supply.

anodize—An electrical surface treatment used for aluminum, often to impart color as well as durability. An anodized skin must be completely removed before welding.

arc—The electrical flame used as a heat source for arc welding and arc cutting. The welding arc is hotter (at nearly 10,000 degrees Fahrenheit) and more concentrated than the 6000-degree oxyacetylene flame. Arc welding is, therefore, generally faster and imposes less heat distortion to the welded workpiece. For most of your home shop arc welding, the arc is directed or aimed by a combination of physical forces and your manipulation of the electrode. The SMAW welding arc possesses a forceful jet from the coated electrode which, to a considerable extent, can be used for control of the underway welding operation. The arc used with the home-compatible wire feed welding processes of GMAW and small diameter FCAW, is controlled mainly by the voltage setting on the power source.

backfire—A sudden extinguishing of a torch flame with a loud cracking sound. This is caused by an overheated tip which, in turn, might be caused by inadequate distance between the tip and the work or by the presence of slag. Backfires can also be caused by a loose tip or connection, allowing air to be sucked into the gas streams. The cause of the backfire should be ascertained and corrected before continuing work. Backfires can cause a dangerous *flashback*. See also flashback.

base metal—The metal joined by welding, brazing, or soldering; also known as parent metal. This is contrasted with filler metal. It is important to know the identity of the base metal in order to use a compatible filler metal. The base metal makes up the weldment.

bead—More correctly, weld bead. This is the deposit upon the base metal made by the welding operation.

blast cleaning—Removal of surface contaminants such as rust, scale, paint, etc., with a high-speed air jet containing sand or some other abrasive. The home welder would usually require this service for repair or restoration work.

brass—A term generally used to describe an alloy of copper and zinc. While brass objects can be fusion welded, soldering and brazing are often more practical.

braze welding—A metal joining process in which the base metal is not melted. The filler metal melts at a temperature higher than 840 degrees Fahrenheit. A chemical cleaner, or flux, is commonly used to remove surface oxides from the base metal. This process differs from brazing or soldering in that the filler metal does not flow between the joint members by capillary force. A distinct weld reinforcement is used with braze welding. In the home shop, braze welding is usually done with an oxyacetylene torch, although a wire feeder loaded with silicon bronze and using C-25 shielding gas achieves a similar effect in less time and with less heat input into the base metal. Braze welding is used extensively in the repair of cast iron.

brazing—A metal joining process similar to braze welding, except that the filler metal flows between the joint members by capillary force.

bronze—A term generally used to describe an alloy of copper and tin. Home welders use low-fuming bronze (AWS R CuZn-c) filler rod for brazing and braze welding. This is an established misnomer because low-fuming bronze contains a high proportion of zinc which makes it a brass. Silicon bronze is another fairly common filler often used with GMAW where low heat input is desired. Brasses and bronzes melt in the range of 1600-1800 degrees Fahrenheit.

cast—The curl that remains in spooled wire after the wire is detached from the spool.

casting—An item manufactured by an ancient but viable process in which a melted substance (metal) is poured or pumped into a mold and allowed to solidify. After cooling, the mold is opened to release the casting within. This way of producing metal objects is in contrast to wrought operations whereby the metal is forced into shape without melting. In a very real sense, a weld deposit is a casting. In the home shop, castings are generally associated with weld repairs. See Chapter 21.

circuit breaker—A safety device inserted into an electrical circuit that automatically opens the cir-

cuit if a specific amperage value is exceeded. Circuit breakers are similar in effect to fuses, although breakers can be reset. To prevent dangerous overloads, every welding power source must be connected to the line electrical source through either a circuit breaker or a fuse.

cold lap—A serious weld defect by which unfused filler metal lays on the surface of the base metal. This is almost always caused by improper and/or untested welding techniques.

connectors—Devices used to join either electrical components or gas apparatus items. In the home shop these are potential sources of electrical energy loss or gas leaks. Some connectors are designed to be frequently assembled and disassembled as easily as an ordinary electrical plug. The home craftsman can often insert connectors into equipment systems to increase portability or general convenience.

contactor—A relay switching device inside a power source that connects the main transformer to the input power. This device permits the safe handling of a welding torch until the operator indicates readiness via a trigger, foot pedal, etc. Contactors are found in nearly all wire feed power sources, and some SMAW units.

demurrage—Rental fees charged for gas cylinders after a specified length of time. This is often a separate fee in addition to the charge for the contents of the cylinder. This expense is sometimes overlooked by the novice welder.

diode—A semiconductor (solid state) electrical device that permits current to flow in only one direction (Fig. A-4). These are sometimes described as electrical "check valves." In many power sources, diodes are arranged in a structure called a rectifier bridge to rectify or change ac into dc. Along with other semiconductor

Fig. A-4. Symbol for a diode (rectifier).

devices, diodes are damaged by heat over 150 degrees Fahrenheit and are therefore located on a heat-radiating heat sink.

distortion—The result of an uneven expansion and contraction of the weldment due to the presence of restraints. See also Chapter 17.

duty cycle—A rating used for welding equipment (power sources, welding guns, etc.) that specifies what percent of a 10-minute period the equipment can be continuously operated at its rated capacity. Example: a 200-amp, 60 percent duty-cycle, power source may be continuously operated at 200 amps for only 6 of every 10 minutes. It must then be allowed to cool for 4 minutes. The home craftsman seldom uses equipment to capacity.

electrical power—Multiplying voltage by amperage equals wattage. This is the familiar unit of electrical heat energy or power used in classifying electrical items from lightbulbs to toasters. The simple volts-times-amps (wattage) relationship does not apply to the welding arc. An increase in welding voltage causes an increase in arc length; this adds resistance and does not always result in increased wattage. The major power (or "heat") control in a welding circuit is amperage.

electrode—Carried in a holder or gun (torch), this is the hottest solid part of the welding circuit located at one end of the arc. The electrode is heated by resistance with the passage of the arc current. The electrode melts and becomes the filler metal with the home welding processes described in this book. Stick electrodes are coated with a flux. Solid wire electrode is generally bare, although it may have a light copper coating, and flux-cored electrode is hollow and filled with flux. Electrodes are classified by their as-deposited tensile strength in psi or kg/sq mm. See also tensile strength.

fasteners—Devices that mechanically join parts of a structure either permanently or temporarily. Examples of fasteners include screws, bolts, rivets, clips, staples, and other special-design devices.

ferrous—A metal that is mainly composed of iron. This includes most of the useful metals—iron and steel. Sometimes stainless steel is classed as non-ferrous, even though it is mainly iron.

filler—The material added to the base metal. This is the electrode itself for FCAW, GMAW, and SMAW. With OAW and other torch processes, it is the hand-fed filler rod. See also weld metal.

fire extinguishers—These familiar devices are classed according to different types of fire. A welding environment has the potential for any type of fire, so multi-purpose extinguishers should be conveniently located. Never use water on burning liquid or electrical fires.

flame cutting—This is a process used mainly for cutting ordinary carbon steel. A special cutting torch or attachment is required. An oxyfuel flame (acetylene, propane, and Mapp are fuels) is used to heat a small area of the metal to a kindling temperature. A stream of pure oxygen then cuts through the plate. Flame cutting is essential for the home welder.

flashback—A potentially dangerous condition whereby the oxyacetylene flame burns back inside the torch tip. If allowed to continue an explosion could easily occur at the regulator and cylinder. Flashbacks are caused by a lack of adequate fuel gas flow through an overheated tip; a backfire can easily set one off. Tip starvation, whereby the gas flow adjustments are inadequate, could cause a flashback—as could a pinched or kinked hose. Should a flashback occur, the first response is to *immediately* shut off torch oxygen valve and then acetylene valve.

flux—A cleaning agent to remove contaminants from metal to permit an effective bond or weld to take place. This term is commonly applied to the coating on the stick electrodes which, besides flux, is also composed of other agents to form slag, permit smooth application of the coating (extruding agents), and other functions. The flux-cored continuous wire electrodes as well as the stick electrodes permit quality welding on slightly contaminated material. A paste or powder flux is often used with oxyacetylene torch work.

forging—A mechanical forming method whereby metal is pounded into a specific shape, often while hot. Forging imparts a refined grain structure to certain alloy steels, which achieves a very high strength. Wrenches, lifting hooks, and connecting rods are some examples of forging. The home welder should usually avoid welding on forgings because heat changes grain structure. In some limited cases, special procedures permit weld repair to forgings.

fuse—As a verb, fuse means to melt; welding fuses metals together. As a noun, a fuse is a safety device placed within an electrical circuit that melts or "blows" at a designed amperage, thereby protecting other circuit components and/or preventing fire. Welding equipment often contains fuses, and the home craftsman should keep the correct size and shape of spares on hand. There are both fast blow and slow blow types of fuses for any given amperage rating. They must never be interchanged. Fast blow fuses protect delicate circuits with an instantaneous interruption of current flow.

galvanize—A process of coating steel with zinc to prevent rust. Many familiar items from garbage cans to rain gutters are galvanized. The home craftsman might send a project or part of a project to a galvanizing shop. Welding on galvanized material requires positive ventilation because the fumes produced are toxic. See Chapter 20.

gas—A state of matter, along with liquid and solid. "Gas welding" is sometimes used to describe oxyacetylene welding as distinct from the arc welding methods. Compressed fuel gases are used with compressed oxygen gas for both cutting and joining operations. Other gases and gas blends are used to protect the weld zone from atmospheric contamination with certain arc welding processes.

gauge—1)A pressure-indicating instrument. 2)A measure of metal thickness: for example, 16-gauge (ga) sheet metal. The higher the gauge number, the thinner the metal. 18 or 20 gauge is the lightest for practical arc welding.

gouging—A metal-removing operation used to remove defective welds or to disassemble a welded structure. Gouging is also sometimes used to prepare weld joints of heavier metal thicknesses. See Air Carbon Arc Cutting (AAC).

heat affected zone—The base metal directly adjacent to the weld that has a grain structure distinct from that of the weld and that of the base metal located farther away from the weld. Simply abbreviated HAZ, this zone is often prone to cracking in metals other than mild steel and is a frequent subject for welding investigators. See HAZ.

heat treat— Processes that modify the grain structure of metal with the controlled application of heat. Hardening, tempering, annealing, and stress relieving are examples of this process. The home craftsman should avoid welding on heat-treated objects. Because welding heat can destroy the effect of heat treatment, a pre-weld and a post-weld sequence of heat treat operations is needed. See also stress relieve.

hood—A protective shield mounted on a head band that deflects ultraviolet and infrared arc radiation. A filter lens carried in a flip visor is used to permit safe observation of the arc. When the visor is lifted, a stationary clear plastic lens mounted on the body of the hood permits safe removal of slag and other chipping operations. See also lenses and Chapter 4.

inclusion—An impurity contained in a weld. These can cause cracking or otherwise weaken a welded structure. Proper cleanliness of the weld area before welding, and removal of slag between passes help prevent inclusions.

induction—Causing an electrical current to flow in a circuit without any direct mechanical connection to a source of voltage. The three essentials for induction are: a magnetic field, a conductor, and relative motion. This action, more specifically, electromagnetic induction is what makes generators, alternators, and transformers work.

inductor—A coil connected in series with the output circuit of a power source that has a stabilizing effect upon the arc.

inert gas—A gas that does not combine with any other substance and is therefore excellent for shielding the weld zone. Argon is the most widely-used true inert gas for the home shop. Carbon dioxide also acts enough like an inert gas in the arc to work as a shield for welding mild steel.

intermittent welds—Also known as "skip welds;" these are welds spaced in short, regular increments.

inverter—Also known as converters, and inverter-converters. A type of power source design that is very compact and efficient compared to a conventional design. Essentially, the frequency of the input to the transformer is raised electronically. This permits using a dramatically smaller transformer. The design had been used in Europe for some time before being introduced to North America. Of particular interest to the home craftsman is the fact that the high level of control possible with these units enables one power source to be used for FCAW, GMAW, SMAW, and even limited GTAW.

ionize—A transformation of the air or other gases between the electrode and the plate in such a manner as to permit the easy formation of an arc. A certain voltage is needed to ionize the air and other gases. Argon gas that is easily ionized enables welding with a lower voltage than is possible with straight carbon dioxide. The stick electrode coatings used with ac contain potassium which, in the smoke cloud that surrounds the arc, enable the easy reignition of the arc which "goes out" 120 times per second.

iron—A metal that is the basic ingredient for all of the steels and the irons. The element iron plus carbon over 2 percent is called iron, less than 2 percent is steel. Used alone, the term often is applied to gray cast iron, the most common of the irons. Gray cast iron is used in the manufacture of engine blocks, cylinder heads, gear boxes, and wood stoves. Gray cast iron in the home shop can be welded, but braze welding is often a better choice. A soldering copper is sometimes called a soldering "iron." See casting and Chapter 21.

joint—The particular arrangement of metal parts to be welded, brazed, or soldered. See also weld joints.

kerf—The slot or groove formed by the removal of material during a cutting operation (Fig. A-5). The kerf zone of a plate is changed to sawdust or slag during the cutting operation. Kerf allowance is taking into account the kerf width before making cuts. This is important to avoid running short of material.

layout—Marking out material in accordance with a blueprint or sketch to indicate cut lines. placement or location lines, center lines, and other essential information. See Chapters 12 and 15.

lenses—Welding lenses are merely flat darkened filter plates designed to protect the welder from infrared rays such as heat, that can be felt, and ultraviolet rays that cannot be felt. These rays are given off by the welding arc and are both harmful to the eye. For oxyacetylene torch work, a Number 5 density lens is usually adequate; for arc welding a Number 10 density lens is required. Individual differences may require the use of somewhat different densities. Generally more amps or larger flames require more filtration. A magnifying lens ("cheater") is also sometimes added to the lens "stack."

line power—The electrical energy delivered to your shop by the utility company. This is single-phase alternating current. Most households in North America have three-wire service connections that allow tapping off either 220 or 110 volts. The amperage capacity for the entire household depends on the size and rating of the service connection. Welding power sources must be compatible with the size of the service connection, otherwise overloads could result. Most home-type power sources have a maximum line draw of 30 to 40 amps. See ac and phase.

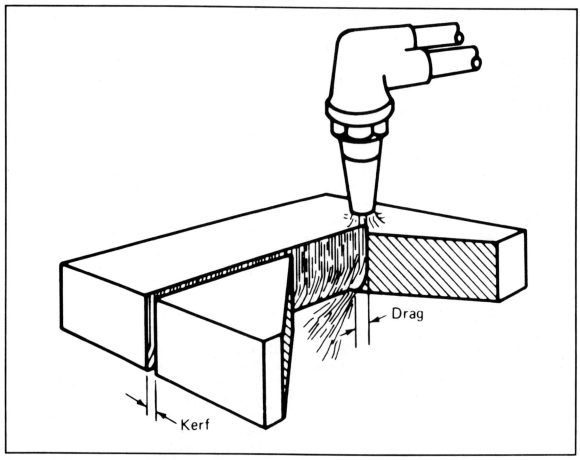

Fig. A-5. Oxyacetylene cutting showing the kerf. (Courtesy of American Welding Society.)

manual—The method of performing a welding operation in which both the application of heat and the addition of filler metal are hand-held operations. Oxyacetylene and SMAW welding are manual processes. See also semiautomatic.

metal transfer—The mode or manner by which filler metal melted from an electrode gets to the plate. There are three common metal transfer modes—short circuit transfer, globular transfer, and spray transfer. Pulsed transfer is not yet common for home applications. See GMAW and Chapters 10 and 11.

mild steel—A popular term to describe steel with a carbon content less than 0.29 percent or 29 "points of carbon." This is the ordinary type of steel used for most steel fabrications. It presents little difficulty in terms of its hardenability and is easy to cut and weld. Steel classification is an extensive subject.

non-ferrous—Containing very little, if any, iron. Common non-ferrous metals include aluminum and copper with its brass and bronze alloys.

Ohm's law—A formula named after George Simon Ohm (d.1854) that describes the relationships of voltage, amperage, and resistance, and is stated as follows: voltage = amperage times resistance in ohms. This is usually written as $E = IR$, where E is electromotive force or voltage, I is intensity of current flow or amps, and R is resistance or ohms. See also electrical power.

parent metal—Same as base metal.

pass—A single continuous weld deposit. While most welds consist of just one pass, larger welds are often made in combined layers or with multiple passes.

phase—The nature of an ac line supply. A single series of wave-like impulses (Fig. A-1) is termed *single-phase*; this is the type of power supplied to residential areas. Single-phase power has distinct zero periods that are usually not noticeable because of the frequency of the waves. A smoother and more efficient manner of delivering power is achieved with three overlapping series of wave-like impulses (Fig. A-6). Called three-phase, this power is usually only available in industrial parts of town. Welding power sources for home use must be designed for single-phase. A three-phase unit cannot be operated from single-phase power (although single-phase units can be operated from three-phase power by "dropping a leg"). Single-phase power sources with dc outputs use filters (inductance coils and capacitors) to smooth over the zero periods. See also ac and Chapters 3 and 10.

plasma—A fourth state of matter; a highly ionized gas stream. A welding arc is a plasma. It con-tinues only as long as energy is input. Plasma Arc Welding and Cutting (PAW and PAC) are processes usually found only in industrial applications. The equipment is expensive. The home craftsman should be aware that cutting aluminum, coppers, and stainless steels is effortless with PAC. Because there is very little heat input, this is one "outside service" that might be worth considering.

plate—A term used to describe the welding workpiece or base metal. Plate is also sometimes used to describe metal in thicknesses greater than 10 or 11 gauge or over 1/8 of an inch. Thicknesses lighter than 1/8 inch are frequently called sheets.

plating—A surface application of a thin layer of a different metal or metals to a base metal. The bond to the base metal is very strong. Plating is done to gain corrosion resistance, improve appearance, and to resist wear. There are a number of plating methods; all require specialized equipment and procedures. Zinc, copper, chromium, cadmium, and tin are just a few of the more common plating metals. Heating plated metals can be hazardous because of the toxic metallic fumes produced; welding often requires special procedures.

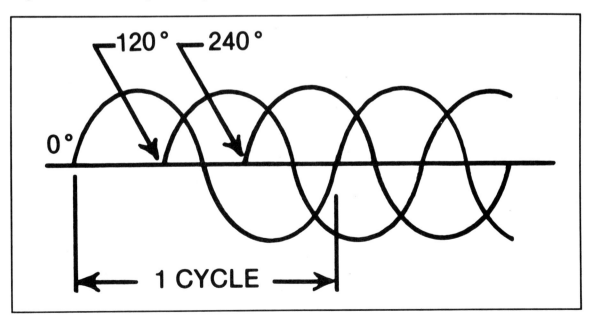

Fig. A-6. Three phase ac. The degrees refer to rotation of a generating device. (Courtesy of Alloy Rods.)

polarity—A term describing the direction of current flow in a dc welding circuit. If the electrode is connected to the positive terminal of the power source, the polarity is direct current electrode positive (DCEP). This is also known as "reverse polarity," and is used for the GMAW processes, the gas-shielded flux-cored process, and for most of the coated stick electrodes. If the electrode is connected to the negative terminal of the power source, the polarity is direct current electrode negative (DCEN). This is also known as "straight polarity," and is used with the self-shielded flux-cored wires and a few of the coated stick electrodes. It is absolutely essential to the nature and quality of the weld that the correct polarity be used. Arc control, bead shape, penetration, deposition rate, and spatter control are some of the factors determined by polarity. With 60 Hz ac, the polarity changes 120 times per second and is simply referred to as ac.

power source—A device to adapt the high-voltage, low-amperage line electrical energy available from the power company into a low-voltage, high-amperage form suitable for arc welding. Sometimes called welding machines. See also CC and CV, line power, phase, and Chapters 3, 4, 10, and 11.

primary—The input coil of a transformer. The input wires of a power source that are plugged into the line; sometimes called the primary cable as distinct from the secondary or welding cables.

prestolite—A trademark belonging to L-TEC Welding and Cutting Systems, also known as Linde Division of Union Carbide Corporation. Prestolite is the air acetylene process often used for soldering and brazing. POL fittings are used to adapt an air acetylene torch to a standard acetylene cylinder. See also Chapter 9.

process—A (welding, brazing, soldering, or cutting) process is a specific method of joining or severing metal using heat. The home craftsman often has more than one process option for a given job. The welding processes covered in this book include OAW, FCAW, GMAW, SMAW. Torch brazing, braze welding, and soldering are also covered. Thermal cutting processes used at home include OAC and AAC (Figs. A-7 and A-8).

protective gear—Items worn by the welder to prevent injury from heat, light, sparks, slag, sharp and heavy objects, noise, and dust. Protective gear includes leather gloves and jackets, welding hoods and goggles, safety glasses, boots, ear plugs, and respirators.

rectifier—A device to change ac to dc. In power sources either the diode or silicon control rectifier (SCR) is used to obtain dc from ac.

scarfing—Removal of metal usually with AAC, same as gouging. A scarf joint is a type of butt joint more familiar to woodworking but sometimes used to maximize joint surface area for soldering or brazing.

secondary—The output coil of a transformer. The secondary circuit refers to the cables and any other conductors connected to the output side of a power source.

semiautomatic—A welding process in which the filler metal is automatically fed but the torch or gun is hand-held. See also manual.

shielding gas—A gas that is directed into the weld zone at a continuous, controlled flow rate to displace the air and prevent vulnerable hot weld metal from combining with oxygen, nitrogen, and other contaminants. For steel welding in the home shop, the C-25 mixed shielding gas produces the best weld. Carbon dioxide is a lower-priced mild steel shielding gas. Shielding gas used for spray transfer of GMAW requires a mixture of at least 85 percent argon. See metal transfer.

short arc—An enduring nickname for short circuit metal transfer along with micro-wire and dip-transfer. See metal transfer, SCR, solid state.

silver brazing—A brazing process that uses a filler metal composed of 25 to 50 percent silver, plus copper, zinc, and cadmium. The melting point is about 1150 to 1200 degrees Fahrenheit. A fluoride/borate-type flux is used. Silver brazing is used to join many different and dissimilar

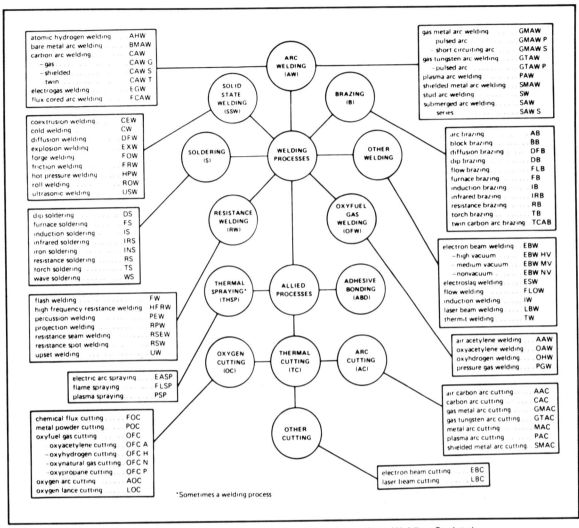

Fig. A-7. Master chart of welding and allied processes. (Courtesy of American Welding Society.)

metals and can be performed with an oxyacetylene torch. Silver brazing is sometimes called "silver" or "hard" soldering; it is brazing, however, because more than 840 degrees Fahrenheit are required. Sil-Fos or "poor man's silver solder" is a low-priced alloy (less silver, more copper and some phosphorous) that works well joining copper to copper without a flux. Ventilation is essential when silver brazing because of toxic fumes. See also brazing.

skip weld—A slang term for intermittent weld. See also weld types.

slag—The sharp iron oxide residue from flame cutting; also the crust that forms over FCAW and SMAW deposits containing impurities floated out from the molten weld deposits. Arc welding slag also protects hot, vulnerable metal from atmospheric contamination and from cooling too rapidly. As a general rule, welding over any kind of slag is poor practice because of the likelihood of weld inclusions.

soldering—A metal joining process that uses a filler metal to bond, not weld, metal parts together. The solder filler metal melts at a tem-

465

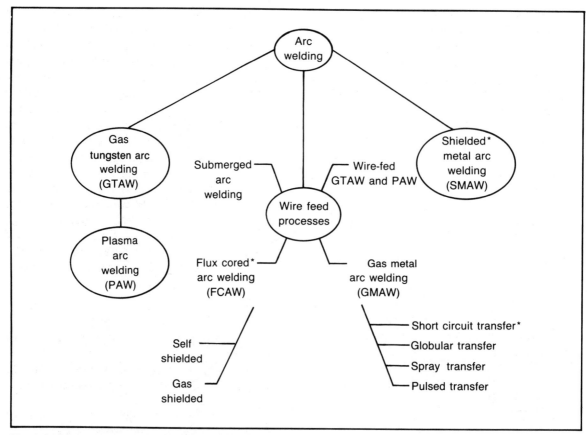

Fig. A-8. Common arc welding processes (* marks processes popular for home use).

perature lower than that of the base metal and always at less than 840 degrees Fahrenheit. Capillary (soaking) action spreads the solder between the members of the solder joint. For best results, large joint surface areas are required, butt joints are generally avoided. Soldering with a tin and lead alloy, also known as "soft soldering," is commonly used to join electrical components and copper alloys. A flux is used to remove surface impurities so that the solder can adhere to the base metal. There are many different solders designed for specific applications. The home welder usually solders with an oxyacetylene torch, a propane torch, a soldering copper (iron), a soldering gun, or a soldering pencil. See also flux, and Chapter 9.

solid state—Electronic components that, during use, exhibit no readily perceived light, heat, mo-

tion, or sound. Also circuits that are mostly made of such components. Examples include diodes, silicon control rectifiers (SCRs), transistors, and integrated circuits (ICs) or "chips." Earlier electronic designs made heavy use of vacuum tubes with their integral heaters, and mechanical switching devices. Solid state components must be kept at temperatures below 150 degrees Fahrenheit. See also diode and SCR.

spot welding—A manner of joining plates, usually in a lap joint, with the weld located within a tightly-defined circular area. There are two types of spot welding: resistance and arc spot. Resistance spot welding works with a high electrical current generating heat by passing through the tightly squeezed small contact area between two copper electrodes. The right combination of heat and pressure is essential and usually controlled automatically. Arc Spot Welding is done with a wire feed process using a specific combination of heat input, wire stickout, and arc-on time. A very tight fitup is essential for arc spot welding. Spot welding should not be confused with plug or slot welding. See also welding joints and weld types.

stainless steel—A group of alloys composed of iron, at least 11.5 percent chromium, about 10 percent nickel, and usually less than 0.15 percent carbon. Stainless steel is used in harsh environments such as piping systems for corrosive fluids. While most stainless alloys can be welded, every effort should be made to identify the base metal for matchup with the correct electrode. In addition, other precautions are also required. See also SS.

steel—A commonplace material with an incredibly complex and fascinating body of structural and behavioral qualities. Steel is mostly composed of iron, plus an exact amount of carbon. Although by definition, steel contains up to 2.0 percent carbon, most steels have well less than 1.0 percent and easily welded steels contain less than 0.29 percent. If carbon is the major additive or alloying element, the material is called carbon steel. If some other element is added in a quantity sufficient to make a significant structural or behavioral difference, the material is called alloy steel. Alloying elements change various properties of steel including hardness and tensile strength. The American Iron and Steel Institute (AISI) along with the Society of Automotive Engineers (SAE) have formulated the AISI/SAE steel classification numbering system. See also CRS and HRS.

stress relieve—As welds cool and the metal contracts, the welded structure usually has internal stresses locked in, due to the limited ability of parts to move with the contraction. These residual stresses can cause a welded object to break if subjected to a sudden and/or heavy load. One of the reasons filler metal is not cheap is that it is designed to yield or stretch upon cooling, thereby relieving some stress and avoiding cracks. If a welded structure is reheated after welding to about 1100 degrees Fahrenheit (a faint red color in a darkened room), held for a few minutes, and cooled gradually, much of the stresses are relieved. Special heat-treating ovens are often used by manufacturers. There are even some more exotic electrical and vibratory methods of stress relief. The home craftsman can stress relieve small welded objects that will be subjected to rough service by first heating them with a torch and then covering them with a fiberglass blanket, or burying them in a bucket of ashes to slow down the cooling rate.

structural shapes—The cross-section of a bar of metal reveals its characteristic shape. Although many shapes are available, the home welder will probably only use a few of the more basic (also more economical) shapes. Among these are:

- Flat bar; measured thickness × width × length.
- Angle bar; measured flange × flange × thickness × length.
- Channel; measured width × height (flanges) × thickness, or by weight per foot × length.
- Rectangular structural tubing; measured height × width × wall thickness × length.
- Beams (I and H); measured by height × flange width × weight per foot × length.
- Round bar; measured by diameter × length.
- Pipe; measurement methods vary.
- Tubing; measured usually by outside dimension and wall thickness.

A proper description of a structural shape must also specify the composition of the material. The home craftsman usually does not need a proper pedigree. Just HRS or CRS is enough information. It is also less expensive to buy remnants. See also CRS, HRS, and Chapter 19.

suitability for design purpose—The adequacy of a welded structure to fulfill its intended function. While every weld should be of at least reasonable quality, an agonizing search for perfection is seldom necessary. See also Chapter 17.

tack weld—A small weld used to hold parts in correct position and alignment prior to application of the final, or production, weld. Tack welds should be carefully made, sloppy tacks are never without consequence. Tack welds affect both the final product appearance and the structural integrity of the product.

tensile strength—The ability of a material to resist being pulled apart. A test sample with a known cross-sectional area is gradually pulled upon with a measurable number of pounds. A breaking point is eventually reached. This is the tensile strength in psi (or kg/sq mm). Ordinary mild steel used by the home welder withstands about 60,000 psi tensile load before breaking.

Filler metal is also rated as to its tensile strength (as deposited); most exceed 70,000 psi. See also electrode, steel, and yield strength.

transformer—A device used to step voltage up or down. A transformer is used in every power source to reduce the high input (line) voltage to a level that permits arc welding. A transformer type power source takes in ac and puts out ac. A simple transformer is composed of a primary and a secondary coil wound onto a special iron core. Welding transformers require no maintenance. See also electrical power, induction, primary, secondary, and Chapter 10.

Troy ounce—480 grains, a "standard" ounce is 437.5 grains. Therefore a Troy ounce is about 1.097 as much as a standard (avoirdupois) ounce. Troy weight is used in weighing silver brazing alloy.

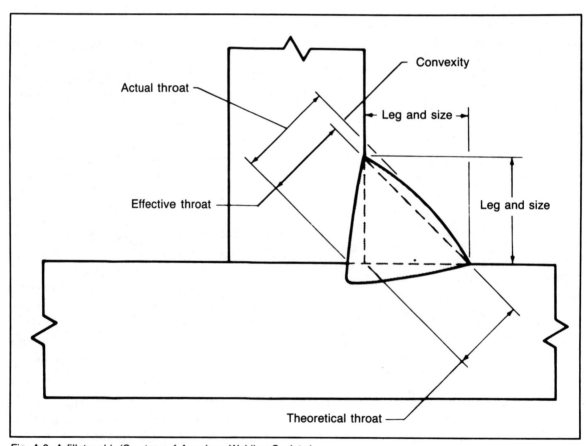

Fig. A-9. A fillet weld. (Courtesy of American Welding Society.)

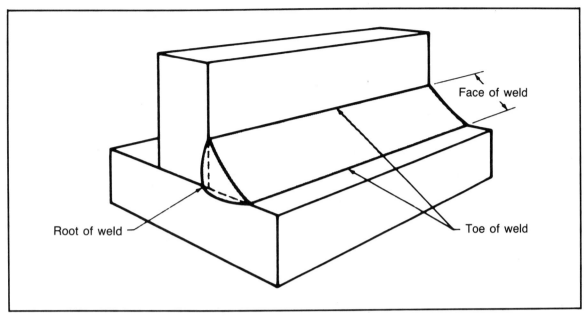

Fig. A-10. Face, root, and toes of a fillet weld. (Courtesy of American Welding Society.)

volt—The unit of electrical pressure named after Alessandro Volta (d.1827). See also OCV, electrical power, line power, Ohm's law, power source, transformer, and Chapter 10.

weld metal—One of the three distinct zones of a welded structure (along with the base metal and heat affected zone). The deposited metal that combines with the assembly parts of the base metal, also the metal deposited onto a surface to provide improved wear, corrosion resistance, or to achieve a special effect. Also known as the weld deposit. See also base metal, filler, and heat affected zone.

weld parts—A weld deposit is cast into a shape defined by the joint. A fillet weld (Figs. A-9 and A-10) is joined to the base metal by its legs. The toes of a fillet are the outside edges of the legs. The face of a fillet can have three possible profiles—flat, convex, or concave. The root of a weld is simply the deepest part of it. The build-up on the face of a groove weld is the reinforcement. See also welding joints, weld size, weld types, and Chapter 17.

weld size—The size of a fillet weld is measured by the length of its legs. This should never be greater than the material thickness of the thinnest member of the joint. A groove weld is either full or partial penetration. The thickness of any excess metal on the face or penetrating through the root is called reinforcement.

weld symbol—A standardized (by AWS) set of marks placed on drawings and blueprints that convey joint preparation, fitup, size of weld, location of weld, and other pertinent information. The home craftsman need not be overly concerned about weld symbols. However, blueprints or plans often contain symbols and so a summary chart is contained in Appendix B.

weld types—Welds are either fillets or grooves. Sometimes beading is counted as a third type of weld for surfacing operations and for welding edge joints (Fig. A-11). Fillets are used on lap, corner, and T joints; groove welds join butt joints and, depending on how you look at it, edge joints.

welding—A process for permanently joining parts of an assembly that involves some melting, mixing and fusing of those parts. Sometimes welding is used as a blanket term and includes related joining operations like brazing. For example, "tack welded" could also mean "tack brazed."

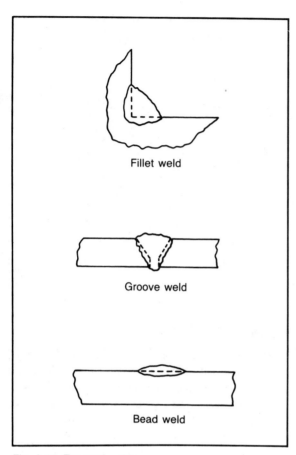

Fillet weld

Groove weld

Bead weld

Fig. A-11. Types of welds.

welding joints—The arrangement of metal parts to be joined. There are five basic types of welding joints—lap, corner, T, butt, and edge (Fig. A-12). The T joint occurs most frequently. There is a rather extensive variety of possible edge preparations and assembly details, most having to do with butt joints in heavier plate thick-

nesses. In general, a welding joint should be arranged to permit a strong weld, with minimal distortion to the structure of the project. See also joint.

weldment—A structure designed for and joined by a welding, brazing, or soldering method. See also base metal.

wheelabrated plate—A method of plate preparation and preservation developed by the Wheelabrator Company. Hot rolled steel is blasted to remove scale and then coated with a colored primer. The home craftsman might encounter remnants with this treatment. Welding on this plate presents little difficulty with FCAW and SMAW other than the added smoke and fumes; however, the GMAW processes often pick up porosity from the reaction with the primer. Flame cutting wheelabrated plate is somewhat more difficult due to the reflective function of the primer.

wrought—Any metal forming operation other than casting; including drawing, rolling, stretching, extruding, and forging. See also cast.

yield strength—The value in psi of tensile load at which a metal or weld takes on a permanent stretch. At the yield point, the metal no longer springs back entirely when the load is removed. As an example—E71 T-1 electrode with a tensile (breaking) strength of 80,000 psi has a yield strength of 69,000 psi. Generally speaking, the lower the yield strength, the more ductile the structure. This can be useful in avoiding weld cracks on highly restrained joints that permit little movement for stress relief. See also tensile strength and Chapter 17.

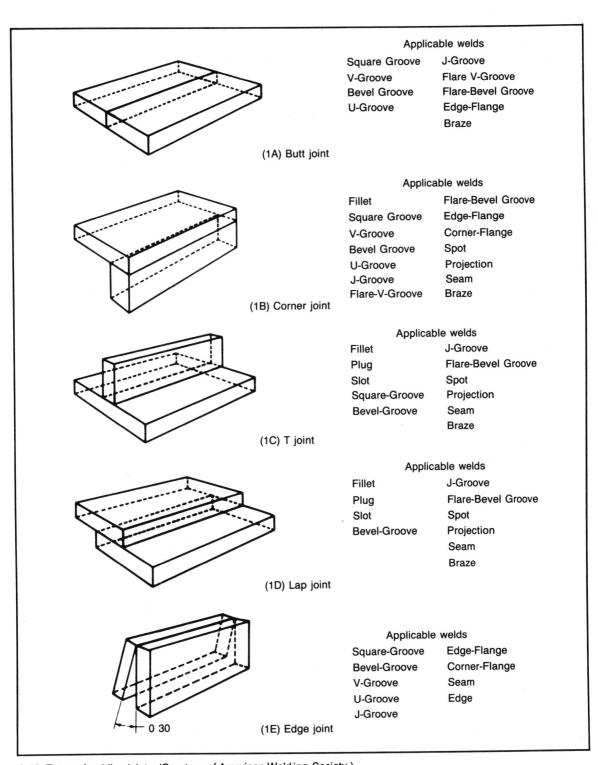

Applicable welds

Square Groove	J-Groove
V-Groove	Flare V-Groove
Bevel Groove	Flare-Bevel Groove
U-Groove	Edge-Flange
	Braze

(1A) Butt joint

Applicable welds

Fillet	Flare-Bevel Groove
Square Groove	Edge-Flange
V-Groove	Corner-Flange
Bevel Groove	Spot
U-Groove	Projection
J-Groove	Seam
Flare-V-Groove	Braze

(1B) Corner joint

Applicable welds

Fillet	J-Groove
Plug	Flare-Bevel Groove
Slot	Spot
Square-Groove	Projection
Bevel-Groove	Seam
	Braze

(1C) T joint

Applicable welds

Fillet	J-Groove
Plug	Flare-Bevel Groove
Slot	Spot
Bevel-Groove	Projection
	Seam
	Braze

(1D) Lap joint

Applicable welds

Square-Groove	Edge-Flange
Bevel-Groove	Corner-Flange
V-Groove	Seam
U-Groove	Edge
J-Groove	

0 30   (1E) Edge joint

A-12. Types of welding joints. (Courtesy of American Welding Society.)

471

Appendix B
Formulas and Tables

Table B-1. Tap Drill Sizes.

A hole is first drilled to a specific diameter called the tap drill size. Threads are then cut in the hole with a corresponding tap. Use the following charts to match a tap drill with a tap. The nominal size states two facts: the diameter and the number of threads per inch. For example 1/4-20 means a 1/4-inch wide thread with 20 threads per inch.

TAP DRILL SIZES
FOR 75% FULL THREAD

NOMINAL SIZE	TAP DRILL SIZE
1/8-40	38
5/32-36	30
3/16-24	26
7/32-24	16
1/4-20	7
1/4-28	3
5/16-18	F
5/16-24	I
3/8-16	5/16
3/8-24	Q
7/16-14	U
7/16-20	25/64
1/2-13	27/64
1/2-20	29/64

9/16-12	31/64
9/16-18	33/64
5/8-11	17/32
5/8-18	37/64
11/16-11	19/32
11/16-16	5/8
3/4-10	21/32
3/4-16	11/16
13/16-10	23/32
7/8-9	49/64
7/8-14	13/16
7/8-18	53/64
15/16-9	53/64
1 -8	7/8
1 -14	15/16
1 1/8-7	63/64

1 1/8-12	1 3/64
1 1/4-7	1 7/64
1 1/4-12	1 11/64
1 3/8-6	1 7/32
1 3/8-12	1 19/64
1 1/2-6	1 11/32
1 1/2-12	1 27/64
1 5/8-5 1/2	1 29/64
1 3/4-5	1 9/16
1 7/8-5	1 11/16
2 -4 1/2	1 25/32
2 1/8-4 1/2	1 29/32
2 1/4-4 1/2	2 1/32
2 3/8-4	2 1/8
2 1/2-4	2 1/4

SIZE OF PIPE	THREADS PER INCH	TAP DRILL SIZE
1/8	27	R
1/4	18	7/16
3/8	18	37/64
1/2	14	23/32
3/4	14	59/64
1	11 1/2	1 5/32

Tap Drill Sizes for American Taper Pipe Taps		
1 1/4	11 1/2	1 1/2
1 1/2	11 1/2	1 47/64
2	11 1/2	2 7/32
2 1/2	8	2 5/8
3	8	3 1/4
3 1/2	8	3 3/4
4	8	4 1/4

TAP DRILL SIZES
FOR 75% FULL THREAD

Machine Screw Sizes A. S. M. E.	
NOMINAL SIZE	TAP DRILL SIZE
0—80	3/64
1—64	53
1—72	53
2—56	50

2—64	50
3—48	47
3—56	45
4—40	43
4—48	42
5—40	38
5—44	37
6—32	36
6—40	33

8—32	29
8—36	29
9—32	26
10—24	25
10—30	22
10—32	21
12—24	16
12—28	14
14—24	7

Table B-2. Decimal Sizes of Number (Wire Gauge) Drills and Letter Drills.

Wire Gage	Decimal Equivalent	Wire Gage	Decimal Equivalent	Wire Gage	Decimal Equivalent
80	.0135	70	.028	60	.040
79	.0145	69	.0293	59	.041
78	.016	68	.031	58	.042
77	.018	67	.032	57	.043
76	.020	66	.033	56	.0465
75	.021	65	.035	55	.052
74	.0225	64	.036	54	.055
73	.024	63	.037	53	.059
72	.025	62	.038	52	.0635
71	.026	61	.039	51	.067

Wire Gage	Decimal Equivalent	Wire Gage	Decimal Equivalent	Wire Gage	Decimal Equivalent
50	.070	40	.098	30	.1285
49	.073	39	.0995	29	.136
48	.076	38	.1015	28	.1405
47	.0785	37	.104	27	.144
46	.081	36	.1065	26	.147
45	.082	35	.110	25	.1495
44	.086	34	.111	24	.152
43	.089	33	.113	23	.154
42	.0935	32	.116	22	.157
41	.096	31	.120	21	.159

Wire Gage	Decimal Equivalent	Wire Gage	Decimal Equivalent	Letter	Decimal Equivalent
20	.161	10	.1935	A	.234
19	.166	9	.196	B	.238
18	.1695	8	.199	C	.242
17	.173	7	.201	D	.246
16	.177	6	.204	E	.250
15	.180	5	.2055	F	.257
14	.182	4	.209	G	.261
13	.185	3	.213	H	.266
12	.189	2	.221	I	.272
11	.191	1	.228	J	.277

Letter	Decimal Equivalent	Letter	Decimal Equivalent
K	.281	U	.368
L	.290	V	.377
M	.295	W	.386
N	.302	X	.397
O	.316	Y	.404
P	.323	Z	.413
Q	.332		
R	.339		
S	.348		
T	.358		

Table B-3. Decimal and Metric Equivalent.

Inches (Fractions)	Inches (Decimals)	Millimeters
1/64	.015625	.397
1/32	.03125	.794
3/64	.046875	1.191
1/16	.0625	1.588
5/64	.078125	1.984
3/32	.09375	2.381
7/64	.109375	2.778
1/8	.125	3.175
9/64	.140625	3.572
5/32	.15625	3.969
11/64	.171875	4.366
3/16	.1875	4.763
13/64	.203125	5.159
7/32	.21875	5.556
15/64	.234375	5.953
1/4	.250	6.350
17/64	.265625	6.747
9/32	.28125	7.144
19/64	.296875	7.541
5/16	.3125	7.938
21/64	.328125	8.334
11/32	.34375	8.731
23/64	.359375	9.128
3/8	.375	9.525
25/64	.390625	9.922
13/32	.40625	10.319
27/64	.421875	10.716
7/16	.4375	11.113
29/64	.453125	11.509
15/32	.46875	11.906
31/64	.484375	12.303
1/2	.500	12.700
33/64	.515625	13.097
17/32	.53125	13.494
35/64	.546875	13.891
9/16	.5625	14.288
37/64	.578125	14.684
19/32	.59375	15.081
39/64	.609375	15.478
5/8	.625	15.875
41/64	.640625	16.272
21/32	.65625	16.669
43/64	.671875	17.066
11/16	.6875	17.463
45/64	.703125	17.859
23/32	.71875	18.256
47/64	.734375	18.653
3/4	.750	19.050
49/64	.765625	19.447
25/32	.78125	19.844
51/64	.796875	20.241
13/16	.8125	20.638
53/64	.828125	21.034
27/32	.84375	21.431
55/64	.859375	21.828
7/8	.875	22.225
57/64	.890625	22.622
29/32	.90625	23.019
59/64	.921875	23.416
15/16	.9375	23.813
61/64	.953125	24.209
31/32	.96875	24.606
63/64	.984375	25.003
1	1.000	25.400

Table B-4. Table of Decimal Equivalents.

1/64	1/32	1/16	1/8	1/4	1/2	Decimal
1/64						.01563
	1/32					.03125
3/64						.04688
		1/16				.0625
5/64						.07813
	3/32					.09375
7/64						.10938
			1/8			.125
9/64						.14063
	5/32					.15625
11/64						.17188
		3/16				.1875
13/64						.20313
	7/32					.21875
15/64						.23438
				1/4		.250
17/64						.26563
	9/32					.28125
19/64						.29688
		5/16				.3125
21/64						.32813
	11/32					.34375
23/64						.35938
			3/8			.375
25/64						.39063
	13/32					.40625
27/64						.42188
		7/16				.4375
29/64						.45313
	15/32					.46875
31/64						.48438
					1/2	.500

33/64	17/32	9/16	5/8	3/4	1	Decimal
33/64						.51563
	17/32					.53125
35/64						.54688
		9/16				.5625
37/64						.57813
	19/32					.59375
39/64						.60938
			5/8			.625
41/64						.64063
	21/32					.65625
43/64						.67188
		11/16				.6875
45/64						.70313
	23/32					.71875
47/64						.73438
				3/4		.750
49/64						.76563
	25/32					.78125
51/64						.79688
		13/16				.8125
53/64						.82813
	27/32					.84375
55/64						.85938
			7/8			.875
57/64						.89063
	29/32					.90625
59/64						.92188
		15/16				.9375
61/64						.95313
	31/32					.96875
63/64						.98438
					1	1.0000

Table B-5. Conversion Tables for Inches to Millimeters.

Inches dec.	mm	Inches dec.	mm	Inches dec.	mm	Inches dec.	mm
0.01	0.2540	0.26	6.6040	0.51	12.9540	0.76	19.3040
0.02	0.5080	0.27	6.8580	0.52	13.2080	0.77	19.5580
0.03	0.7620	0.28	7.1120	0.53	13.4620	0.78	19.8120
0.04	1.0160	0.29	7.3660	0.54	13.7160	0.79	20.0660
0.05	1.2700	0.30	7.6200	0.55	13.9700	0.80	20.3200
0.06	1.5240	0.31	7.8740	0.56	14.2240	0.81	20.5740
0.07	1.7780	0.32	8.1280	0.57	14.4780	0.82	20.8280
0.08	2.0320	0.33	8.3820	0.58	14.7320	0.83	21.0820
0.09	2.2860	0.34	8.6360	0.59	14.9860	0.84	21.3360
0.10	2.5400	0.35	8.8900	0.60	15.2400	0.85	21.5900
0.11	2.7940	0.36	9.1440	0.61	15.4940	0.86	21.8440
0.12	3.0480	0.37	9.3980	0.62	15.7480	0.87	22.0980
0.13	3.3020	0.38	9.6520	0.63	16.0020	0.88	22.3520
0.14	3.5560	0.39	9.9060	0.64	16.2560	0.89	22.6060
0.15	3.8100	0.40	10.1600	0.65	16.5100	0.90	22.8600
0.16	4.0640	0.41	10.4140	0.66	16.7640	0.91	23.1140
0.17	4.3180	0.42	10.6680	0.67	17.0180	0.92	23.3680
0.18	4.5720	0.43	10.9220	0.68	17.2720	0.93	23.6220
0.19	4.8260	0.44	11.1760	0.69	17.5260	0.94	23.8760
0.20	5.0800	0.45	11.4300	0.70	17.7800	0.95	24.1300
0.21	5.3340	0.46	11.6840	0.71	18.0340	0.96	24.3840
0.22	5.5880	0.47	11.9380	0.72	18.2880	0.97	24.6380
0.23	5.8420	0.48	12.1920	0.73	18.5420	0.98	24.8920
0.24	6.0960	0.49	12.4460	0.74	18.7960	0.99	25.1460
0.25	6.3500	0.50	12.7000	0.75	19.0500	1.00	25.4000

474

Table B-6. Some Key Steel Metal.

The larger the gauge number, the thinner the plate. The usual thin limit for practical arc welding is 16 gauge. In most cases, the thickness of steel heavier than 11 gauge is specified in fractions of an inch.

Gauge No.	Thickness in Inches
3	0.250 (1/4)
7	0.188 (3/16)
11	0.125 (1/8)
14	0.0781
16	0.0625 (1/16)
18	0.050

Table B-7. Recommended Metric Tap Drill Sizes.

BASED ON APPROX. 75% THREAD			
Theoretical drill size mm		Closest American drill	
		size	mm equiv.
1.1590		57	1.0922
1.2590		3/64	1.1913
1.4590		54	1.3970
1.5612		53	1.5113
1.6102		1/16	1.5875
1.7615		51	1.7018
1.9102		49	1.8542
2.0615		46	2.0574
2.1615		45	2.0828
2.4155		3/32	2.3825
2.5184		40	2.4892
2.9155		33	2.8702
3.2693		30	3.2639
3.3180		30	3.2639
3.7693		26	3.7338
4.0257		22	3.9878
4.1235		20	4.0894
4.2204		19	4.2164
4.6235		14	4.6228
5.0257		9	4.9784
5.2693		5	5.2197
6.0257		15/64	5.9537
6.2693		D	6.2484
6.7823		H	6.7564
7.0257		I	6.9088
7.7823		N	7.6708
8.0257		5/16	7.9375
8.5385		Q	8.4328
8.7823		11/32	8.7325
9.0257		S	8.8392
9.5385		3/8	9.5250
10.2950		Y	10.2616
10.5385		Z	10.4902
10.7823		27/64	10.7162
12.0516		15/32	11.9075
12.5385		31/64	12.3037
12.7832		1/2	12.700
13.5385		17/32	13.4950
14.0516		35/64	13.8912
14.5385		9/16	14.2875

BASED ON APPROX. 75% THREAD			
Theoretical drill size mm		Closest American drill	
		size	mm equiv.
15.5385		39/64	15.4787
15.5643		39/64	15.4787
16.0516		5/8	15.8750
16.5385		41/64	16.2712
16.5643		41/64	16.2712
17.5643		11/16	17.5625
18.0516		45/64	17.8587
18.5385		23/32	18.2557
19.5643		49/64	19.4462
20.0516		25/32	19.8425
20.5385		51/64	20.2413
21.0773		53/64	21.0337
22.0516		55/64	21.8288
22.5385		7/8	22.2250
23.0516		29/32	23.0175
23.5385		59/64	23.4163
23.0773		29/32	23.0175
24.0773		15/16	23.8125
25.0516		63/64	25.0038
25.0773		63/64	25.0038
26.0516		1 1/64	25.7962
26.5900		1 1/32	26.1925
27.0773		1 1/16	26.9875
28.0516		1 3/32	27.7825
28.5900		1 1/8	28.5750
30.0516		1 11/64	29.7663
29.5900		1 5/32	29.3675
30.0773		1 11/64	29.7663
31.0516		1 7/32	30.9575
30.5900		1 13/64	30.5587
32.1028		1 17/64	31.7500
33.0733		1 19/64	32.9413
34.0516		1 21/64	33.7337
34.1028		1 21/64	33.7337
35.1028		1 3/8	34.9250
36.0773		1 13/32	35.7175
37.0516		1 29/64	36.9087

Table B-8. Basic Welding Symbols and Their Location Significance.

Location Significance	Fillet	Plug or Slot	Spot or Projection	Seam	Back or Backing	Surfacing	Scarf for Brazed Joint	Flange Edge
Arrow Side	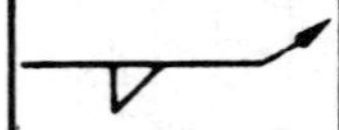	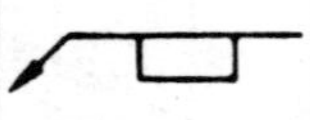	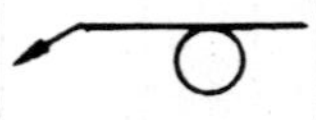	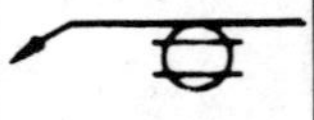	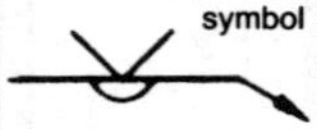 Groove weld symbol		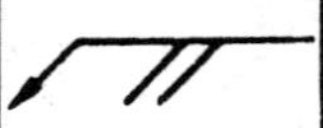	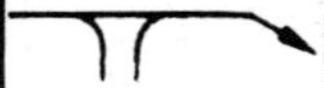
Other Side	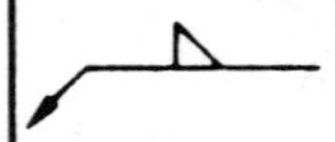			 Not used	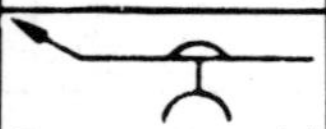 Groove weld symbol	Not used		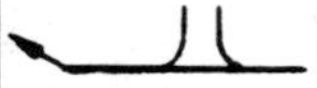
Both Sides	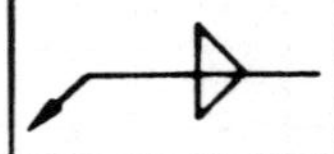 Not used	Not used	Not used	Not used	Not used	Not used		Not used
No Arrow Side or Other Side Significance	Not used	Not used			Not used	Not used	Not used	Not used

Supplementary Symbols Used with Welding Symbols

Convex Contour Symbol

Convex contour symbol indicates face of weld to be finished to convex contour

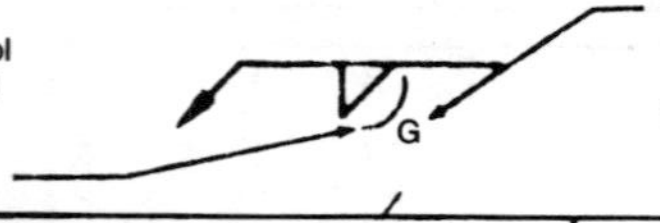

Finish symbol (user's standard) indicates method of obtaining specified contour but not degree of finish

Weld-All-Around Symbol

Weld-all-around symbol indicates that weld extends completely around the joint

Joint with Backing

With groove weld symbol

See note

Note: Material and dimensions of backing as specified

Joint with Spacer

With modified groove weld symbol

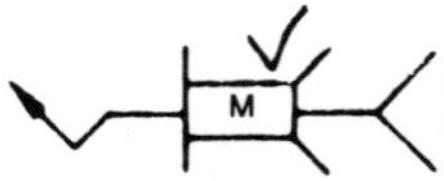

See note

Double bevel groove

Note: Material and dimensions of spacer as specified

Melt-Thru Symbol

Any applicable weld symbol

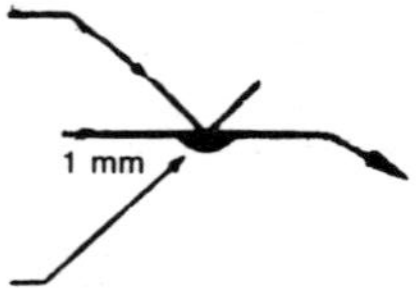

1 mm

Melt-thru symbol is not dimensioned (except height)

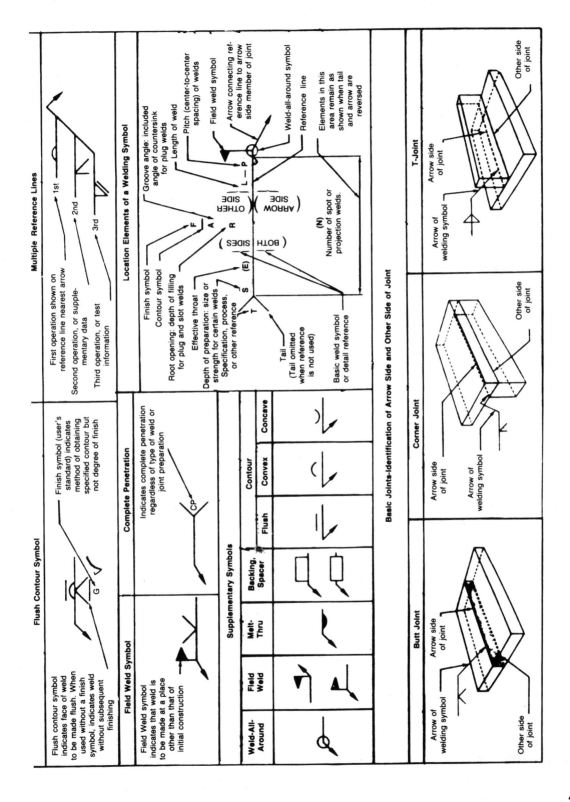

Flush Contour Symbol

Flush contour symbol indicates face of weld to be made flush. When used without a finish symbol, indicates weld without subsequent finishing

Finish symbol (user's standard) indicates method of obtaining specified contour but not degree of finish

G

Complete Penetration

Indicates complete penetration regardless of type of weld or joint preparation

CP

Field Weld Symbol

Field Weld symbol indicates that weld is to be made at a place other than that of initial construction

Multiple Reference Lines

1st — First operation shown on reference line nearest arrow

2nd — Second operation, or supple-mentary data

3rd — Third operation, or test information

Location Elements of a Welding Symbol

Groove angle: included angle of countersink for plug welds

Length of weld

Pitch (center-to-center spacing) of welds

Arrow connecting ref-erence line to arrow side member of joint

Field weld symbol

Weld-all-around symbol

Reference line

Elements in this area remain as shown when tail and arrow are reversed

Finish symbol

Contour symbol

Root opening; depth of filling for plug and slot welds

Effective throat

Depth of preparation: size or strength for certain welds

Specification, process, or other reference

Tail (Tail omitted when reference is not used)

Basic weld symbol or detail reference

Number of spot or projection welds

(N)

L — P

OTHER SIDE

ARROW SIDE

(BOTH SIDES)

(E)

S

T

R

A

F

Supplementary Symbols

Weld-All-Around	Field Weld	Melt-Thru	Backing, Spacer	Contour		
				Flush	Convex	Concave

Basic Joints–Identification of Arrow Side and Other Side of Joint

Butt Joint

Arrow of welding symbol

Arrow side of joint

Other side of joint

Corner Joint

Arrow side of joint

Arrow of welding symbol

Other side of joint

T-Joint

Arrow side of joint

Arrow of welding symbol

Other side of joint

477

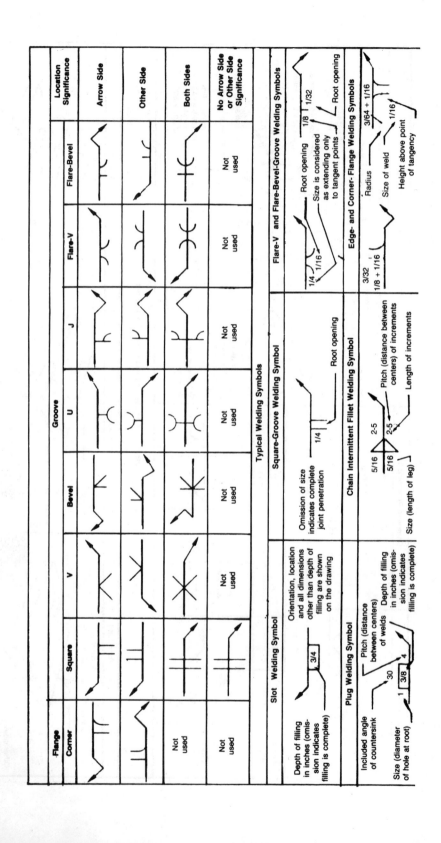

The following table shows welding symbols:

Flange		Groove							Location Significance
Corner	Square	V	Bevel	U	J	Flare-V	Flare-Bevel		
									Arrow Side
									Other Side
Not used									Both Sides
Not used	Not used	Not used	Not used	Not used	Not used	Not used	Not used		No Arrow Side or Other Side Significance

Typical Welding Symbols

Slot Welding Symbol

Orientation, location and all dimensions other than depth of filling are shown on the drawing

Depth of filling in inches (omission indicates filling is complete) — 3/4

Square-Groove Welding Symbol

Omission of size indicates complete joint penetration — 1/4

Root opening

Flare-V and Flare-Bevel-Groove Welding Symbols

Root opening — 1/8 T 1/32
Size is considered as extending only to tangent points
1/4
1/16

Root opening

Plug Welding Symbol

Pitch (distance between centers) of welds
Depth of filling in inches (omission indicates filling is complete)
Included angle of countersink — 30
Size (diameter of hole at root) — 1 3/8 4

Chain Intermittent Fillet Welding Symbol

5/16 2-5
5/16 2-5
Pitch (distance between centers) of increments
Length of increments
Size (length of leg)

Edge- and Corner-Flange Welding Symbols

3/64 + 1/16
3/32
1/8 + 1/16
Radius
Size of weld
1/16
Height above point of tangency

478

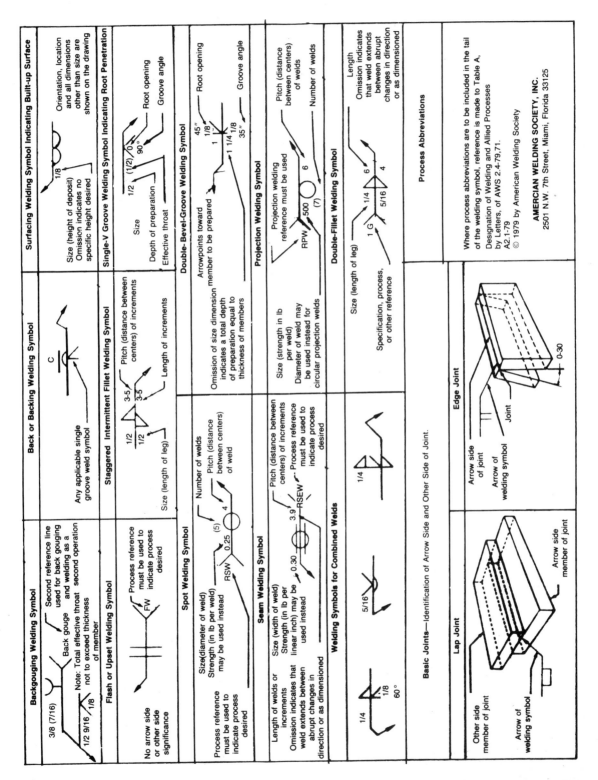

Backgouging Welding Symbol

3/8 (7/16)
Second reference line used for back gouging and welding as a
Back gouge
Note: Total effective throat second operation not to exceed thickness of member
1/2 9/16 1/8

Flash or Upset Welding Symbol

Process reference must be used to indicate process desired
FW
No arrow side or other side significance

Spot Welding Symbol

Size(diameter of weld)
Strength (in lb per weld) may be used instead
Process reference must be used to indicate process desired
RSW 0.25
(5) Number of welds
4 Pitch (distance between centers) of weld

Seam Welding Symbol

Size (width of weld)
Strength (in lb per linear inch) may be used instead
Length of welds or increments Omission indicates that weld extends between abrupt changes in direction or as dimensioned
RSEW 0.30
3.9 Pitch (distance between centers) of increments
Process reference must be used to indicate process desired

Welding Symbols for Combined Welds

5/16
1/4
1/4
1/8
60°

Lap Joint

Other side member of joint
Arrow of welding symbol
Arrow side member of joint

Back or Backing Welding Symbol

C
Any applicable single groove weld symbol

Staggered Intermittent Fillet Welding Symbol

1/2
1/2
3-5 Pitch (distance between centers) of increments
3-5 Length of increments
Size (length of leg)

Double-Bevel-Groove Welding Symbol

Arrowpoints toward member to be prepared
Omission of size dimension indicates a total depth of preparation equal to thickness of members

Basic Joints—Identification of Arrow Side and Other Side of Joint.

Edge Joint

Arrow side of joint
Arrow of welding symbol
Joint
0-30

Surfacing Welding Symbol Indicating Built-up Surface

1/8
Orientation, location and all dimensions other than size are shown on the drawing
Size (height of deposit) Omission indicates no specific height desired

Single-V Groove Welding Symbol Indicating Root Penetration

1/2 (1/2) 0
90°
Size
Depth of preparation
Effective throat
Root opening
Groove angle

45°
1/8
1
1 1/4 1/8
35°
Root opening
Groove angle

Projection Welding Symbol

Projection welding reference must be used
RPW 500
(7)
6 Pitch (distance between centers) of welds
Size (strength in lb per weld)
Diameter of weld may be used instead for circular projection welds
Number of welds

Double-Fillet Welding Symbol

1/4
5/16
6
4
1 G
Size (length of leg)
Specification, process, or other reference
Length Omission indicates that weld extends between abrupt changes in direction or as dimensioned

Process Abbreviations

Where process abbreviations are to be included in the tail of the welding symbol, reference is made to Table A, Designation of Welding and Allied Processes by Letters, of AWS 2.4-79.71.
A2.1-79
© 1979 by American Welding Society

AMERCIAN WELDING SOCIETY, INC.
2501 N.W. 7th Street, Miami, Florida 33125

Table B-9. Usability, Performance and Impact Values of Flux-Cored Electrode Designations.

AWS classification[a]	External shielding gas	Current and polarity	V-notch impact
EXXT-1 (Multiple-pass)	CO_2[b]	dc, electrode positive	20 ft-lb @ 0°F
EXXT-2 (Single-pass)	CO_2[b]	dc, electrode positive	None
EXXT-3 (Single-pass)	None	dc, electrode positive	None
EXXT-4 (Multiple-pass)	None	dc, electrode positive	None
EXXT-5 (Multiple-pass)	CO_2[b]	dc, electrode positive	20 ft-lb @ -20°F
EXXT-6 (Multiple-pass)	None	dc, electrode positive	20 ft-lb @ -20°F
EXXT-7 (Multiple-pass)	None	dc, electrode negative	None
EXXT-8 (Multiple-pass)	None	dc, electrode negative	20 ft-lb @ -20°F
EXXT-10 (Single-pass)	None	dc, electrode negative	None
EXXT-11 (Multiple-pass)	None	dc, electrode negative	None
EXXT-G (Multiple-pass)	c	c	None
EXXT-GS (Single-pass)	c	c	None

a. EXXT-9 not used.
b. Argon-CO_2 mixtures may also be used
c. As agreed by supplier and user.

Table B-10. Chemical Composition—Carbon Steel Bare Wires.

AWS Class	Major Alloying Elements - % By Weight					
	Carbon	Manganese	Silicon	Titanium	Zirconium	Aluminum
ER70S-2	0.07	0.90-1.40	0.40-0.70	0.05-0.15	0.02-0.12	0.05-0.15
ER70S-3	0.06-0.15	0.90-1.40	0.45-0.70	—	—	—
ER70S-4	0.07-0.15	1.00-1.50	0.65-0.85	—	—	—
ER70S-5	0.07-0.19	0.90-1.40	0.30-0.60	—	—	0.50-0.90
ER70S-6	0.07-0.15	1.40-1.85	0.80-1.15	—	—	—
ER70S-7	0.07-0.15	1.50-2.00	0.50-0.80	—	—	—
ER70S-G	No Chemical Requirements					

Further Reading

These are a few of the better readily available books on welding theory and practice. Some earlier books are now out of print but still can be found in the collections of larger libraries.

REFERENCE BOOKS

Althouse, Andrew et al. *Modern Welding*. Goodheart-Wilcox Company, Inc. 1984.

Cary, Howard. *Modern Welding Technology*. Prentice-Hall Inc., 1979.

Daniele, Joseph. *Early American Metal Projects*. McKnight and McKnight Publishing Company.

O'Con, R. and Carr, R. *Metal Fabrication: A Practical Guide*. Prentice-Hall, Inc. 1985.

Hobart Brothers Company. *Pocket Welding Guide*. 1983.

Jeffus, L. and Johnson, H. *Welding Principles and Applications*. Delmar Publishers Inc. 1984.

Lincoln Electric Company. *The Procedure Handbook of Arc Welding*. 1986.

Linde Division of Union Carbide Corporation. *The Oxy-Acetylene Handbook*.

Linde Division of Union Carbide Corporation. *MIG Welding Handbook*.

Linde Division of Union Carbide Corporation. *Welding Power Handbook*.

Miller Electric Manufacturing Company. *Principles of Arc Welding*.

Miller Electric Manufacturing Company. *Basic Electricity*.

Smith, Dave. *Welding Skills and Technology*. McGraw-Hill Book Company. 1984.

Walker, John. *Metal Projects*. Goodheart-Wilcox.

The Welding Encyclopedia. 1986. Monticello Books Inc., Box 19, Lake Zurich, IL 60047.

PERIODICALS

The Farm Journal. (Monthly).

Hot Rod. (Monthly).

The Mother Earth News. (Bimonthly).

Stabilizer. The Lincoln Electric Company. 22782 St. Clair Ave. Cleveland, Ohio 44117. (Quarterly).

Welding Design and Fabrication. Penton/IPC, Inc. (Monthly).

NOTE: There are many ideas to be discovered in the metalworking section of public libraries. These sources are simply too numerous to list.

Index

Edited by Cherie R. Blazer

Other Bestsellers of Related Interest